Chas. Harris

# SETH THOMAS
# Clocks & Movements

## A Guide to
## Identification and Prices



*Tran Duy Ly*

At last there is a comprehensive guide to honor the clocks of Seth Thomas, the renowned American clockmaker...a single, unique source for collectors and dealers to use in identifying and placing values on these extraordinary clocks.

Private homes and public centers throughout the world have long moved to the reliable beat of Seth Thomas clocks. Now, Tran Duy Ly, well known collector and publisher, has compiled a comprehensive collection of these timepieces, including those rare and exceptional models that are anticipated to appreciate in value and are most desired by collectors.

The clocks and movements in this guide were assembled from an extensive collection of rare, original Seth Thomas catalogs. The illustrations, descriptions and prices that are provided will prove indispensable to all who seek to further their understanding and appreciation of these outstanding clocks.

- **EASY TO USE:**
  - — Separate sections for clocks and movements.
  - — All models completely indexed.
- **CONVENIENT REFERENCE GUIDE:**
  - — Clocks of the same type grouped together.
  - — Clocks with similar appearance grouped next to each other.
  - — Unique Movements Table.
- **SUGGESTED RETAIL PRICES:**
  - — Realistic up-to-date 1985 values.

INDEX COMPILED BY
John Cammarata, P.E., CMC

**Library of Congress Cataloging in Publication Data**

Ly, Tran Duy, 1933-
Seth Thomas clocks & movements.

1. Clocks and watches—Catalogs. 2. Clocks and watches—Collectors and collecting. 3. Seth Thomas Clock Company—Catalogs. I. Title. II.Title: Seth Thomas clocks and movements.
TS549.L9 1985 681.1'13'0973 85-6029

ISBN 0-930163-26-5



**First Published in USA 1985 by Arlington Book Company**
**P.O. Box 327, Arlington, Virginia 22210-0327 USA**

# INTRODUCTION

When the many individual collectors of clocks decided to start a national association for their common good (National Association of Watch and Clock Collectors), their thirst for information was insatiable. For a definitive origin of clocks, the only authentic references were the original sales catalogs. These often were so rare that they brought prices rivaling the clocks they pictured.

Soon, the enterprising collectors had the original catalogs reprinted — but then they found themselves with a plethora of individual catalogs. They also discovered they did not have the time necessary to conduct repetitious searches through each, often with frustrating and unsuccessful results.

To solve these problems, the Editor-Publisher of this Guide has gathered a vast and comprehensive collection of original Seth Thomas catalogs — those which best serve to present the widest selection of this maker's products. The clocks and movements pictured and described within these covers were taken from those catalogs, and thus may be considered a very wide, representative sampling of this prolific maker's clocks and movements.

Most past reproductions of company catalogs have illustrated the flat dimensional appearance of the clock and dial. Now, in this volume, movements are treated with equal importance, occupying over one hundred pages of illustrations and detailed descriptions. This provides a valuable and important additional aid to the collector and restorer of Seth Thomas clocks.

The Editor-Publisher has created a format which makes it easier for the reader to obtain information without tiring, wasteful groping through endless pages. The inclusion of prices, assiduously researched, makes it a valuable price guide as well as an indicator of the economic trends in clock collecting. Also, for the first time, the inclusion of a cross-referenced movements table makes it possible to quickly locate different clocks which were produced with the same model movement. Thus, the format makes itself useful to the greatest segment of collectors, restorers and dealers. Yet it does so without overloading its size or cost.

Now that you have reached this point, please take a moment to quickly scan the remaining pages. I am sure the many illustrations and descriptions will reward you with pleasure, and you will depart with a greatly improved appreciation of the many fine timepieces manufactured by Seth Thomas.

**Henry B. Fried**

# ACKNOWLEDGMENTS

**I want to take this opportunity to express my sincere appreciation to my following friends who provided me with valuable material, information and up-to-date prices without which this publication could not have been accomplished.**

**Tran Duy Ly**

**STACY WOOD, JR.**
**FNAWCC**
*NAWCC Headquarters Administrator and Museum Director, Columbia, Pa.*

**JEROLD BOXENHORN**
*Rosedale, N.Y.*

**STEVEN BERGER**
*Algonquin, Ill.*

**HENRY B. FRIED,**
**FNAWCC**
*Flushing, N.Y.*

**C. THOMAS McCALLY**
*Rockville, Md.*

**JIM HOUSTON**
*Damacus, Md.*

**DICK NORRIS**
*Annandale, Va.*

**ROBERT S. EDWARDS**
*Reston, Va.*

**A. BJORNESTAD**
*Glendale, Ca.*

**PAUL V. HEFFNER**
**FNAWCC**
*Waynesville, Oh.*

**H. KIRK FALLIN**
**FNAWCC**
*Fort Myers, Fla.*

**BRENDA BRUSH**
*Washington, D.C.*

**MARGARETHA WILLIAMS**
*Fairfax, Va.*

**CHRIS BAILEY**
**FNAWCC**
*Director-Curator of American Clock and Watch Museum, Bristol, Conn.*

**DONALD SUMMAR**
*Librarian*
*NAWCC Museum, Inc.*
*Columbia, Pa.*

**MARTY RUDDOCK, M.D.**
*Bedford, Oh.*

**JOHN CAMMARATA**
*Syosset, N.Y.*

**DR. STEVE PETRUCELLI**
*Director, Adams Brown Co.*
*Cranbury, N.J.*

**DR. HARVEY SCHMIDT**
*Flushing, N.Y.*

**DANA J. BLACKWELL**
**FNAWCC**
*Naugatuck, Conn.*

**DR. DOUGLAS WRIGHT**
*Wethersfield, Conn.*

**DR. WESLEY G. HARDING**
*Rockville, Md.*

**KENNETH D. ROBERTS**
**FNAWCC**
*Fitzwilliam, N.H.*

**DR. SNOWDEN TAYLOR**
**FNAWCC**
*Tappan, N.Y.*

**J. W. SIMMONS,**
**FNAWCC**
*Eagle Rock, Va.*

**JAMES W. KRAUSE**
*Huntington Beach, CA.*

**R. & D. GAENGER**
**FNAWCC**
*E. Williston, N.Y.*

**JACK SWEISSFORD**
**FNAWCC**
*Royersford, Pa.*

**IAN K. ROOME**
*Costa Mesa, Ca.*

**PHIL McDERMOTT**
*St. Louis, Mo.*

**JIM COULSON**
*Knoxville, Tn.*

**ROBERT HANSEN**
*New Hampton, N.Y.*

**JOSEPH BOURELL**
*Chicago, Ill.*

**JUDY CRAMER**
*Alexandria, Va.*

**MARCIA D'ARCANGELO**
*Washington, D.C.*

**Dr. RALSTON READ**
*Vienna, Va.*

**HOWARD CROUTHAMEL**
*Upper Marlboro, Md.*

**JOE SERNAK**
*Union, N.J.*

**JAMES ELLER**
*Greensboro, N.C.*

**ELMER CRUM**
*Skokie, Ill.*

**MARTIN SWETSKY**
*Brooklyn, N.Y.*

**GENE O'BRIEN**
*St. Louis, Mo.*

**TOM MARREN**
*St. Louis, Mo.*

**R. LEON O'BRIANT**
*Raleigh, N.C.*

**DON REINHART**
*Cockeysville, Md.*

**ROBERT BROWN**
*Millington, N.J.*

**LYNN GALLO**
*Lindenhurst, N.Y.*

**ARLO GREER**
*Takoma Park, Md.*

**BETTINA LY**
*Bristol, Va.*

**M.H. RABENECK**
*Kansas City, Mo.*

**JOHN WYATT**
*Herndon, Va.*

# CONTENTS



## SECTION I: History



## SECTION II: Domestic Clocks



## SECTION III: Public Clocks



## SECTION IV: Movements

**Seth Thomas**

# SECTION I

## HISTORY

# SETH THOMAS

**Seth Thomas** is a name that has represented quality and craftsmanship in American clockmaking since the day, one-hundred and seventy-two years ago, when Seth Thomas opened his factory in what was then known as the town of Plymouth Hollow, in Connecticut. The name Seth Thomas has the respect of collectors the world over, and it is representative of the oldest and finest timepieces ever produced in Connecticut.

Born in Wolcott Connecticut on August 19th, 1785, just nine years after the United States won its independence, Seth Thomas acquired a desire for honest, hard work that was and is typical of the farmers and small businessmen in that part of the country. Shortly after completing an apprenticeship as a carpenter, he formed an association with the clockmaker Eli Terry in 1807, and in 1810 he joined with Silas Hoadley to purchase Terry's Factory. Seth Thomas sold his interest in this factory in 1813.

Intent on owning his own business, he shortly thereafter purchased another clock factory in Plymouth Hollow. This business soon prospered and has since produced the fine clocks for which the Seth Thomas Clock Company is so well known. On May 3, 1853, nearing old age, Seth Thomas organized his business into the Seth Thomas Clock Company, a joint stock corporation, in order to assure continuity of operation after his death. He died at the age of seventy-three in Plymouth Hollow, on January 28, 1859. In 1865, six years after his death, the name of the town where he had distinguished himself as a leader and a successful businessman was changed to Thomaston in honor of his memory.

Thomas' sons managed the company after his death, and greatly expanded the business, introducing many new models of clocks. In 1931, the company became a division of General Time Instrument Corporation, subsequently renamed General Time Corporation. In 1970 this corporation became a part of Tally Industries Company. Members of the Seth Thomas family retained leadership of the Seth Thomas Company until the death of Seth E. Thomas, Jr., great grandson of the Company's founder, on June 5, 1932.

# CHRONOLOGY

| | |
|---|---|
| 1785 | August 19: Seth Thomas is born in Wolcott, Connecticut. |
| 1799 | Is apprenticed to learn carpenter and joiner trade. |
| 1807 | Assists clockmaker Eli Terry in Terry's factory in Plymouth. |
| 1808 | April 20: Marries Phelinda Tuttle. |
| 1810 | May 12: Phelinda Tuttle dies. |
| 1810 | July 7: Thomas and Silas Hoadley buy Eli Terry's clock factory. |
| 1813 | December 4: Thomas sells his interest in the clock factory to Hoadley. |
| 1813 | December 9: Thomas buys another clock factory in Plymouth Hollow, which town becomes the site of all future expansion. |
| 1818 | October 2: Enters agreement with Eli Terry to manufacture Terry's new wood movement shelf clock. |
| 1830 | Changes from manufacture of pillar and scroll shelf clock to bronze looking-glass clock. |
| 1842 | Begins producing brass movements. |
| 1845 | Discontinues manufacture of wood movements. |
| 1853 | May 3: Anticipating death, establishes Seth Thomas Clock Commpany, a joint stock corporation. |
| 1859 | January 17: Youngest son, Aaron, becomes president of the company. |
| 1859 | January 28th: Seth Thomas dies at Plymouth Hollow. |
| Ca. 1855-1860 | Spring driven clocks are introduced. |
| Ca. 1862 | Calendar clocks are introduced. |
| 1865 | Plymouth Hollow renamed Thomaston to honor Seth Thomas. |
| 1865 | October 17: Seth Thomas' Sons & Company is organized to manufacture marine or lever movements as well as high quality brass pendulum movements. |
| Ca. 1869 | Begins producing tower clocks. |
| 1879 | Seth Thomas' Sons & Company consolidated with parent company. |
| 1879 | Sharp gothic or steeple clock is introduced (discontinued in 1917). |
| 1879 | Regulator No. 10 is introduced. |
| 1880 | Regulators No. 12, No. 14 and No. 15 are introduced (discontinued 1909-1911). |
| 1884 | Pocket watches are introduced (discontinued in 1915). |
| 1884 | Ship's bell lever clock is introduced. |
| 1884 | Walnut cased "kitchen clocks" are introduced (discontinued in 1909). |
| 1887 | Solid marble cased clocks are introduced (discontinued around 1895). |

1888 Brass movement hall clocks are introduced (most are discontinued by 1911).

1888 High precision observatory clock is introduced (discontinued in 1913).

1890 Black enameled wooden cased clocks introduced (discontinued around 1917).

1890 Oak wood "kitchen clocks" are introduced (discontinued in 1915).

1892 Black enameled iron case clocks are introduced (discontinued around 1895).

1892 William Thomas Woodruff, nephew of Aaron Thomas, becomes president of the company.

1898 Porcelain cased clocks are introduced (discontinued by 1902).

1898 Imitation French "crystal" clocks with beveled glass sides are introduced.

1900 Secondary Master Clocks using electric current are introduced.

1904 Tambour clock is introduced.

1913 New Regulator No. 1 is introduced—Lower glass painted in a Maroon with Gold Gilt Lines.

1913 New Regulator No. 70 is introduced.

1914 Sonora Chimes are introduced.

1917 Last Perpetual Calendar model clock discontinued.

1919 Seth E. Thomas, Jr., Seth Thomas' grandson, becomes president of the company.

1924 Banjo clocks are introduced.

1928 Electric clocks with alternating current are introduced.

1931 January 1: Seth Thomas Clock Company becomes a division of General Time Instruments Corporation, later renamed General Time Corporation.

1932 June 5: Seth E. Thomas, Jr. dies, and the Seth Thomas family no longer has control of the company for the first time since it was founded.

1970 General Time becomes a division of Talley Industries.

1982 In May Seth Thomas plant closes at Thomaston, Connecticut. Operation moves to Georgia.

Factories: THOMASTON, CONN.

# GUIDE USE INSTRUCTIONS

Two years ago, when I began the task of getting together materials for this Guide, I did not anticipate the pleasure my search would bring me, nor did I fully appreciate the enormity of the job before me. I have now reached the point where it is essential to publish the materials gathered in this volume, and to postpone any further supplements to a later date. The next volume on Seth Thomas will contain watches as well as many additional clocks and movements, with particular emphasis on those which are contained in catalogs not represented in this publication. It will also contain the fine clocks manufactured by Seth Thomas' Sons & Company, which was merged with Seth Thomas Clock Company in 1879.

In addition to the illustrations, descriptions and current prices provided in this Guide, I have also included a unique Movements Table. This table will enable you to establish movement numbers for different models, to identify cases which should be used for authentic case/movement matches, and to identify suspected "marriages" of mismatched cases and movements.

The clocks and movements pictured in this Guide have been compiled from original Seth Thomas catalogs, including many rare editions of the 19th century. Use of catalog illustrations have permitted the display of hundreds of different clocks and movements, many more than would be possible if it were necessary to rely solely upon photographs of clocks in private collections. Use of catalog illustrations have also permitted display of modifications and changes in the appearance of many models, with certainty that the changes shown were made by the Company and are not the result of modification or restoration occurring after the clock left the factory.

Illustrated catalogs were issued by Seth Thomas as a means of letting the public know when new models were being introduced. Separate movement catalogs were distributed to dealers, repairmen, and salesmen and also to individual clockmakers who made or bought their own cases and completed assemblies for sale under their private label. These catalogs contain technical specifications and illustrations which are invaluable to the restorer, collector and dealer as a guide in determining model and correctness of a clock or movement in question.

The Guide will confirm model and movement combination and permit you to seek out the proper movement for your case. Before doing this, you may find it helpful to flip through the four sections just to get an idea of the kinds of material contained therein. Particular attention should be given to the Movements Table provided at the beginning of Section 4.

When using the Guide, remember that it is divided into four sections, each containing different types of information:

**Section 1** contains a brief history of Seth Thomas and the Seth Thomas Clock Company. It also contains comments on the use of the Guide.

**Section 2** contains illustrations, descriptions and current prices of Seth Thomas domestic clocks used in the home, office and factory.

**Section 3** contains illustrations, descriptions and current prices of Seth Thomas Public Clocks. This section, unlike Sections 2 and 4, includes both clocks and movements.

**Section 4** contains illustrations, descriptions and current prices of Seth Thomas movements, as well as a Movements Table, which is designed to facilitate comparisons of clocks and their movements.

Note: The dates listed at the top of each page indicate a range of production years of the clocks shown on each page. In most instances a single model was not made during that entire period.

To assist you in finding a particular clock model, Section 2 is presented in alphabetical order, by type (e.g. Alarm Clocks, Banjo Clocks, etc.). Within each type, the models are further identified in chronological order (e.g. 1886-1910, 1895-1906, etc.). Descriptions accompanying the illustrations are the original catalog descriptions, and many include the list price of the model at the time the catalog was issued.

Section 3, Public Clocks, is separated into three categories: Bracket Clocks, Street Clocks, and Tower Clocks. Due to the large number of Tower Clocks presented, the pages containing these clocks are further arranged by model number.

Section 4 begins with the Movements Table. Following this are the movements, grouped chronologically, except that electric and miscellaneous movements are each grouped together.

The 1985 prices contained in the Guide have been compiled from the top horologists in the country and are professional estimates for antique clocks and movements in mint original condition. Please do not confuse the 1985 prices with the originai list prices accompanying many of the illustrations.

Occasionally, you will encounter an early Seth Thomas clock or movement which cannot be identified using this Guide or any other reference materials. Seth Thomas clock production has been essentially continuous since the Company was founded, and factory records for this entire period are not available. Furthermore, catalogs were either not printed or cannot be located for many of the models that were produced; also, a number of clocks were made in small runs or as special order models and were never shown in Trade Catalogs. There are, however, a number of aids to identification of the earlier clock models:

- Brass movements were not introduced until after 1842. Wooden movements indicate the clock was produced before that date.
- Spring driven movements were introduced between 1855 and 1860. Therefore, the presence of a spring driven movement indicates the clock was produced after 1855. The earliest of these have brass springs.
- The name of the town printed on the label with the Seth Thomas name changed over time. "Plymouth" appeared on the earliest clocks, but this changed to "Plymouth Hollow" sometime during the period 1840-45. (The exact date of this transition is not yet known, but it was perhaps a part of the change to brass movements in 1844-45, and it is treated as such herein.). Subsequently there was a change to a third name—from "Plymouth Hollow" to "Thomaston" in 1865.
- In addition to the changes in town names, the name and address of the label makers changed frequently, and since these also appear on the labels, they can be used to date the clock.

I am deeply appreciative of the assistance rendered to me by Chris Bailey, Esq., Director-Curator of the American Clock and Watch Museum and by Paul Heffner, a respected collector and researcher, in connection with this article.

*Tran Duy Ly*

# SETH THOMAS PRODUCTION CHART

| LABEL PRINTER | SETH THOMAS LOCATION | MOVEMENT TYPE | DATES PRODUCED |
|---|---|---|---|
| Hudson & Skinner | Plymouth | Wood | before 1839 |
| L. Skinner | Plymouth<br>Plymouth Hollow | Wood | 1839 - 1845 |
| T. M. Newson | Plymouth Hollow | Brass | 1845 - 1846 |
| Elihu Geer<br>26 State Street | Plymouth Hollow | Brass | 1846 - 1847 |
| Elihu Geer<br>1 State Street | Plymouth Hollow | Brass | 1847 - 1849 |
| Geo. D. Jewett | Plymouth Hollow | Brass | 1852 |
| Elihu Geer<br>10 State Street | Plymouth Hollow | Brass | 1850 - 1855 |
| Case, Tiffany & Co. | Plymouth Hollow | Brass | 1854 - 1858 |
| Elihu Geer<br>16 State Street | Plymouth Hollow | Brass | 1856 - before 1865 |
| Hartford Steam Printing Co. | Plymouth Hollow | Brass | 1860 - before 1865 |
| Case, Lockwood & Co. | Plymouth Hollow<br>Thomaston | Brass | 1859 - after 1865 |
| Francis & Loutrel | Thomaston | Brass | 1863 - unknown |
| Case, Lockwood, Brainard & Co. | Thomaston | Brass | after 1865 |
| Thomaston Express Print | Thomaston | Brass | after 1865 |
| American Printing Co. | Thomaston | Brass | after 1865 |

Use this chart to determine approximate date of manufacture for clocks produced before 1865. (Note differences in printer's address, where included on the chart, since this can be used for dating purposes.)

# SECTION II

## DOMESTIC CLOCKS

# 1886-1910 SETH THOMAS ALARM CLOCKS

$45.00

**ECHO ALARM**

Nickel Plated. With Switch to Shut off Alarm. 4½ inch Dial.

Time .....................List, $1.54
Time Alarm ..............List, 1.60
Time Alarm, 50 in a box, Special Price, List, 1.54

$65.00

**LONG ALARM**

Metal Case, Oxidized Copper Finish. Time Alarm. 4½ inch Dial. Height, 9 inches. The alarm rings about fifteen minutes with one winding, but can be switched off at pleasure.

List, $4.00

$60.00

**TOCSIN ALARM**

Metal Case. Gun Metal, with polished brass embossed front. Time Alarm. 4½ inch Dial. Height, 7½ inches. The alarm rings about five minutes with one winding, but can be switched off at pleasure.

List, $3.00

$50.00

**MODE ALARM**

Embossed Metal Case. Made in two finishes, Gun Metal with Polished Brass Flowers, or Rich Gold Plated and Lacquered. One Day, Lever, Time Alarm. 3 inch Dial. Height, 6½ inches.
Gun Metal, List, $2.50
Gold Plated, List, $3.20

$85.00

1886

MIKADO LEVER

GOLD GILT FRONT AND HANDLE. NICKEL PLATED FRAME AND GLASS SIDES.

1 Day, Time, Alarm.
1 Day, Strike.
3 inch Dial.
Hight, 7 inches.

$45.00

**PRIM**

Rich Gold. One Day, Lever. 3 inch Dial.
Height, 6 inches.
Time, List, $3.50
Time Alarm, List, $3.80

**NUTMEG LEVER, ALARM.**

$70.00

1886

1-day, Time, Alarm, 3-inch Dial........................$2 40

**THE "PILOT" ALARM**

(Patented)

$165.00

**THE "PILOT" ALARM**

A Reminder Alarm Clock for pilot houses or such places where an accurate alarm reminder is desired. Movable dial rim arranged to show at a glance the exact running time for steamer's course. The alarm set from the face, rings at the expiration of the time as set. Easily adjusted from one minute to any number of minutes, or hours and minutes.

Eight inches wide. List, $20.00
Descriptive circular furnished.

# 1895 - 1906 SETH THOMAS ALARM CLOCKS

## LONG ALARM.

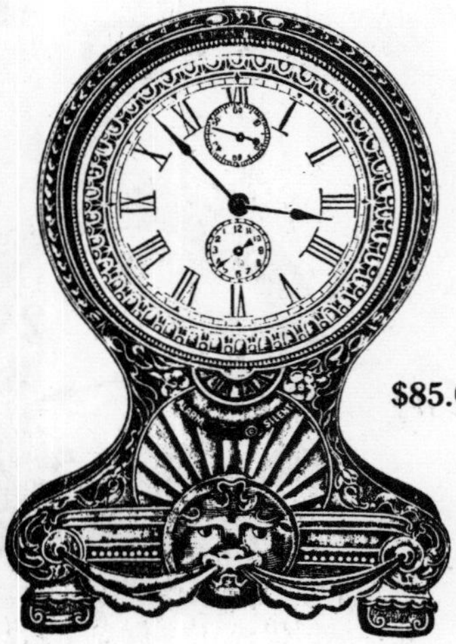

$85.00

1 Day, Time, Alarm.
4½ inch Dial. Hight, 9¼ inches.

The alarm rings twenty minutes with one winding, but can be switched off at pleasure.

## JOKER LEVER.

Gold Gilt Front and Handle. 3-inch Dial.
Nickel Frame and Glass Sides.

$85.00

1 Day, Lever, Time, Alarm. 1 Day, Lever, Strike.
3 inch Dial. Hight, 7 inches.

## ARTIST LEVER.

$85.00

1 Day, Lever, Time, Alarm. 1 Day, Lever, Strike.
3 inch Dial. Hight, 7 inches.

## CRYSTAL

$85.00

1 Day, Lever, Time, Alarm. 1 Day, Lever, Strike.
3 inch Dial. Hight, 7 inches.

$150.00

4½ inch Dial.

## ANVIL, LEVER, ALARM.

NICKEL PLATED.

1-Day Time, Alarm.................$2 40
When the time side only is wound the man is seen at work, the winding of alarm puts him at rest.

## ELK, ALARM.

NICKEL PLATED.

$100.00

1 Day, Lever, Time, Alarm.
4½ inch Dial. Hight, 7½ inches.

$75.00

1906

## ODETTE ALARM

Metal Case. Rich Gold Finish. One Day, Lever, Time Alarm. 3 inch Dial. Height, 10 inches.

Price, $7.00

## LODGE LEVER.

Height, 7 inches. Metal Case. Gold Gilt Front.
Nickel Frame and Sides. 3-inch Dial.

$100.00

1 Day, Lever, Time. 1 Day, Lever, Time, Alarm.
1 Day, Lever, Strike. 3 inch Dial. Hight, 7 inches.

## STUDENT LEVER.

$75.00

GOLD GILT FRONT, NICKEL PLATED FRAME, SIDES AND BELL.

1 Day, Lever, Time Alarm.
3 inch Dial. Hight, 7 inches.

Price, $2.25

# 1910-1913 SETH THOMAS ALARM CLOCKS

## AUTOMATIC EIGHT DAY LONG ALARM CLOCKS

Requires winding but once a week. Does not need any attention at night. Automatically alarms each morning, ringing until you get up and stop it.

**VIGIL**

Metal Case of Artistic Design.
Height, 12½ inches. Dial, 5 inches.
Fine Bronze Finish.
Time-Alarm ............List, $9.50 **$115.00**
Strike and Alarm .......List. 11.50 **$125.00**
Rich Gold Finish Burnished and Lacquered.
Time-Alarm ............List, $11.50
Strike and Alarm .......List, 13.50

**ADVANCE**

Black Adamantine Finish, Green Marbleized Top and Base Moldings, the best imitation of marble ever produced. Guaranteed not to crack, chip or peel. Gold Plated Feet and Ornaments.

Height, 11 inches. Dial, 5 inches.
Time-Alarm ........................List, $7.00 **$75.00**
Strike and Alarm......................List. 9.00 **$100.00**

**$60.00**

**MODERN**

Old Oak and Mahogany Finish.
Height, 9½ inches. Dial, 5 inches.
Time Alarm, List, $5.40

**PROGRESS**

Quartered Oak and Mahogany, Highly Finished and Polished.
Height, 12½ inches. Dial, 5 inches.
Time Alarm.............List, $7.00
Strike and Alarm........List, 9.00

**$75.00**

**$100.00**

**$55.00**

*1913*

**No. 2 EIGHT-DAY AUTOMATIC ALARM MAHOGANY CASE**

Height, 6½ inches. Width, 6½ inches.
Same description as regular Automatic Alarm.
List Price, $13.50.
Also fitted with Radium Treated Dials and Hands.
List Price, $14.75.

**$100.00**

**GRAND**

Metal Case and Dial, Verde Antique Base. Gun Metal Top and Hands. 7 inch Gold Dial, Silver Center and Raised Black Numerals and Minute Dots.
Height, 10 inches. Base, 7½ inches.
Time-Alarm, List, $8.50

*Only clocks in ORIGINAL MINT condition and in good running order will command top prices. A replaced or damaged dial, missing parts or chipped beveled glass will reduce prices accordingly.*

# 1928-1929 SETH THOMAS BANJO CLOCKS

**$100.00**

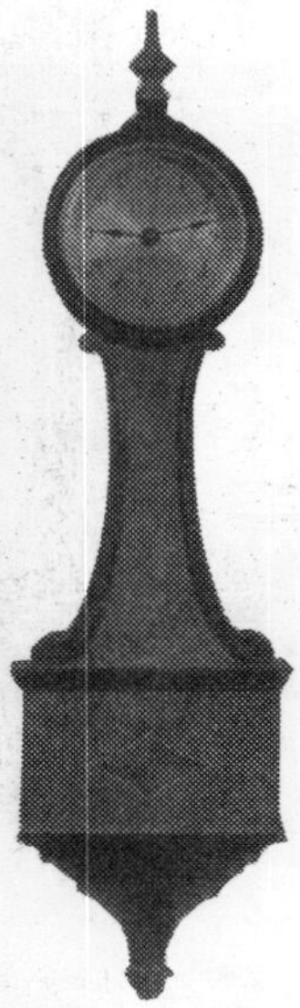

**Pomfret**

Mahogany case with burled mahogany panels. 3-inch silvered dial. Four jeweled, eight-day time movement. Height 19½ inches. Width 5 3/16 inches.

**$175.00**

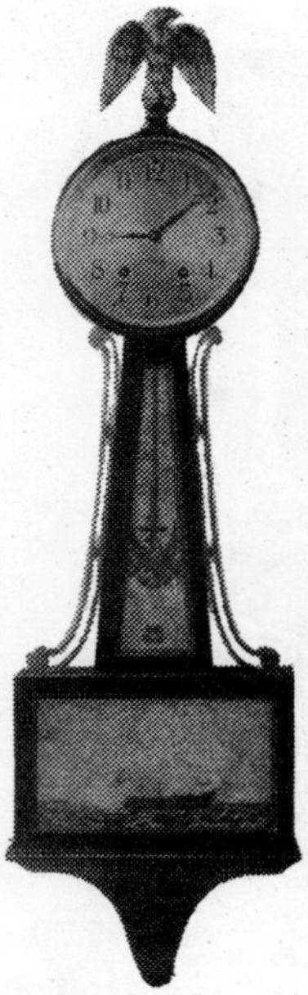

**Banjo No. 2**

***Rod Strike***

Mahogany case with decorated panels. Polished brass side ornaments. 5-inch silvered dial. Eight-day, four-jeweled movement No. 125. Strikes hour and half hour on vertical rod. Height 29 inches. Width 8¾ inches.

**$100.00**

**Deerfield**

Mahogany case. Burled mahogany panel. 3-inch silvered dial. Four-jeweled, eight-day time movement No. 103-A. Height 15½ inches. Width 5 inches.

**$150.00**

**Banjo 5**

Mahogany case. Polished brass side ornaments. 6-inch dial. Eight-day pendulum time movement No. 123. Height 30 inches. Width 9½ inches.

**$150.00**

**Banjo No. 3**

Mahogany case with decorated panels. Polished brass side ornaments. Eight-day pendulum time movement No. 123. 6-inch dial. Height 30 inches. Width 9½ inches. Panel in blue with gold ornamentation.

**$75.00**

**Danvers**

Mahogany case. Gold border and colored flowers in panels. 3-inch silver dial. Eight-day, 4-jewel lever movement No. 103-A. Width 6 inches. Height 18¾ inches.

**Color Banjo Clocks**

*Blackstone* (black case)
*Grenville* (green case)
*Redress* (red case)
Flowers in panels in four colors and gold border. Height 18¾ inches. 3-inch silvered dial. Eight-day, jeweled No. 103-A time movement.

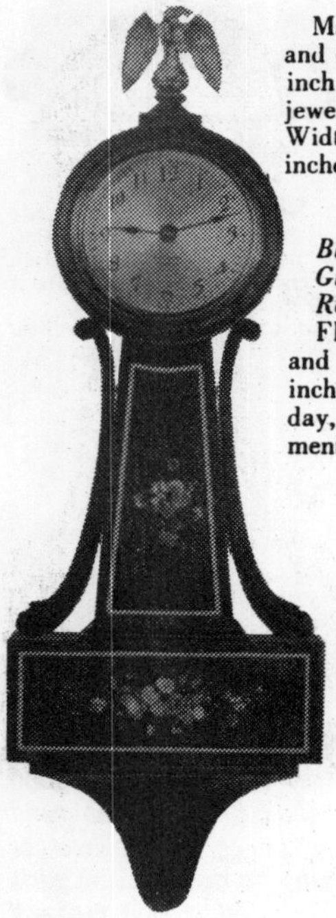

DANVERS

DUFFIELD

**DUFFIELD . . . . . . $40.00**

With self-starting electric time movement

**$75.00**

**BOSTON . . . . . . $50.00**

Same with 8-day, seven jeweled, lever time movement

The design of this fine banjo clock reflects the craftsmanship of Simon Willard. The case is mahogany and has Boston Tea Party panels in full color. 5⅛-inch silverplated dial with Roman numerals in black. Height 30½-inches. Width 8⅞-inches. Thickness 4-inches. Packed singly. Shipping weight 18 pounds.

**$125.00**

**CRANDALL . . . . $27.50**

With 8-day pendulum movement No. 120
Strikes hours and half hours

Reflects the warmth, character and charm invariably associated with Early American architecture. Mahogany case with George Washington-Mount Vernon panels in full color. Polished brass ornaments. 5⅛-inch silverplated dial. Height 28¾-inches. Width 9¾-inches. Thickness 3¾-inches. Packed singly. Shipping weight 11 pounds.

**$175.00**

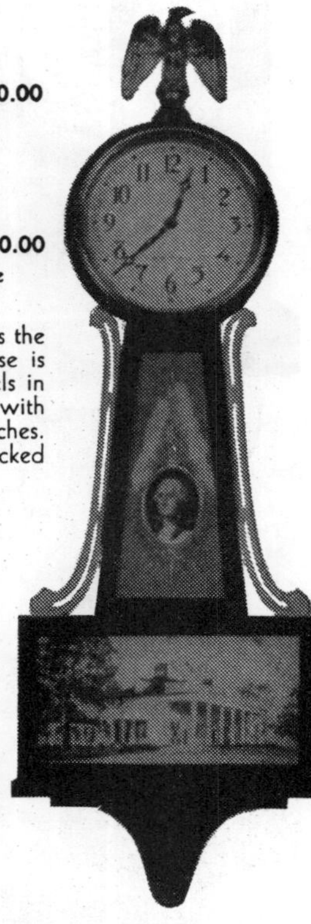

CRANDALL

# 1928-1936 SETH THOMAS BANJO CLOCKS

$450.00

**Banjo No. 1**

Mahogany case and decorated panels. Eight-day pendulum time movement No. 41. Polished brass side ornaments. 7-inch dial. Height 37½ inches. Width 10½ inches. Also fitted with No. 117 full 7-jeweled movement.

$425.00

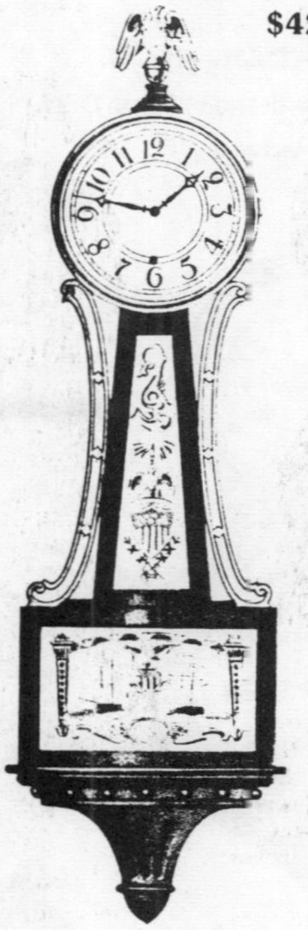

**BANJO No. 1**

Reproduction of original Willard Banjo clock. Mahogany case and decorated panels. Eight day pendulum time movement No. 123. Also fitted with No. 117 fine seven-jeweled lever time movement. 7 inch dial. Height 37½ inches. Width 10½ inches.

$125.00

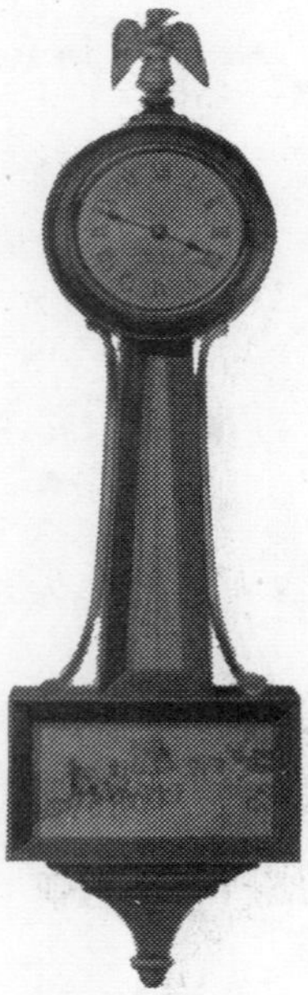

**Ramsgate**

Mahogany case and decorated panels. Polished brass side ornaments. 3-inch silvered dial. Height 21 inches. Width 6 inches. Eight-day, 4-jewel lever time movement No. 103-A.

$75.00

**Electric Banjo 24**

Mahogany case with decorated panels. 5-inch silvered dial. Polished brass side ornaments. Electric time movement. Height 29 inches. Width 8¾ inches.

**MANSFIELD . . . . . $19.95**

With self-starting electric time movement

**$75.00**

**DELAWARE . . . . . $19.95**

Same with 8-day lever time movement No. 5100

Mahogany case with George Washington-Mount Vernon panels in full color. Polished brass ornaments. 3⅜-inch silverplated dial with Roman numerals. Height 23-inches. Width 6⅜-inches. Thickness 2¾-inches. Packed singly. Shipping weight 7 pounds.

$125.00

**$125.00**

**BROOKFIELD . . . . $35.00**

With self-starting electric movement. Strikes hours and half hours

A lovely Colonial clock ideal for both Early American and conventional rooms. Mahogany case with George Washington-Mount Vernon panels in full color. Top and side ornaments in polished brass. 5⅛-inch silverplated dial. Height 28¾-inches. Width 9¾-inches. Thickness 3¾-inches. Packed singly. Shipping weight 11 pounds.

$125.00

**CORNWALL**

*Time only*

| | |
|---|---|
| Case | Mahogany. Marine panels. Polished brass ornaments |
| Dial | 5 inch silverplated |
| Movement | 8-day pendulum time |
| Height | 29 inches |
| Width | 9¾ inches |
| **Retail** . . . . | **$27.50** |

# 1863-1906 SETH THOMAS CALENDAR CLOCKS

$1500.00

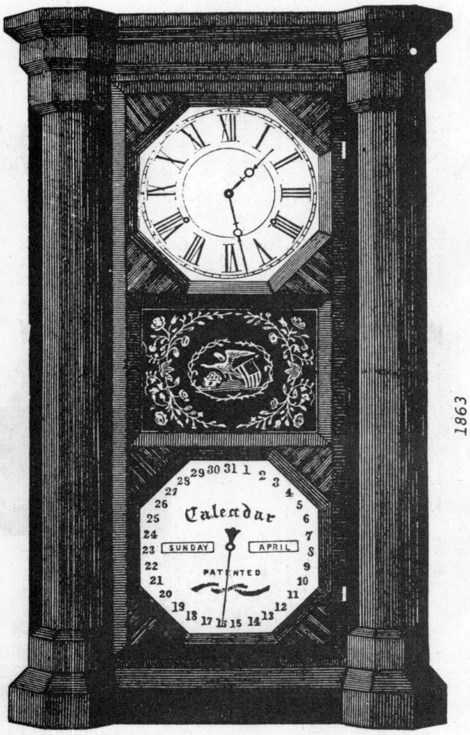


1863

**8 DAY PARLOR CALENDAR.**
Strike - Weight
Height of Clock
30½ inches

$3200.00

**OFFICE CALENDAR.—No. 3.**

8 DAY—SPRING—TIME—6 INCH TIME DIAL.

Hight, 24 inches.

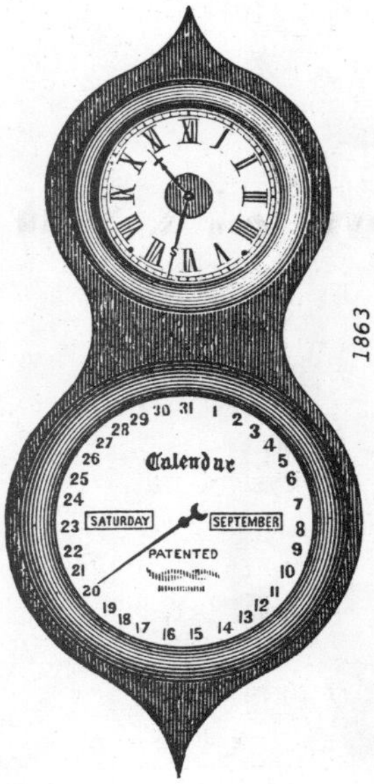


1863

$2750.00

PARLOR CALENDAR No. 10.

WALNUT CASE

1900

8 Day, Weight, Strike, Cathedral Bell.
PERPETUAL CALENDAR
Height, 36 inches. 10 inch Dials.

$2250.00

PARLOR CALENDAR No. 11.

OLD OAK

1900

8 Day, Strike, Cathedral Bell.
Height, 30 inches. 8 inch Dials.
PERPETUAL CALENDAR.

1902

$2500.00

**OFFICE CALENDAR**
**No. 12**

WALNUT, CHERRY OR OAK.

8 Day, Weight, Time.
Hight, 48 inches. 10 inch Dial.
PERPETUAL CALENDAR.
Price, $42.00

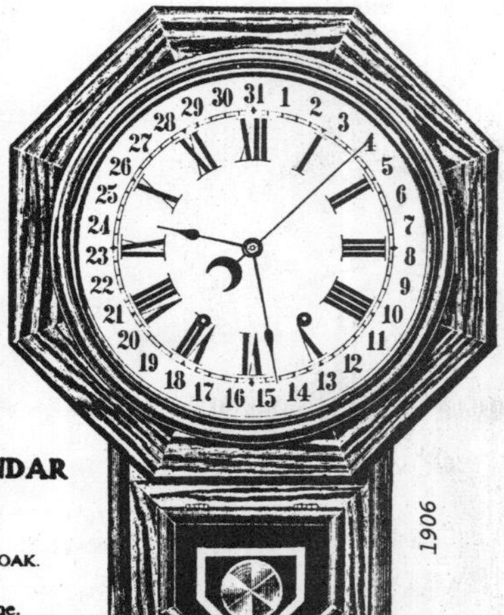

1906

**DROP OCTAGON CALENDAR**
Mahogany Veneer and Old Oak, Polished. 12-inch Dial.
Height, 23½ inches.

| | | |
|---|---|---|
| Time ....................List, $8.30 | | $350.00 |
| Strike ....................List, 8.90 | | $375.00 |

# 1875-1886 SETH THOMAS CALENDAR CLOCKS

**$950.00**

PARLOR CALENDAR No. 1

ROSEWOOD OR WALNUT VENEER, POLISHED.

1875

8 Day, Weight.
Strikes the hour.
Hight, 32 inches.
10 inch Dials.

Price, $25.00. Without Calendar, $18.00.

**$700.00**

PARLOR CALENDAR No. 3

ROSEWOOD AND WALNUT VENEER.

8 Day, Spring, Strike. 8 inch Dials.
Height, 27 inches................$15 00

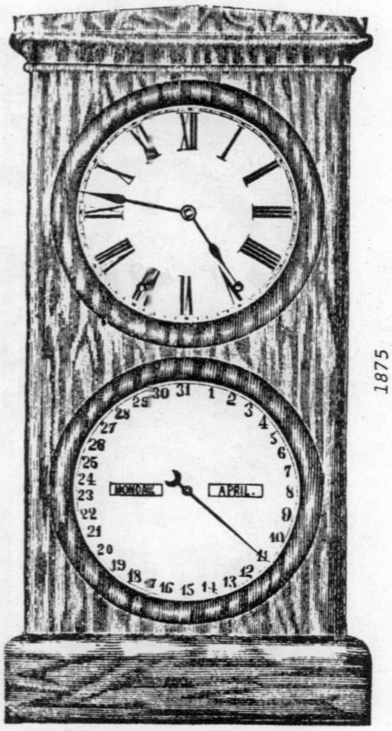


1875

**$1200.00**

**OFFICE CALENDAR—No. 2.**

8 DAY. TIME. WEIGHT. 14 INCH DIAL.
Hight 42 1-2 inches.

1875

PARLOR CALENDAR No. 4.

WALNUT VENEER, POLISHED.

**$1000.00**

8 Day, Spring.
Strikes the hour.
Hight, 25 inches.
8 inch Dials.

Price, $20.00.

OFFICE CALENDAR No. 5.

Walnut case, with polished French walnut trimmings.

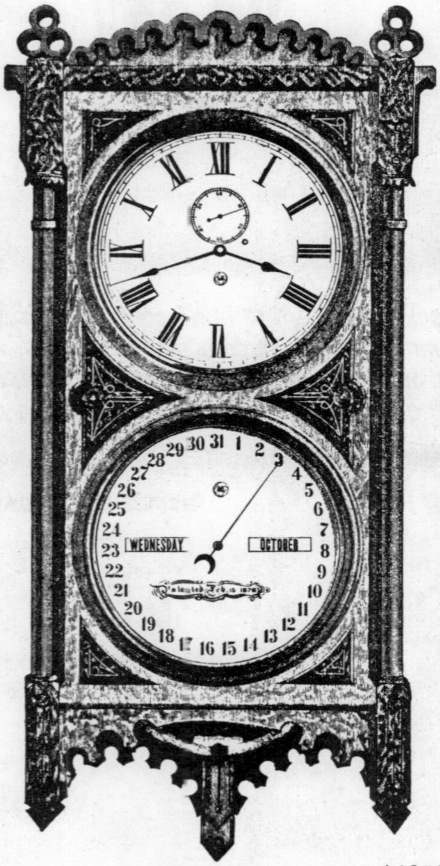


8 Day, Weight, Silent. Hight, 50 inches. 14 inch Dials.

Price, $50.00.

**$4250.00**

**$750.00**

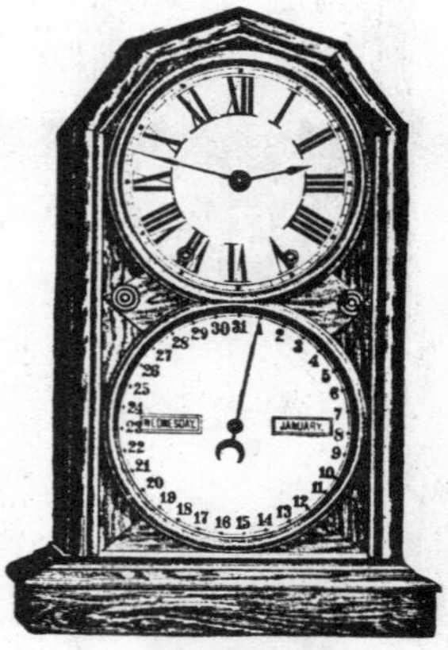

**PARLOR CALENDAR NO. 5.**

POLISHED WALNUT VENEER.

8 Day, Spring, Strike. 8 inch Dials.
Height, 20 inches................$12 75

# 1875-1904 SETH THOMAS CALENDAR CLOCKS

**$2000.00**

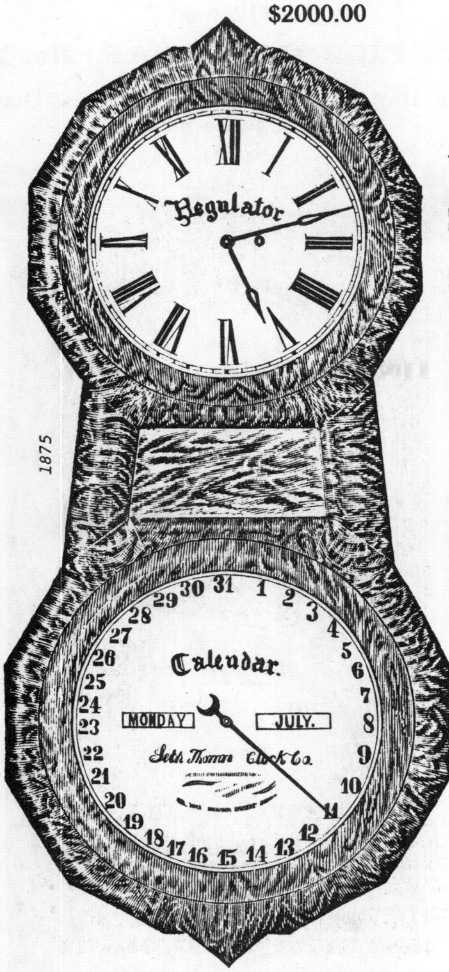


1875

**OFFICE CALENDAR.—No. 1.**

3 Day. Time. Weight. 12 Inch Time Dial.

Hight, 40 inches.

**$9000.00**

OFFICE CALENDAR No. 8.

POLISHED WALNUT CASE.

PERPETUAL CALENDAR.

8 Day, Weight, Time. Beats Seconds.

14 inch Dials. Hight, 66 inches.

**$8000.00**

OFFICE CALENDAR No. 11.

MADE IN WALNUT, OAK AND CHERRY.

8 Day, Weight, Time. Beats Seconds.

14 inch Dials. Hight, 68½ inches.

PERPETUAL CALENDAR.

1904

***Only clocks in ORIGINAL MINT condition and in good running order will command top prices. A replaced or damaged dial, missing parts or chipped beveled glass will reduce prices accordingly.***

OFFICE CALENDAR No. 10.

MADE IN WALNUT, OAK AND CHERRY.

8 Day, Weight, Time.
10 inch Dials.
Hight, 49 inches.

PERPETUAL CALENDAR.

**$2500.00**

**OFFICE CALENDAR No. 10**

WALNUT, CHERRY, OAK OR OLD OAK.

8 Day, Weight, Time.

Hight, 49 inches. 10 inch Dial.

PERPETUAL CALENDAR.

Price, $40.00

**$2500.00**

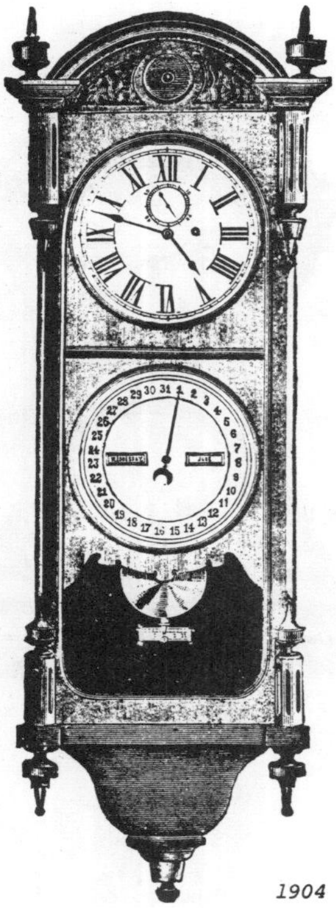

1904

# 1875-1904 SETH THOMAS CALENDAR CLOCKS

**$2250.00**

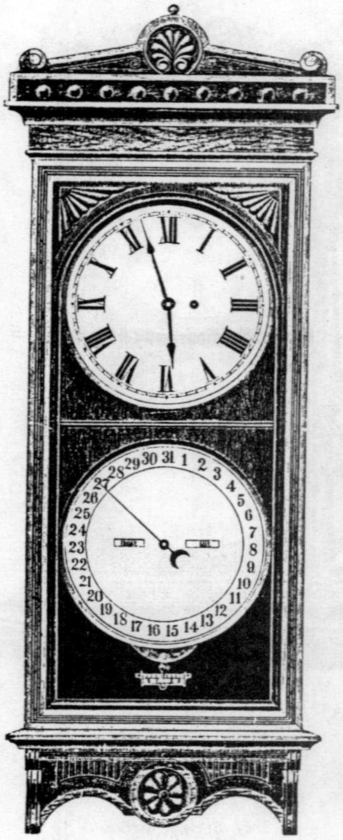

OFFICE CALENDAR No. 13

WALNUT, CHERRY, OAK
OR OLD OAK.

8 Day, Weight, Time.

Hight, 49 inches. 12 inch Dial.

Price, $44.00

**$1100.00**

PARLOR CALENDAR No.7
Hand Carved, Walnut Case
Height 29 inches
8 -inch Dial

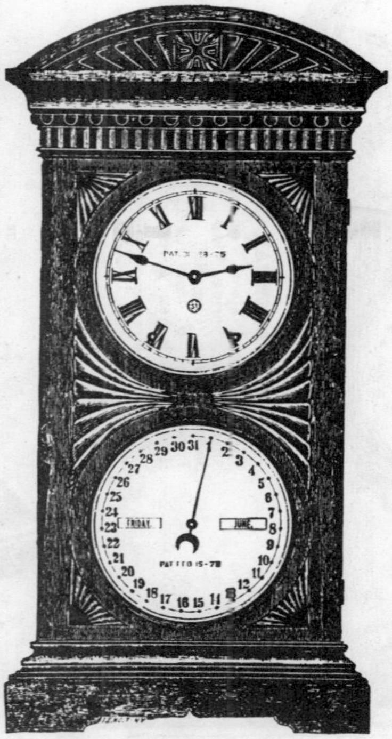

8 Day, Spring, Strike
Perpetual Calendar

*NOTE: Most of the Fashion Calendars were manufactured by the Southern Calendar Clock Company of St. Louis, Missouri. They have been included here because Seth Thomas Clock Company supplied them with the clock movement and the calendar mechanism.*

**$1875.00**

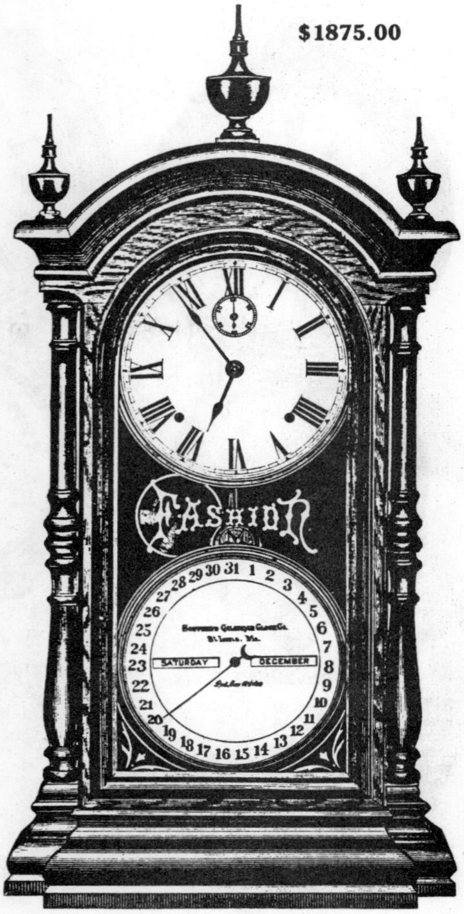


FASHION No.3
Walnut, 8 Day, 7½ inch
Dial. Short Pendulum. Spring
Strike. Hight 32 inches.

**$2700.00**

OFFICE CALENDAR No. 9

WALNUT, CHERRY OR OAK.

8 Day, Weight, Time. Beats Seconds.

Hight, 68 inches. 14 inch Dial.

PERPETUAL CALENDAR.

Price, $83.50

**$2000.00**

FASHION No.4
Walnut, 8 Day, 7½ inch
Dial. Short Pendulum. Spring
Strike. Hight 32 inches

**$2250.00**

FASHION No.5
Walnut, 8 Day, 7½ inch
Dial. Long Pendulum. Spring
Strike. Hight 32 inches

# 1886-1910 SETH THOMAS CALENDAR CLOCKS

**OFFICE CALENDAR No. 6**

Rosewood or Walnut Veneer. Polished. 8 Day, Spring, Time. 8 Day, Spring, Strike, Cup Bell. 12 inch Time Dial. 10 inch Cald. Dial. Height, 32 in. Perpetual Cald.

Time ................List, $22.00 $850.00.
Strike ................List, 24.00 $950.00

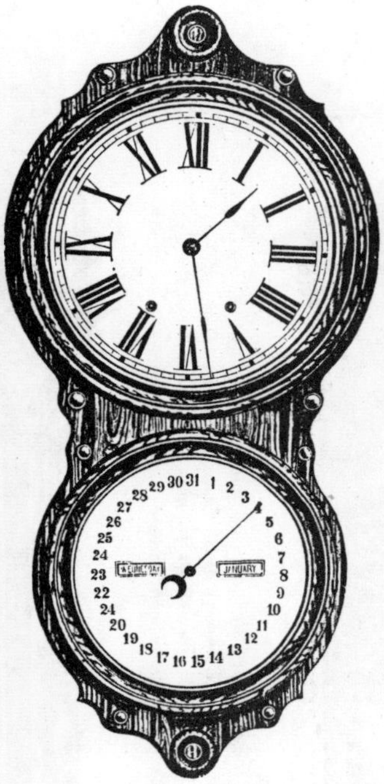

**OFFICE CALENDAR No. 7**

$950.00
$1000.00

**OFFICE CALENDAR No. 7**

Rosewood or Walnut Veneer, Polished. 8 Day, Spring, Time. 8 Day Spring, Strike, Cup Bell. 10 inch Time Dial. 8 inch Cald. Dial. Height, 26½ inches. Perpetual Cald.

Time ................List, $20.00
Strike ................List, 22.00

$2750.00

PARLOR CALENDAR No. 8.

PERPETUAL CALENDAR.

WALNUT CASE.

MAPLE TRIMMINGS.

8 Day, Spring, Strike.

Cathedral Bell.

8 inch Dials.

Hight, 27½ inches.

$875.00

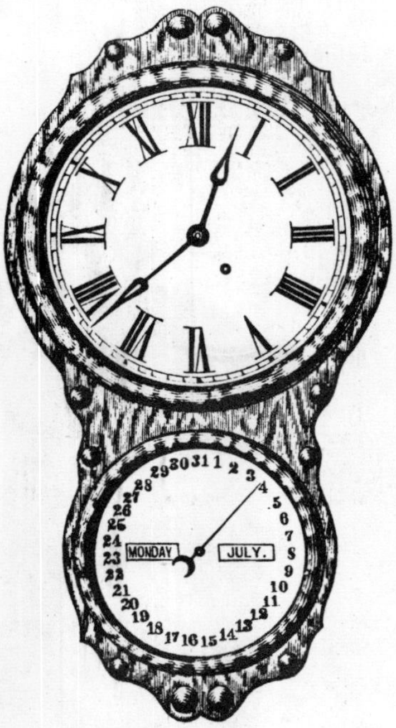


**OFFICE CALENDAR. No. 4.**

12 inch dial. Height, 28 inches.
EIGHT DAY, TIME.

$3500.00

PARLOR CALENDAR No. 9.

WALNUT CASE.

8 Day, Spring, Strike.

Cathedral Bell.

8 inch Dials.

Hight, 29½ inches.

PERPETUAL CALENDAR.

$950.00

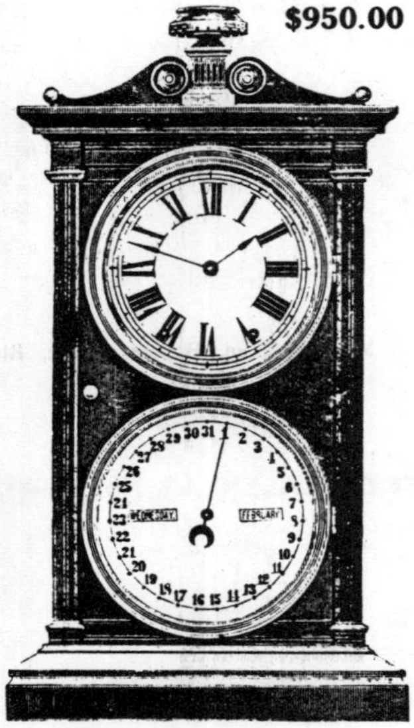

**PARLOR CALENDAR No. 6**

Walnut Veneer, Polished. Eight Day. Spring Strike, Cup Bell. 8 inch Dials. Height, 27 inches. Perpetual Calendar.

List, $20.00

# 1906 -1910 SETH THOMAS CARRIGAGE CLOCKS

**$425.00**

**CARRIAGE No. 1**

Gilt Metal Case, with Beveled Glass Sides and Top. Eight Day, Lever, Time, Porcelain Dial, Fine Polished Cut Steel Pinion Movement with Compensating Horizontal Balance; Jeweled Escapement. Supplied, fitted in Leatherette Case 5½ inches high, 4 inches wide, 4 inches deep. An accurate timepiece. Height of Clock, with handle up, 6 inches.

Price, **$22.70**

**$185.00**

**CARRIAGE** No. 6

Polished Gold.
Eight Day Time. 3 inch Etched Metal Silvered Dial.
Height, including handle, 9 inches.
List. **$35.00**

**$150.00**

**CARRIAGE No. 5**

Rich Gold Metal Case, with Beveled Glass and Top. Eight Day, Lever, Time, Porcelain Dial, Fine Polished Cut Steel Pinion Movement with Compensating Horizontal Balance; Jeweled Escapement. An accurate timepiece. Height of Clock, with handle up, 7¼ inches.

Price, **$31.50**

**COMPANION**

A noiseless timepiece, consisting of a nickel Seth Thomas seven-jeweled watch movement, having Breguet hairspring and cut compensating balance, fitted in a metal block, held in a leather case, made up in the best possible manner. Stem winding and setting, Porcelain dial, Roman or Arabic, with heavy numerals and hands. (Arabic will be sent unless Roman specified.) An accurate timepiece.

Made in Red, Brown, Black, Blue and Light Green Genuine Turkey Morocco, and English Pig.

List, **$8.00**

Made in Brown, Black and Dark Green Genuine Seal and Red Levant.

List, **$9.00**

Made in Genuine Alligator and Lizard.

List, **$10.00**

**COMPANION**

**$65.00**

**$185.00**

**CARRIAGE** No. 7

Rich Gold.
Eight Day Time. 3 inch Etched Metal Silvered Dial.
Height, including handle, 9 inches.
List, **$40.00**

# 1910—1915 SETH THOMAS CHIME CLOCKS

## SONORA CHIME CLOCKS

*Only clocks in ORIGINAL MINT condition and in good running order will command top prices. A replaced or damaged dial, missing parts or chipped beveled glass will reduce prices accordingly.*

**$325.00**

**CHIME CLOCK No. 1**

Adamantine Mahogany Finish.

A durable surface that can be polished when necessary, and will always look like fine figured mahogany. Gold gilt metal columns, brass feet.

Height, 12½ inches. Base, 16 inches. List, $25.00

**$350.00**

**CHIME CLOCK No. 4**

Adamantine Mahogany Finish.

A durable surface that can be polished when necessary, and will always look like fine figured mahogany. Large celluloid onyx columns, gold gilt metal caps, bases and feet.

Height, 14 inches. Base, 19 inches. List, $30.00

**$350.00**

**CHIME CLOCK No. 7**

Adamantine Black Marble Finish.

A durable surface that can be polished when necessary, and will always look like fine marble. Adamantine fancy marble finish blocks, celluloid onyx columns, bronze finish metal caps, bases and feet.

Height, 13 inches. Base, 19 inches. List, $30.00

**$350.00**

**CHIME CLOCK No. 10**

Adamantine Black Marble Finish.

A durable surface that can be polished when necessary, and will always look like fine marble. Adamantine fancy marble finish top, and front blocks. Adamantine black marble finish side mouldings and dome tops. Large celluloid onyx columns in front, and two large columns on each side. All columns have gold gilt metal caps and bases. Gold gilt large ornamental feet.

Height, 12½ inches. Base, 20 inches. List, $35.00

**$425.00**

**CHIME CLOCK No. 12**

Early English Oak.

Eight inch, Convex, Porcelain Dial, Cast Sash, with Convex Beveled Glass.

Height, 16 inches. Depth, 8½ inches. List, $35.00

**$275.00**

**CHIME CLOCK No. 2**

Large deep-toned bells; 6-inch Silvered Dial; Convex Beveled, Glass; Cast Brass Sash, Mat and Feet.

Height, 14½ inches.
List, $30.00

**$300.00**

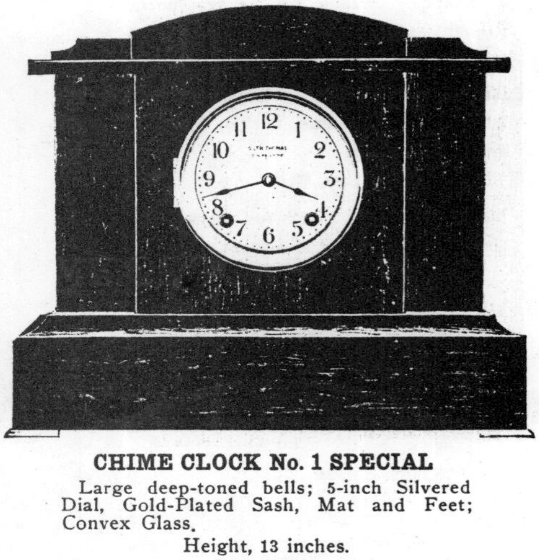

**CHIME CLOCK No. 1 SPECIAL**

Large deep-toned bells; 5-inch Silvered Dial, Gold-Plated Sash, Mat and Feet; Convex Glass.

Height, 13 inches.
List, $25.00

# 1913 SETH THOMAS CHIME CLOCKS

## WESTMINSTER CHIME CLOCKS
## AND WHITTINGTON AND WESTMINSTER CHIME CLOCKS

**ON PATENTED "SONORA" RESONATED BELLS.**

Nos. 17, 19 and 21 play the Westminster Chime every 15 minutes. Nos. 217, 219 and 221 are arranged to play either the Whittington or Westminster every 15 minutes. Both Chimes can be silenced, leaving the hour only striking. All these Cases are fitted with Convex Silvered Dials with Raised Bronze Numerals and Cut Serpentine Hands.

**CHIME CLOCK No. 21** **$425.00**

Real Bronze Case.
8-inch Dial.
Height, 12½ inches.
Westminster Quarter-Hour Chime on four large deep-toned Bells.
List, $80.00

**CHIME CLOCK No. 221** **$650.00**

Same Case as No. 21.
Whittington and Westminster Quarter-Hour Chime on eight large deep-toned Bells.
List, $100.00

*4 Bell:* **$475.00**
*5 Bell:* **$550.00**
*8 Bell:* **$800.00**

**CHIME CLOCK No. 19** **$475.00**

Real Bronze Case.
6-inch Dial. Height, 15 inches.
Westminster Quarter-Hour Chime.
Five Bells. List, $65.00

**CHIME CLOCK No. 219** **$800.00**

Same Case as No. 19.
Whittington and Westminster
Quarter-Hour Chime.
Eight Bells. List, $85.00

**$475.00**

**CHIME CLOCK No. 17**

Real Bronze Case.
6-inch Dial. Height, 13 inches.
Westminster Quarter-Hour Chime.
Five Bells. List, $65.00

**CHIME CLOCK No. 217** **$800.00**

Same Case as No. 17.
Whittington and Westminster
Quarter-Hour Chime.
Eight Bells. List, $85.00

*Only clocks in ORIGINAL MINT condition and in good running order will command top prices. A replaced or damaged dial, missing parts or chipped beveled glass will reduce prices accordingly.*

**$375.00**

**CHIME CLOCK No. 16**

Mahogany, Fine Figured Pattern.
Eight inch, Convex Porcelain Dial, Cast Sash, with Convex Beveled Glass.
Height, 18 inches. Depth, 8½ inches. List, $40.00

**$450.00**

**CHIME CLOCK No. 18**

Real Bronze.
Eight inch, Convex, Silvered Dial, with Raised Numerals and Minute Dots. Cast Sash, with Convex, Beveled Glass.
Height, 16 inches. Depth, 8½ inches. List, $75.00

**$475.00**

**CHIME CLOCK No. 20**

Real Bronze.
Eight inch, Convex, Silvered Dial, with Raised Numerals and Minute Dots. Cast Sash, with Convex Beveled Glass.
Height, 18 inches. Depth, 8½ inches. List, $75.00

# 1913—1923 SETH THOMAS CHIME CLOCKS

$500.00

**CHIME CLOCK No. 103**
8-inch Dial Colored in Antique Finish with Ornamental Corners. Fine Mahogany Burl Panel.
Length, 31 inches. Width, 12 inches.
Depth, 8 inches.
List Price, **$100.00.**

$350.00

**CHIME CLOCK No. 101**
8¼-inch Silvered Dial.
Length, 28 inches. Width, 13 inches.
Depth, 7½ inches.
List Price, **$75.00.**

$375.00

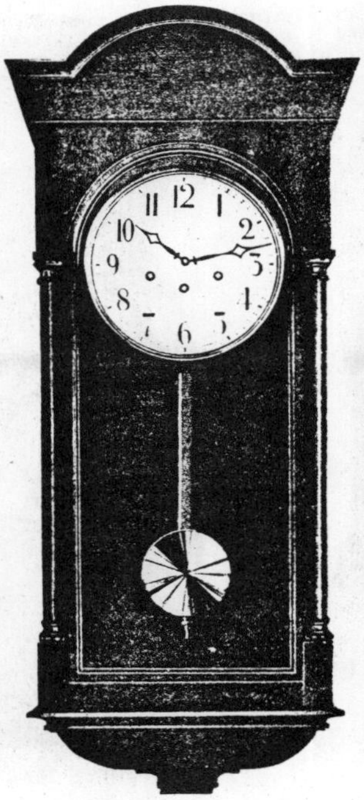

**CHIME CLOCK No. 102**
8-inch Silvered Dial.
Length, 31 inches. Width, 13 inches.
Depth, 7½ inches.
List Price, **$90.00.**

$400.00

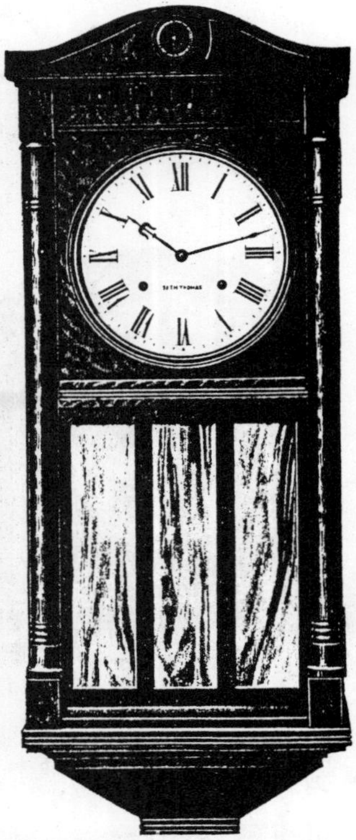

**HANGING CHIME CLOCK No. 110**
Westminster Quarter-Hour Chime. Mahogany and Golden Oak. 8-inch White Dial. Colored Glass in door. Brass Grill Work in sides of case.
Height, 33 inches. List, **$50.00**

**CHIME CLOCK No. 210**
Same Case as No. 110.
Whittington and Westminster Quarter-Hour Chime on 8 Bells.
List, **$75.00**

$600.00

## WESTMINSTER CHIME CLOCKS

Mahogany Cabinet Cases and Highest Grade 3-Train Movement No. 113 Chiming each Quarter and Striking the Hour on Five Perfectly Tuned Rods.

$150.00

Chime 96
*Westminster Chime*
Mahogany case with inlaid marquetry. No. 124 chime movement. Raised bronze numerals on 5 inch silvered dial. Height 10⅜ inches. Base 8¼ inches. Depth 5⅝ inches.

$150.00

**CHIME No. 95**
Two-tone Mahogany Cabinet Case. Old English design. 5-inch silvered dial. Also with raised bronze numerals on either silvered or gold dial. Height 11 inches. Base 7¼ inches.

$110.00

**CHIME CLOCK No. 59**
Mahogany Cabinet Case. Six-inch Convex Silvered Dial, Cast Brass Sash with Silvered Mat and Convex Beveled Glass.
Height, 11½ inches. Length, 18¼ inches.
List Price, **$65.00.**

$125.00

Chime 100
*Westminster Chime*
American Walnut case with burled panels. No. 124 chime movement. Raised bronze numerals on 5-inch silvered dial. Height 12 inches. Base 10 inches. Depth 6⅛ inches.

# 1913-1928 SETH THOMAS CHIME CLOCKS

$350.00

**CHIME CLOCK No. 66**
Westminster Quarter-Hour Chime on 4 deep-toned Bells. 8-inch Dial.
Height, 18 inches. List, **$50.00**

**CHIME CLOCK No. 266** **$750.00**
Same Case at No. 66.
Whittington and Westminster Quarter-Hour Chime, 8 deep-toned bells. List, **$75.00**

$185.00

**CHIME CLOCK No. 6**
5-inch Silvered Dial, Gold-Plated Sash, Mat, Caps, Bases, Feet and gilt lines in columns; Convex Glass.
Height, 14 inches.
List, **$25.00**

$375.00

**CHIME CLOCK No. 50**
Westminster Quarter-Hour Chime. Five Bells. Mahogany Fine-Figured Pattern. Hand-Carved, Highest Grade Workmanship. 6-inch Dial.
Height, 15½ inches.
List, **$45.00**

$250.00

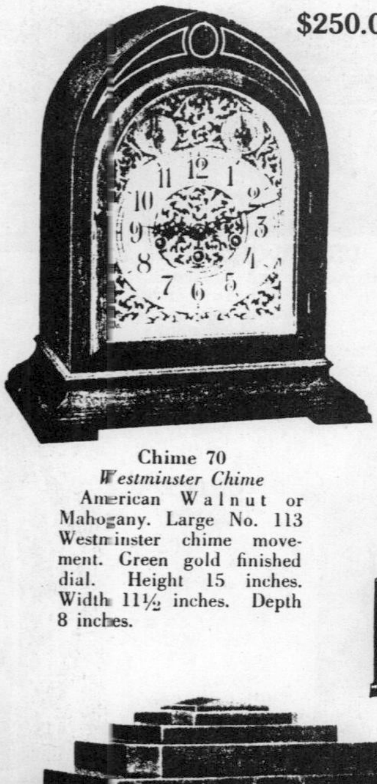

Chime 70
*Westminster Chime*
American Walnut or Mahogany. Large No. 113 Westminster chime movement. Green gold finished dial. Height 15 inches. Width 11½ inches. Depth 8 inches.

$250.00

**CHIME CLOCK No. 51**
Westminster Quarter-Hour Chime on 4 large deep-toned Bells. Golden Oak, Fumed Oak and Early English Oak. 6-inch Dial.
Height, 14 inches.
List, **$32.50**

$375.00

**SULGRAVE**
*Westminster Chime*

**There is a substantial atmosphere about this attractively designed model. The case is in walnut and with the twisted columns and carved grill work it presents an imposing appearance. Silver dial with gold decorations. Finest grade No. 113 chime movement. Height 14⅜ inches, Base 13¼ inches. Depth 8¼ inches. $145.00**

$300.00

**ASTOR**
*Westminster Chime*

**Present day tendency in architecture is evident in this massive chime clock. American Walnut case. Silver dial. Finest grade No. 113 movement. Deep throated rods announce each quarter hour. Height 16½ inches. Base 17½ inches. Depth 8¾ inches. $135.00**

# 1915 SETH THOMAS CHIME CLOCKS

## WHITTINGTON AND WESTMINSTER CHIME CLOCKS

**ON EIGHT LARGE DEEP-TONED "SONORA" PATENTED RESONATED BELLS.**

**$900.00**

*To receive top prices, clocks must be in original mint condition and in good running order — A replacement dial or missing parts will reduce prices accordingly.*

***4 Bell:* $525.00**

***5 Bell:* $635.00**

***8 Bell:* $900.00**

Mahogany cases, fine figured grain. Highest grade cabinet work. Dull rubbed finish.

Chimes every fifteen minutes and can be switched to play either Whittington or Westminster by turning an indicator over dial. There is another indicator over dial to silence strike if desired. Chimes can be silenced by a switch at side of case.

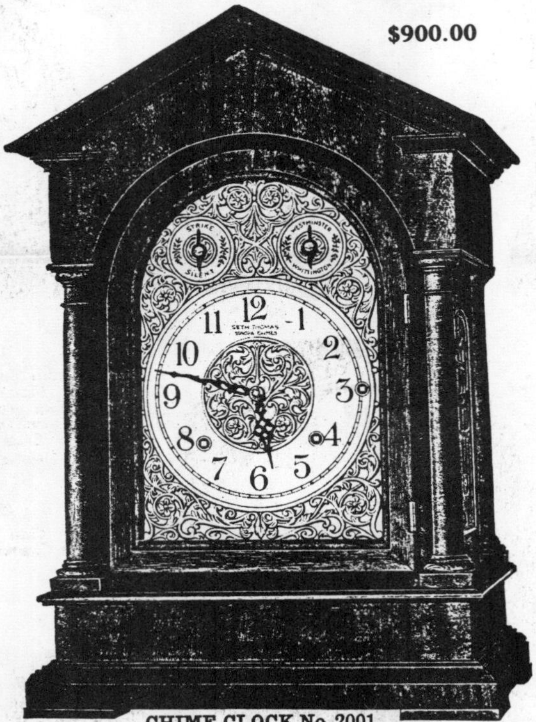

The dials have 5½-inch etched minute circle and dull gold-plated cast pierced open work on a solid metal plate that is dull silver finished, making a pleasing contrast.

The doors are fitted with heavy beveled plate glass.

All patterns have Mahogany grille work at sides, with the exception of the No. 2003, which has metal grilled sides gold-plated, as shown in cut.

**CHIME CLOCK No. 2001**
Whittington and Westminster Chimes.
Height, 18½ inches.
Width, 12½ inches.
Depth, 9 inches.
Plain Silvered Dial..List, $70.00
Fancy Pierced Dial..List, 80.00

**$900.00**

**CHIME CLOCK No. 2000**
Whittington and Westminster Chimes.
Height, 16 inches. Width, 12 inches. Depth, 9 inches.
Plain Silvered Dial..List, $70.00
Fancy Pierced Dial..List. 80.00

**CHIME CLOCK No. 2002**

Same as No. 2003, but Plain Mahogany, with Mahogany Grill Sides.
Plain Silvered Dial.
List, $65.00
Fancy Pierced Dial.
List, $75.00

**$800.00**

**$900.00**

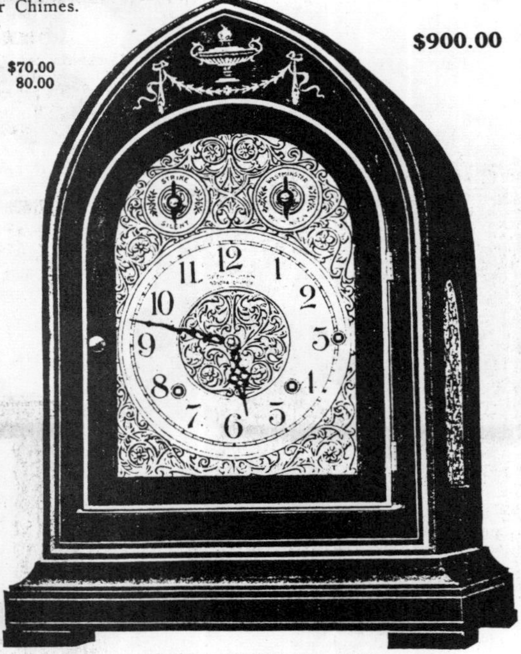

**CHIME CLOCK No. 2003**
Whittington and Westminster Chimes.
Height, 18 inches. Width, 12½ inches. Depth, 9½ inches.
Gold-Plated Metal Grill Sides.
Plain Silvered Dial..List, $70.00
Fancy Pierced Dial..List. 80.00

# 1915 SETH THOMAS CHIME CLOCKS

**WESTMINSTER CHIME CLOCKS AND WHITTINGTON AND WESTMINSTER CHIME CLOCKS**

ON PATENTED "SONORA" RESONATED BELLS.

Cabinet Cases, Dull Rubbed Finish.

**$325.00**

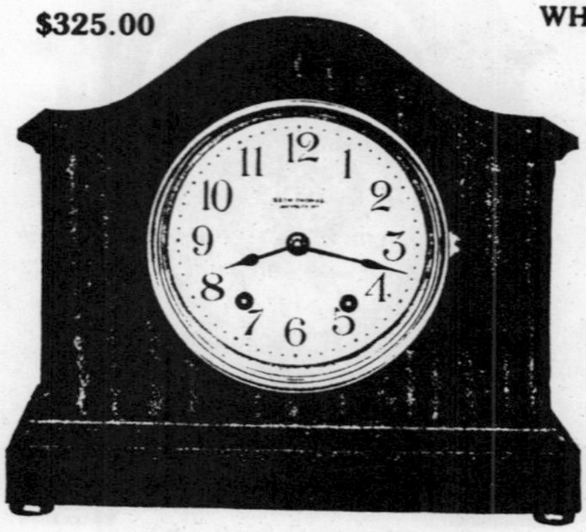

**$500.00**

**CHIME CLOCK No. 55**

Westminster Quarter-Hour Chime on 4 Bells. 5-inch Dial. Brass Sash with Convexed Beveled Glass.

Height, 9½ inches.
Base, 10½ inches.
List, **$30.00**

**CHIME CLOCK No. 255**

Same Case as No. 55.
Whittington and Westminster Quarter-Hour Chime. 8 Bells.
List, **$50.00**

**$650.00**

**CHIME CLOCK No. 58**

Westminster Quarter-Hour Chime. 5 Bells. Mahogany Fine-Figured Pattern. Hand-Carved Case of perfect proportions. 8-inch Convex Silver Dial with Raised Bronze Numerals and Cut Serpentine Hands. Brass Sash with Convex Beveled Glass.

Height, 14 inches. Base, 32½ inches.
List, **$100.00**

**CHIME CLOCK No. 258**

Same Case as No. 58. Whittington and Westminster Quarter-Hour Chime on 8 Bells.
List, **$125.00**

**$850.00**

**CHIME CLOCK No. 57** **$350.00**

Westminster Quarter-Hour Chime on 4 Bells. 6-inch Convex Silvered Dial. Brass Sash with Convex Beveled Glass. Mahogany Fine Figured Pattern.

Height, 11 inches.
Base, 17½ inches.
List, **$35.00**

**CHIME CLOCK No. 257** **$650.00**

Same Case as No. 57. Whittington and Westminster Quarter-Hour Chime on 8 Bells.

List, **$55.00**

**$285.00**

**CHIME CLOCK No. 5**

5-inch Silvered Dial, Gold-Plated Sash, Mat and Feet; Mahogany Adamantine Caps and Columns with black lines in columns; Convex Glass.

Height, 14 inches.
List, **$25.00**

**$325.00**

**CHIME CLOCK No. 00**

5-inch Silvered Dial. Gilt Ornamental Corners, Gilt Feet.

Height, 9½ inches. Length, 11½ inches.
List, **$20.00**

**$275.00**

**CHIME CLOCK No. 7**

5-inch Silvered Dial, Gold-Plated Sash, Mat and Feet; Mahogany Adamantine Caps and Columns with black lines in columns; Convex Glass.

Height, 13 inches.
List, **$25.00**

# 1920 - 1922 SETH THOMAS CHIME CLOCKS

**$325.00**

**CHIME CLOCK No. 71**
Mahogany Cabinet Case. Silvered Dial.
Height, 13⅜ inches. Width, 10¼ inches.
Depth, 7¼ inches.
With Silvered Dial. List Price, $80.00.

**$325.00**

**CHIME CLOCK No. 72**
Mahogany Cabinet Case. Silvered Dial.
Height, 14¾ inches. Width, 10¼ inches. Depth, 7¼ inches.
With Silvered Dial. List Price, $80.00.

**$450.00**

**CHIME CLOCK No. 73**
Mahogany Cabinet Case. Silvered Dial.
Height, 13⅞ inches. Width, 10¾ inches.
Depth, 7⅛ inches.
With Silvered Dial. List Price, $110.00.

The above three patterns also furnished with fancy Pierced Dial, Hand-Cut, Hand-Engraved and Gold Plated as shown on Chime No. 73. List extra, $20.00.

Convexed Etched Silvered Dials. Brass Sash with Convexed Beveled Glass.

**CHIME CLOCK No. 61**
Westminster Quarter-Hour Chime. 5 Bells. Mahogany, Shaded Marqueterie. 6-inch Dial. Height 13¼ inches. List, $38.00

**CHIME CLOCK No. 261**
Same Case as No. 61.
Whittington and Westminster Quarter-Hour Chime on 8 Bells. List, $58.00

**$400.00**

**$750.00**

## WESTMINSTER CHIME CLOCKS AND WHITTINGTON AND WESTMINSTER CHIME CLOCKS

**ON PATENTED "SONORA" RESONATED BELLS.**

**$350.00**

**CHIME CLOCK No. 14**
Westminster Quarter-Hour Chime. 5 Bells. Mahogany Fine-Figured Pattern, 6-inch Dial.
Height, 14 inches. List, $35.00

**CHIME CLOCK No. 214**
Same Case at No. 14.
Whittington and Westminster Quarter-Hour Chime on 8 Bells. List, $55.00

**$700.00**

**$400.00**

**CHIME CLOCK No. 64**
Westminster Quarter-Hour Chime. 5 Bells. Mahogany, Shaded Marqueterie. 6-inch Dial. Height, 14 inches. List, $38.00

**CHIME CLOCK No. 264**
Same Case as No. 64.
Whittington and Westminster Quarter-Hour Chime on 8 Bells. List, $58.00

**$750.00**

Nos. 11, 14, 61 and 64 play the Westminster Chime every 15 minutes. Nos. 211, 214 261 and 264 are arranged to play either the Whittington or Westminster Chime every 15 minutes Both chimes can be silenced, leaving the hour only striking.

**Cabinet Cases. Dull Rubbed Finish.**

**$350.00**

**CHIME CLOCK No. 11**
Westminster Quarter-Hour Chime. 5 Bells. Mahogany Fine-Figured Pattern. 6-inch Dial.
Height, 13¼ inches. List, $35.00

**CHIME CLOCK No. 211**
Same Case as No. 11.
Whittington and Westminster Quarter-Hour Chime on 8 Bells. List, $55.00

**$700.00**

# 1922-1928 SETH THOMAS CHIME CLOCKS

**Highest** Grade Mahogany Tambour Cases, 6-inch Silvered Convex Dial, Dull Gold Cast Sash, Silvered Mat, Bevelled Convex Glass. Also with Raised Bronze Numerals. Three-Train Chime Movement No. 113 on Perfectly Tuned Rods.

*Westminster Chime Clocks*

**$125.00**

**CHIME CLOCK No. 74**
Height, 10 inches. Base, 20¾ inches.
Depth, 6¾ inches.
List Price, Silvered Dial, **$85.00.**
With Raised Bronze Numerals, **$91.00.**

**$125.00**

Chime 99
*Westminster Chime*
American Walnut or Mahogany. 5-inch silvered dial with raised bronze numerals. Base 20½ inches. Height 9½ inches. Depth 6⅛ inches.

**$125.00**

**CHIME CLOCK No. 75**
Height, 9¾ inches. Base, 22½ inches.
Depth, 6⅞ inches.
List Price, Silvered Dial, **$90.00.**
With Raised Bronze Numerals, **$96.00.**

**$125.00**

Chime 98
*Westminster Chime*
Mahogany. No. 124 chime movement. 5 inch silvered dial with raised bronze numerals. Base 20½ inches. Height 9½ inches. Depth 5⅞ inches.

**$125.00**

**CHIME CLOCK No. 77**
Height, 10 inches. Base, 20½ inches.
Depth, 6¾ inches.
List Price, Silvered Dial, **$75.00.**
With Raised Bronze Numerals, **$81.00.**

**$125.00**

Chime 80
*Westminster Chime*
Mahogany. Large No. 113 Westminster chime movement. 6-inch silvered dial. Also with raised bronze numerals. Height 10¼ inches. Base 20¾ inches. Depth 7 inches.

**$125.00**

Chime 90
*Westminster Chime*
Mahogany. Westminster chime movement No. 124. Raised bronze numerals on 5 inch silvered dial. Height 9 inches. Base 21¼ inches. Depth 5⅝ inches.

**$125.00**

Chime 79
*Westminster Chime*
Mahogany case with contrasting wood panels. Large No. 113 movement. 6-inch silvered dial with raised bronze numerals. Base 20¼ inches. Height 9¼ inches. Depth 6½ inches.

**$125.00**

Chime 97
*Westminster Chime*
Mahogany. No. 124 chime movement. 5 inch silvered dial with raised bronze numerals. Base 19¼ inches. Height 8¾ inches. Depth 6 inches.

# 1875-1894 SETH THOMAS CITY CLOCKS

$275.00

NORMANDY.

CABINET CASE, WALNUT.

15 Day, Half Hour Strike.
Cathedral Bell.
4½ inch Porcelain Dial.
French Sash, with Beveled Glass.
Hight, 15 inches.

$275.00

OMAHA.

MADE IN WALNUT, EBONY AND OAK.

8 Day, Spring, Strike.
Cathedral Bell.
6 inch Dial.
Hight, 19 inches.

$275.00

**CINCINNATI.**

8 DAY. SPRING. STRIKE.
Hight, 16 inches.

VICTORIA.

V. P.

8 Day $200.00
30 Hour $140.00

8 Day, Spring, Strike.
Hight, 18 inches.

CLINTON. $165.00

POLISHED OLD OAK, BRONZED METAL TRIMMINGS.

5 inch Porcelain Dial.
Hight, 10 inches.
8 Day, Half Hour Strike,
Cathedral Bell.

**VICTORIA.**

8 Day $185.00
30 Hour $135.00

8 DAY OR 30 HOUR.
SPRING. STRIKE.
Hight, 15 1-2 inches.

*8 Day* $215.00
*30 Hour* $150.00

ATHENS.

WALNUT CASE.

1 Day, Spring, Strike.
8 Day, Spring, Strike.
Wire Bell.
6 inch Dial.
Hight, 17 inches.

$375.00

**PITTSBURGH, V. P.**

8 Day. Strike. Spring.
Hight, 23 inches

SALEM. $165.00

POLISHED OLD OAK, BRONZED METAL TRIMMINGS.

5 inch Porcelain Dial.
8 Day, Half Hour Strike,
Cathedral Bell.
Hight, 11½ inches.

$175.00

**ST. LOUIS.**

8 Day $250.00

8 DAY OR 30 HOUR.
SPRING. STRIKE.
Hight, 15 1-2 inches.

# 1875-1906 SETH THOMAS CITY CLOCKS

**$165.00**

**TUXEDO.**

Polished Oak, Gilt Metal Trimmings. Height, 10¾ inches, 5-inch Porcelain Dial.

[DEDICATE], $12 00.

**CANDLESTICK**

IN GOLDEN OAK OR MAHOGANY. BRASS HOLDER AND BASE. STAND RESTS ON GILT METAL FEET.

**$35.00**

Hight, 9¾ inches.

Price, $4.5C per pair.

**NAPLES**

POLISHED CABINET CASE. GOLDEN OAK OR DARK MAHOGANY, WITH HAND CARVINGS.

**$185.00**

8 Day, Half Hour Strike, Cathedral Bell.
4 in. Cream Porcelain Dial,
French Sash, with Beveled Glass.
Hight, 12 inches. Base, 9¼ inches.
Price $12.50

**CANDLESTICK**

IN GOLDEN OAK OR MAHOGANY. BRASS HOLDER AND BASE, STAND RESTS ON GILT METAL FEET

**$35.00**

Hight, 9¾ inches.

Price, $4.50 per pair.

**SEVILLE**

**MENTONE**

POLISHED CABINET CASES. MADE IN MAHOGANY OR GOLDEN OAK. METAL ORNAMENTS IN BRONZE FINISH.

**$175.00**

**$175.00**

8 Day, Half Hour Strike, Cathedral Bell, 4 inch Cream Porcelain Dial, French Sash with Beveled Glass.

Hight, 11¼ inches. Base, 10 inches.
Price, $10.50

Hight, 11¼ inches. Base, 10 inches.
Price, $9.40

LISBON.

**$175.00**

LIBSON

Height, 10½ inches, Base 9,3/4 inches. 8-Day - Strike, 4-inch Dial,Cathedral Bell,List $8.65

COLUMBUS V.P.
Height 24½ inches.

**$300.00**

8 Day Spring, Strike
Cathedral Bell.

*8 Day* **$225.00**

*30 Hour* **$185.00**

DETROIT
Height, 19½ inches.
6, Inch Dial

1 Day, Spring, Strike.
8 Day, Spring, Strike.

**$375.00**

MINSTER
Made in Mahogany,
Oak, Elbony and Olive

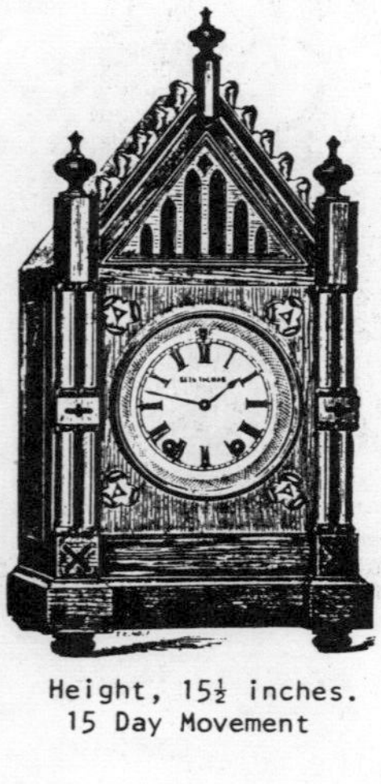

Height, 15½ inches.
15 Day Movement

**$275.00**

BOSTON V.P.
Height, 21 inches

8 Day Spring, Strike
Cathedral Bell.

# 1880 - 1910 SETH THOMAS CITY CLOCKS

**$300.00**

*To receive top prices, clocks must be in original mint condition and in good running order — A replacement dial or missing parts will reduce prices accordingly.*

NASHVILLE.

**$150.00** 1 Day, Spring, Strike.

**$225.00** 8 Day, Spring, Strike.

5 inch Dial.

Hight, 16 inches.

NEW ORLEANS.
V. P.

**$140.00** 1 Day, Spring, Strike.

**$200.00** 8 Day, Spring, Strike.

Hight, 16 inches.

LOUISVILLE.

V. P.

8 Day, Spring, Strike.

**Hight, 22 inches.**

**$225.00**

**CORNELL**

**$225.00**

**HARVARD**

**$225.00**

**YALE**

*1910* **COLLEGE SERIES** *1910*

**Eight Day, Half Hour Strike, Cathedral Bell. 6 inch Dials. 23 inches high.**

**$450.00**

CASTLE.

WALNUT CASE.

1 Day, Weight, Strike.

8 inch Dial.

Hight, 27 inches.

Wood Rod.

**$225.00**

**OXFORD**

**$225.00**

**NEW YORK**

**$225.00**

**CAMBRIDGE**

# 1881-1886 SETH THOMAS CITY CLOCKS

**$350.00**

OREGON.

CHERRY AND COCOBOLA FINISH.

MARQUETERIE TOP.

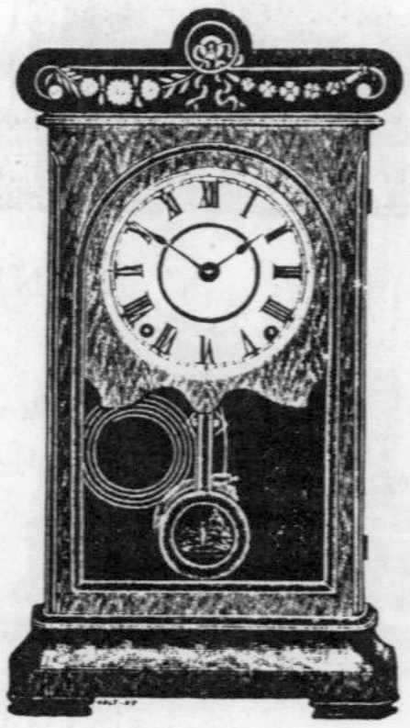

8 Day, Spring, Strike.

Cathedral Bell.

6 inch Dial.

Hight, 19½ inches.

**$350.00**

GREEK.

MADE IN WALNUT, EBONY, OAK AND CHERRY.

8 Day, Spring, Strike.

Cathedral Bell.

6 inch Dial.

Hight, 24 inches.

**$375.00**

ST. PAUL.

WALNUT VENEER, POLISHED.

8 Day, Spring, Strike.

Cathedral Bell.

6 inch Dial.

Hight, 21 inches.

**$225.00**

DENVER.

WALNUT VENEER, POLISHED.

8 Day, Spring, Strike.

Wire Bell.

6 inch Dial.

Hight, 16½ inches.

**$225.00**

NORFOLK.

V. P.

**Spring - Walnut Case.**

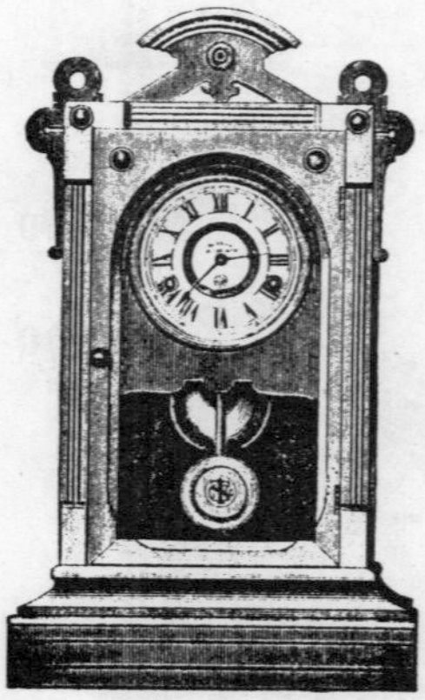

8 Day, Spring.

Strikes the hour.

Hight, 19½ inches.

Price, $7.50.

**$225.00**

NORFOLK.

WALNUT CASE.

1 Day, Spring, Strike.

8 Day, Spring, Strike.

Wire Bell.

6 inch Dial.

Hight, 19½ inches.

**$250.00**

**CHICAGO,**

**V. P.**

8 Day. Strike. Spring.

Hight, 17 inches.

**$250.00**

CHICAGO.

ROSEWOOD AND WALNUT VENEER, POLISHED.

8 Day, Spring, Strike.

6 inch Dial.

Hight, 17 inches.

# 1886 - 1887 SETH THOMAS CITY CLOCKS

**$350.00**

DALLAS.

METAL ORNAMENTS AT BASE.

MADE IN WALNUT, OAK AND CHERRY.

8 Day, Spring, Strike.
Striking half hours on
Cup Bell and hours on
Cathedral Bell.
4½ inch Porcelain Dial.
French Sash, with Beveled Glass.
Hight, 13 inches.

**$350.00**

CARSON.

SHEET BRASS CASE, NICKELED BODY.

GOLD GILT ORNAMENTS AT HEAD AND BASE.

BRONZED ALLIGATOR PANEL.

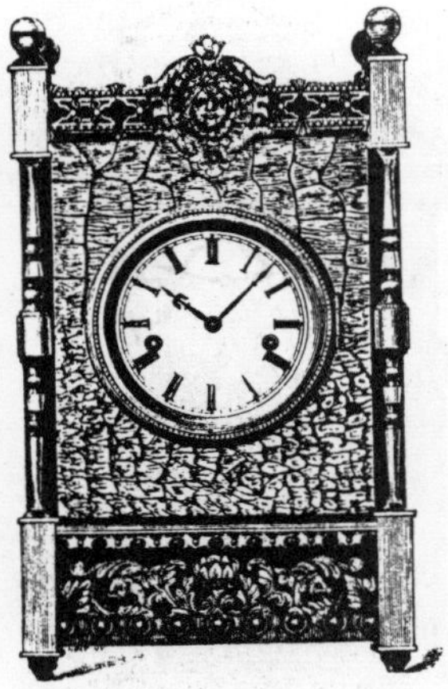

8 Day, Spring.
Striking half hours on a
Cup Bell and hours on
Cathedral Bell.
4½ inch Porcelain Dial.
French Sash, with Beveled Glass.
Hight, 14½ inches.

**$350.00**

AUSTIN.

METAL ORNAMENTS AT HEAD AND BASE.

MADE IN WALNUT, OAK AND CHERRY.

8 Day, Spring, Strike.
Cup Bell and hours on
Striking half hours on
Cathedral Bell.
4½ inch Porcelain Dial.
French Sash, with Beveled Glass.
Hight, 15½ inches.

ARCH TOP.

**$300.00**

FRENCH WALNUT VENEER, POLISHED.

8 Day, Spring, Strike.
6 inch Dial.
Hight, 16 inches.

***Only clocks in ORIGINAL MINT condition and in good running order will command top prices. A replaced or damaged dial, missing parts or chipped beveled glass will reduce prices accordingly.***

**$185.00**

**MILAN**

Height, 10¼ inches. Base, 9½ inches.
Price, **$15.20**

ALBANY.

**$275.00**

WALNUT VENEER, POLISHED.

8 Day, Spring, Strike.
Cathedral Bell.
6 inch Dial.
Hight, 20 inches.

**$225.00**

TACOMA.

OAK CASE.

6 inch Dial. Hight, 23 inches.
8 Day, Spring, Half Hour Strike, Wire Bell.
WITH AND WITHOUT ALARM.

**$165.00**

**CORDOVA**

Gold Plated Metal Caps and Bases.
Height, 10¾ inches. Base, 9¾ inches.
Price, **$13.30**

GENOA.

**$175.00**

8 Day, Half Hour Strike, Cathedral Bell
4 in. White Porcelain Dial.
French Sash, with Beveled Glass.
Hight, 9⅜ in. Base, 9⅞ in.
Price, $9.30.

# 1886 - 1887 SETH THOMAS CITY CLOCKS

**$575.00**

ATLAS.

QUARTER STRIKE.

MADE IN WALNUT, OAK AND CHERRY.

8 Day, Spring.

Striking quarter hours on two Cup Bells and hours on Cathedral Bell.

6 inch Dial.

Hight, 22½ inches.

**$165.00**

DAYTON.

MADE IN POLISHED WALNUT, OAK AND CHERRY.

8 Day, Spring, Strike.

Cathedral Bell.

6 inch Dial.

Hight, 15½ inches.

**$600.00**

HECLA.

QUARTER STRIKE.

MADE IN WALNUT, OAK AND CHERRY.

8 Day, Spring.

Striking quarter hours on two Cup Bells and hours on Cathedral Bell.

6 inch Dial.

Hight, 22½ inches.

**$165.00**

TOLEDO.

MADE IN POLISHED WALNUT, OAK AND CHERRY.

8 Day, Spring, Strike.

Cathedral Bell.

6 inch Dial.

Hight, 15½ inches.

**$350.00**

STERLING.

CABINET CASE.

MADE IN WALNUT, OAK AND MAHOGANY.

15 Day, Half Hour Strike.

Cathedral Bell.

4½ inch Porcelain Dial.

French Sash, with Beveled Glass.

Hight, 12 inches.

**$300.00**

BUFFALO.

WALNUT CASE.

BRASS PILLARS.

8 Day, Spring, Strike.

Cathedral Bell.

6 inch Dial.

Hight, 20½ inches.

**$300.00**

PRINCETON.

WALNUT CASE.

8 Day, Spring, Strike.

Cathedral Bell.

6 inch Dial.

Hight, 23½ inches.

**$250.00**

CAMDEN.

WALNUT CASE.

8 Day, Strike.

Wire Bell.

6 inch Dial.

Hight, 22½ inches.

# 1886 - 1887 SETH THOMAS CITY CLOCKS

**$350.00**

CONCORD.

WALNUT CASE.

8 Day, Spring, Strike.

Cathedral Bell.

6 inch Dial.

Hight, 22½ inches.

**$250.00**

OGDEN.

WALNUT CASE.

8 Day, Spring, Strike.

Cathedral Bell.

6 inch Dial.

Hight, 21½ inches.

**$250.00**

MOBILE.

WALNUT CASE.

8 Day, Spring, Strike.

Cathedral Bell.

6 inch Dial.

Hight, 22 inches.

**$250.00**

RENO.

WALNUT CASE.

8 Day, Spring, Strike.

Wire Bell.

6 inch Dial.

Hight, 19½ inches.

**$300.00**

NEWARK.

WALNUT CASE.

8 Day, Spring, Strike.

Wire Bell.

6 inch Dial.

Hight, 22 inches.

**$400.00**

UTICA.

WALNUT CASE.

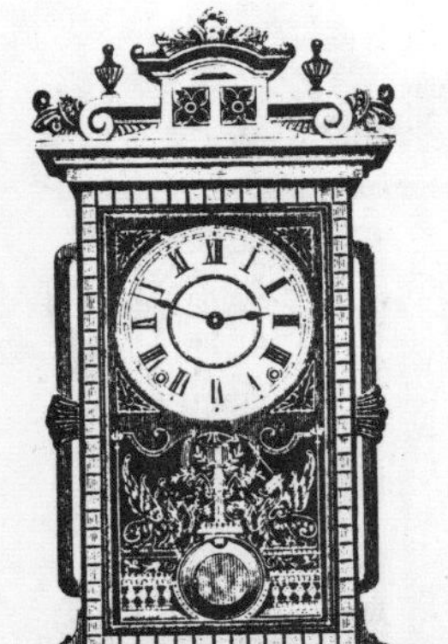

8 Day, Spring, Strike.

Cathedral Bell.

6 inch Dial.

Hight, 22 inches.

**$400.00**

SANTA FE.

WALNUT CASE.

8 Day, Spring, Strike.

Cathedral Bell.

6 inch Dial.

Hight, 22½ inches.

**$275.00**

MADISON.

WALNUT CASE.

8 Day, Strike.

Wire Bell.

6 inch Dial.

Hight, 24 inches.

# 1890-1904 SETH THOMAS CITY CLOCKS

**$850.00**

**LINCOLN.**

WALNUT, OAK OR OLD OAK.

Hight, 27 inches.

**$900.00**

**GARFIELD.**

WALNUT, OAK OR OLD OAK.

Hight, 29 inches.

8 Day, Weight, Strike, Cathedral Bell. 8 inch Dial. Brass Weights, Wooden Rod.

**SUMMIT.** **$165.00**

Height, 18⅛ inches; 6-inch Dial.

8-day, Spring, Half-hour Strike........$ 7.20
8-day, " " " Alarm.. 7.88

**$185.00**

**MIDLAND.**

Height, 22 inches. 6-inch Dial.

8-day, Spring, Strike..................$ 8.00
8-day, " " Alarm............ 8.68

ALBERT.
Height, 16½ inches.

1 Day Spring Strike. **$135.00**
8 Day Spring Strike. **$185.00**

**QUINCY.**

Height, 21 inches; 6-inch Dial.

**$175.00**

8-day, Spring, Half-hour Strike........$ 8.00
8-day, " " " Alarm.. 8.66

**$150.00**

**DERBY.**

Height, 20 inches; 6-inch Dial.

8-day, Strike........................$6.00
8-day, " Alarm.................. 6.68

**$185.00**

**KIRKWOOD.**

Height, 21 inches; 6-inch Dial.

8-day, Spring, Half-hour Strike..........................$8.00
8-day, Spring, Half-hour Strike, Alarm.................. 8.66

**MARLIN.**

Height, 21 inches; 6-inch Dial.

**$150.00**

1-day, Spring, Strike..............................$ 6.46
1-day, " " Alarm.......................... 7.12

**MACON.**

Height, 21 inches; 6-inch Dial.

**$175.00**

8-day, Spring, Half-hour Strike........$ 8.00
8-day, " " " Alarm.. 8.66

# 1891 SETH THOMAS CITY CLOCKS

**$165.00**

**DOVER.**
Height, 20 inches; 6-inch Dial.

8-day, Strike........................................$ 6.54
8-day, " Alarm........................................ 7.20

**$150.00**

**CHESTER.**
WALNUT OR OLD OAK.

8 Day, Half Hour Strike, Cathedral Bell.
5 inch Dial, Roman or Arabic. Hight, 16 inches.

**$175.00**

**BANGOR.**
Height, 20 inches; 6-inch Dial.

8-day, Strike........................................$ 7.00
8-day, " Alarm........................................ 7.68

**$175.00**

**SELMA.**
WALNUT OR OLD OAK.

8 Day, Half Hour Strike, Cathedral Bell.
5 inch Dial. Hight, 16½ inches.

**$175.00**

**AKRON.**
Height, 21½ inches; 6-inch Dial.

8-day, Spring, Strike, Cathedral Bell..............$ 9.00
8-day, " " Alarm, Cathedral Bell....... 9.66

**$285.00**

**ROME.**
ROSEWOOD, WALNUT, OAK OR OLD OAK.

**$285.00**

**LYONS.**
ROSEWOOD, WALNUT, OAK OR OLD OAK VENEER.

Hight, 20½ inches. Hight, 21½ inches.
8 Day, Spring, Half Hour Strike, Cathedral Bell. 6 inch Dial.
WITH AND WITHOUT ALARM.

**$275.00**

**TAMPA.**
Height, 22 inches; 6-inch Dial.

8-day, Spring, Strike, Cathedral Bell............$ 9.54
8-day, " " " " Alarm.... 10.20

*To receive top prices, clocks must be in original mint condition and in good running order — A replacement dial or missing parts will reduce prices accordingly.*

# 1891 SETH THOMAS CITY CLOCKS

*To receive top prices, clocks must be in original mint condition and in good running order — A replacement dial or missing parts will reduce prices accordingly.*

$135.00

**RUSTIC.**
Height 10 inches.

1-day, time.......................$2 80
1-day, time, alarm..................3 50

*8 Day* **$165.00**

*30 Hour* **$120.00**

**ATLANTA**
ROSEWOOD OR WALNUT VENEER, POLISHED.

$235.00

8 Day, Spring, Half Hour Strike, Cathedral Bell.
6 inch Dial. Hight, 19½ inches.
WITH AND WITHOUT ALARM.
Price, Strike, $6.75; Strike, Alarm, $7.25

$275.00

BEE.

CABINET CASE, MARQUETERIE PANEL.

MADE IN WALNUT, EBONY AND OAK.

15 Day, Half Hour Strike.

Cathedral Bell.

4½ inch Porcelain Dial.

French Sash, with Beveled Glass.

Hight, 14 inches.

$185.00

**PONTIAC.**
Height, 22 inches; 6-inch Dial.

8-day, Spring, Strike, Cathedral Bell......................$ 9.34
8-day, Spring, Strike, Alarm, Cathedral Bell..............10.00

$175.00

**PORTLAND.**
WALNUT OR OLD OAK.

8 Day, Half Hour Strike, Cathedral Bell.
5 inch Dial. Hight, 17 inches.

$185.00

**CARTHAGE.**
Height, 22 inches; 6-inch Dial.

8-day, Spring, Strike, Cathedral Bell......................$ 9.60
8-day, Spring, Strike, Alarm, Cathedral Bell..............10.28

$175.00

**LENOX.**
Height, 22 inches; 6-inch Dial.

8-day, Spring, Half-hour Strike..........................$8.54
8-day, Spring, Half-hour Strike, Alarm....................9.20

**CALAIS.**
OLD OAK OR MAHOGANY.

$225.00

8 Day, Half Hour Strike, Cathedral Bell.
5 inch Dial. Hight, 13½ inches.

# 1910-1928 SETH THOMAS CRYSTAL REGULATOR CLOCKS

**$200.00**

**CRYSTAL DORIC No. 0**
4-inch Dial. Height, 9½ inches.
List, $65.00

**$200.00**

**CRYSTAL GOTHIC No. 0**
4-inch Dial. Height, 10¼ inches.
List, $65.00

**$225.00**

**EMPIRE No. 0**
Polished Gold Finish. Round Corners.
Height, 10¾ inches.
List, $27.50

**$250.00**

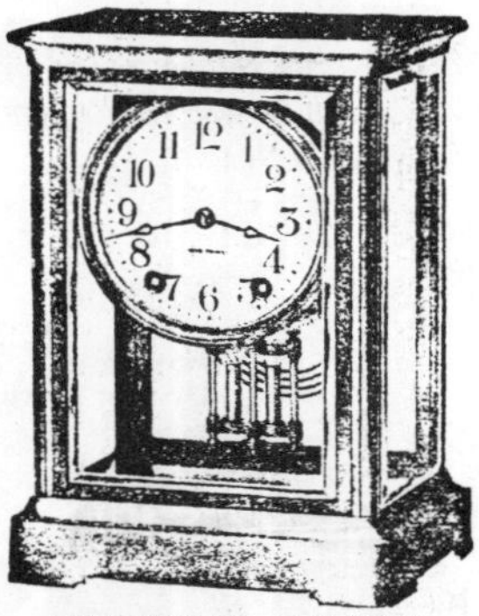

**EMPIRE No. 0 EXTRA**
Rich Gold Finish. Round Corners. Height,10 3/4 inches

**$225.00**

**EMPIRE No. 00**
Polished Gold Finish.
Eight Day. 4 inch Dial.
Height, 12¾ inches. Base, 7¾ inches.
List, $38.00

*Most white metal and brass clock cases are gilded or silver plated. Great care must be taken when cleaning these surfaces. The thin layer of gilding or silver plating will be easily removed by cleaning solutions and strong rubbing. If you believe cleaning is necessary, hire the services of a professional and avoid the expense of re-gilding or replating.*

**$250.00**

**EMPIRE No. 00 EXTRA**
Rich Gold Finish.
List, $42.00

**$375.00**

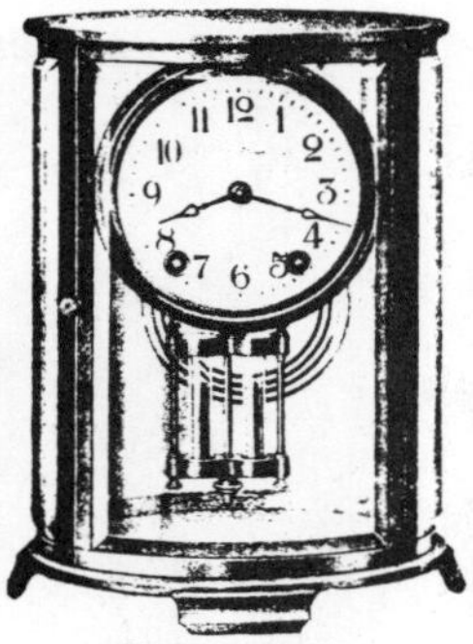

**ORCHID No. 0**
Rich Gold Finish. Oval Base and Top fitted with Convex Beveled Glass on all sides.
Height, 9½ inches.
List, $40.00

**$225.00**

**CRYSTAL DORIC No. 1**
5-inch Dial. Height, 11 inches. List, $75.00

**$235.00**

**CRYSTAL DORIC No. 2**
6-inch Dial. Height, 11 inches. List, $75.00

**$225.00**

**CRYSTAL GOTHIC No. 1**
5-inch Dial. Height, 11¾ inches.
List, $75.00

**$225.00**

**EMPIRE No. 1 EXTRA**
Rich Gold Finish. Sharp Corners.
Height, 10¾ inches.
List, $29.00

**$250.00**

**ALL BRONZE EMPIRE No. 1**

Dull Finish. Eight Day. 4 inch Metal Silvered Dial with Raised Numerals and Minute Dots.
Height, 10¾ inches.
List, $33.70

*Only clocks in ORIGINAL MINT condition and in good running order will command top prices. A replaced or damaged dial, missing parts or chipped beveled glass will reduce prices accordingly.*

**$275.00**

**ALL BRONZE EMPIRE No. 0**

Polished Finish. Round Corners on Cap and Base. Otherwise same as above.

# 1900 - 1910 SETH THOMAS CRYSTAL REGULATOR CLOCKS

Empire No. 1

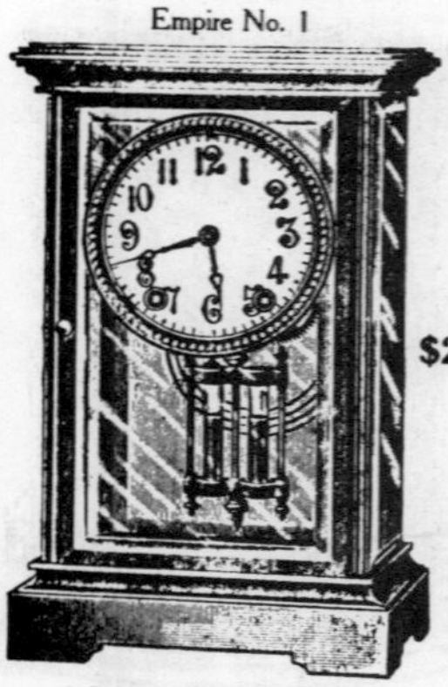

$235.00

8 Day, Half Hour Strike.
3¾ inch Dial.
Hight, 11 inches. Base, 6¾ inches.

Price, $24.00

ORCHID No. 1.

$250.00

15 Day, Half Hour Strike 3 3/4 inch Dial. Height, 10¼ inches. Base 7 3/4 inches.

Empire No. 2

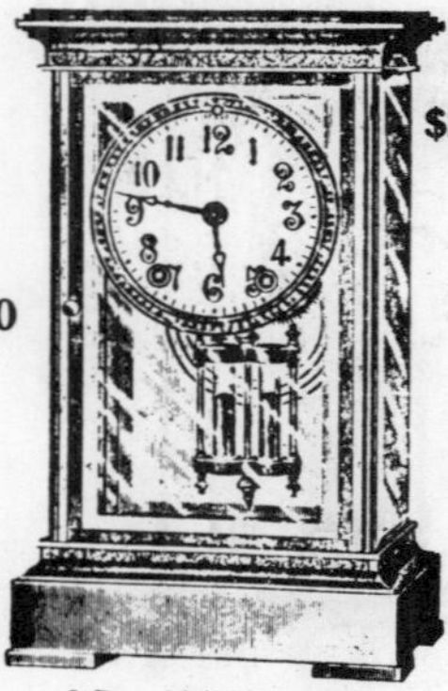

$275.00

8 Day, Half Hour Strike.
3¾ inch Dial.
Hight, 11¾ inches. Base, 7 inches.

Price, $30.00

ORCHID No. 2.

$225.00

15 Day, Half Hour Strike 3 3/4 inch Dial. Height, 9 7/8 inches. Base 6 3/4 inches. Price, $27.00.

$285.00

EMPIRE No. 3

Mahogany and Old Oak Base and Top.
Height, 12¼ inches. Base, 7 inches.
List, $41.00

Orchid No. 3 $225.00

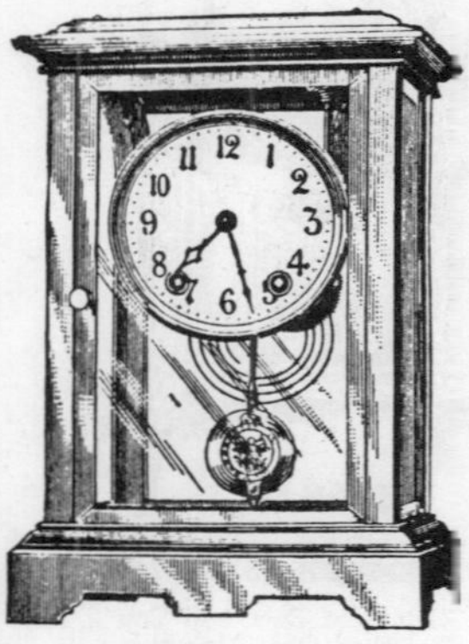

Hight, 10¼ inches. Base, 6¾ inches.

Empire No. 4 $285.00
EMBOSSED

3¾ inch Dial.
Hight, 11 inches. Base, 6¼ inches.
Price, $25.00

$275.00

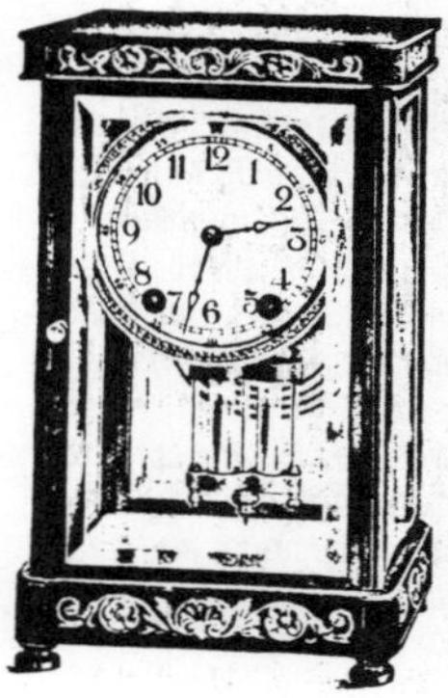

EMPIRE No. 4

Bronze Top and Base, Raised Gold Ornaments, Gold Body. Height, 11 inches. Base, 6¼ inches.
List, $30.00

$225.00

Orchid No. 4

Hight, 10¾ inches. Base, 6¾ inches.
Price, $30.00

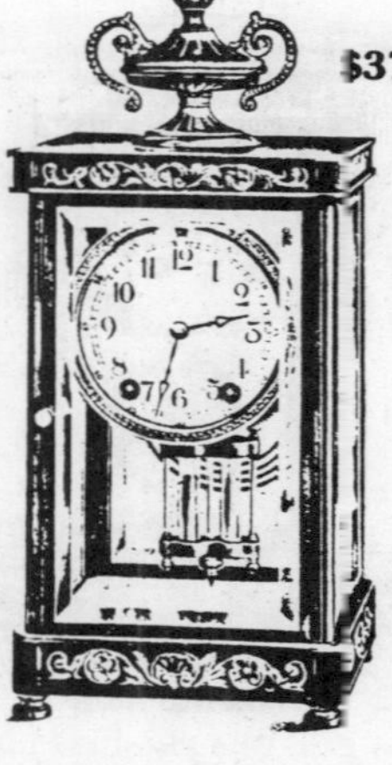

$375.00

EMPIRE No. 5

Bronze Top and Base, Raised Gold Ornaments, Gold Urn and Body. Height, 14¼ inches. Base, 6¼ inches.
List, $36.00

Empire No. 5

$375.00

3¼ inch Dial. Hight, 14 inches, Base, 6¼ inches.
Price, $26.50

$250.00

Orchid No. 5

Hight, 11 inches. Base, 6¾ inches.
Price, $31.00

# 1900 - 1910 SETH THOMAS CRYSTAL REGULATOR CLOCKS

**$300.00**

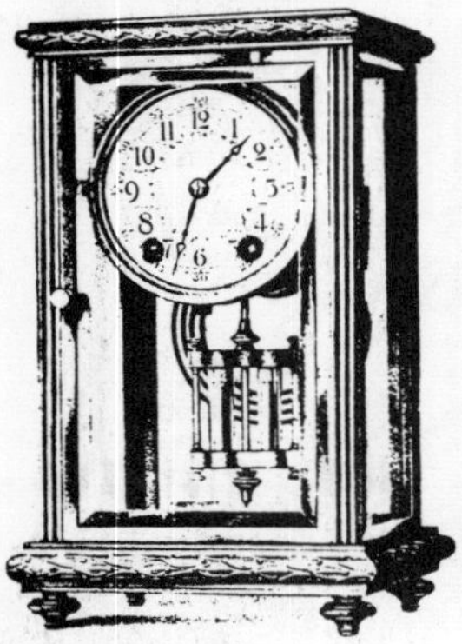

**EMPIRE No. 6**

Rich Gold Finish.
Height, 10½ inches.
Base, 6½ inches.
List, $30.00

Empire No. 6
EMBOSSED

**$350.00**

3¾ inch Dial.
Hight, 11¼ inches. Base, 7¼ inches.
Price, $30.00

Orchid No. 6

**$380.00**

Hight, 11 inches. Base, 8½ inches.
Price, $40.00

**$400.00**

**ORCHID No. 6 EXTRA**

Rich Gold Finish. Oval Base and Top, fitted with Convex Beveled Glass on all sides. Fifteen-Day High Grade Movement. 3½ inches Dial. Height, 11 inches. Base 8½ inches. List, $45.00

*Most white metal and brass clock cases are gilded or silver plated. Great care must be taken when cleaning these surfaces. The thin layer of gilding or silver plating will be easily removed by cleaning solutions and strong rubbing. If you believe cleaning is necessary, hire the services of a professional and avoid the expense of re-gilding or replating.*

Empire No. 7

**$375.00**

3¾ inch Dial.
Hight, 14¼ inches. Base, 7¼ inches.
Price, $32.00

**$250.00**

**EMPIRE No. 7**

Rich Gold Base and Top.
Polished Gold Uprights.
Height, 12½ inches. Base, 8½ inches.
List, $30.00

**$275.00**

**EMPIRE No. 7 WITH URN**

Rich Gold Base, Top and Urn.
Polished Gold Uprights.
Height, 15½ inches. Base, 8½ inches.
List, $32.00

**$325.00**

**EMPIRE No. 7 WITH BUST**

Rich Gold Base, Top and Bust.
Polished Gold Upright.
Height, 17 inches. Base, 8½ inches.
List, $33.00

**$350.00**

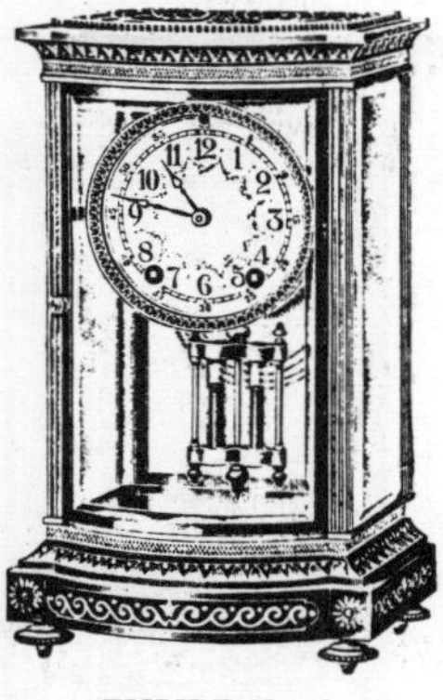

**EMPIRE No. 8**

Top and Base made in Rich Gold, and Bronze Finishes. Convex Beveled Glass Front, Decorated Dial. Height, 12½ inches. Base, 7 inches.
List, $40.80

**$300.00**

**ORCHID No. 8**

Brass Case, Gold Plated and Lacquered. Fifteen Day. Movement Cased in a Cylinder. Cup Bell. 3½ inch Porcelain Dial. French Sash, Convex Beveled Glass. Height, 11 inches. Base, 8 inches.

List, $37.80

**$350.00**

**EMPIRE No. 9**

Rich Gold Finish, Decorated Dial.
Height, 13 inches. Base, 8 inches.
List, $38.00

# 1900 - 1910 SETH THOMAS CRYSTAL REGULATOR CLOCKS

Empire No. 10 **$375.00**

Hight, 15¼ inches. Base, 8 inches.
Price, $35.00

ORCHID No. 10 **$500.00**

**ORCHID No. 10**
Metal Case, Gold Plated, Burnished and Lacquered, Rich Gold Trimmings. Fifteen Day. Movement Cased in a Cylinder. Cup Bell. 3½ inch Porcelain Dial. French Sash, Convex Beveled Glass. Height, 18 inches. Base, 11 inches.
List, $56.00

**$375.00**

**EMPIRE No. 11**
Rich Gold Finish.
Height, 14 inches. Base, 8 inches.
List, $31.50

Empire No 12 **$300.00**

Hight, 14 inches. Base, 8 inches.
Price, $35.00

**$350.00**

**EMPIRE No. 13**
Rich Gold Finish...List, $26.00
Bronze Finish ......List, 25.00
Height, 11½ inches.

**$425.00**

**EMPIRE No. 13**

**EMPIRE No. 13 and LION**
Rich Gold Finish...List, $28.00
Bronze Finish .....List, 27.00
Height Clock with Lion,
15¾ inches.

*Most white metal and brass clock cases are gilded or silver plated. Great care must be taken when cleaning these surfaces. The thin layer of gilding or silver plating will be easily removed by cleaning solutions and strong rubbing. If you believe cleaning is necessary, hire the services of a professional and avoid the expense of re-gilding or replating.*

**$400.00**

Empire No. 14

Rich Gold Finish.
Height, 16 inches. Base, 8 inches.
List, $43.00

**$325.00**

**EMPIRE No. 15**
Rich Gold Finish.
Height, 15½ inches. Base, 8 inches.
List, $31.50

Empire No. 16 **$425.00**

Hight, 16 inches. Base, 8 inches. Price, $31.50

**$400.00**

**EMPIRE No. 17**
Rich Gold Base and Top Burnished.
Polished Gold Uprights.
Height, 17 inches. Base, 8½ inches.
List, $34.00

**$380.00**

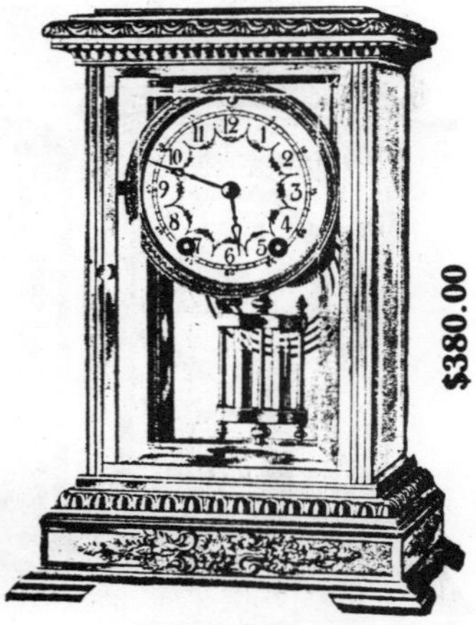

**EMPIRE No. 18**
Rich Gold Base and Top Burnished.
Polished Gold Uprights.
Height, 12 inches. Base, 8½ inches.
List, $31.50

# 1904-1915 SETH THOMAS CRYSTAL REGULATOR CLOCKS

$425.00

**EMPIRE No. 19**
Rich Gold Base and Top Burnished.
Polished Gold Uprights.
Height, 15½ inches. Base, 8½ inches.
List, $33.00

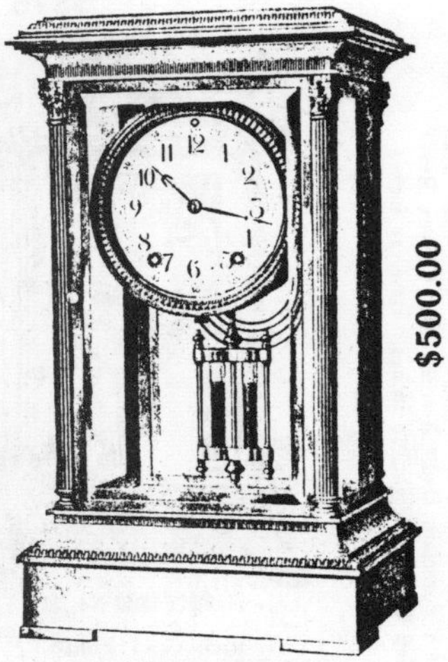

$500.00

EMPIRE No. 20
Rich Gold Finish.
A Fine Pattern and Extra Fine Finish.
Height, 14 inches. Base, 8 inches.
List, $61.00

**EMPIRE No. 21**
MAHOGANY TOP AND BASE

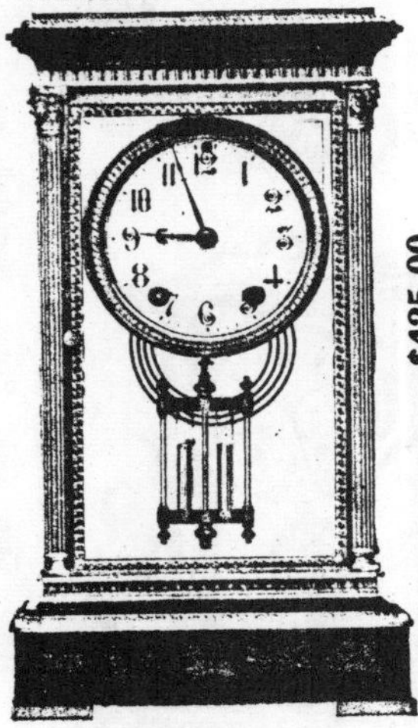

$425.00

4 inch Dial. Hight, 14
inches. Base 7, 7/8
inches. Price, $42.00

Empire No. 22

$300.00

8 Day, Half Hour Strike.
Hight, 13⅜ inches. Base, 7¾ inches.

Price, $37.00

Empire No. 23
BRAZILIAN ONYX TOP AND BASE.

$500.00

15 Day, Half Hour Strike.
Fine Polished Movement.
Hight, 12¾ inches. Base, 8½ inches.
Price, $54.00

$225.00

**EMPIRE No. 23**
Rich Gold Finish, Concave Base.
Height, 11½ inches. Base, 7¼ inches.
Porcelain Dial.
List, $38.00
Metal Dial Silvered, with Raised Numerals as shown on Empire No. 63.
List, $41.50

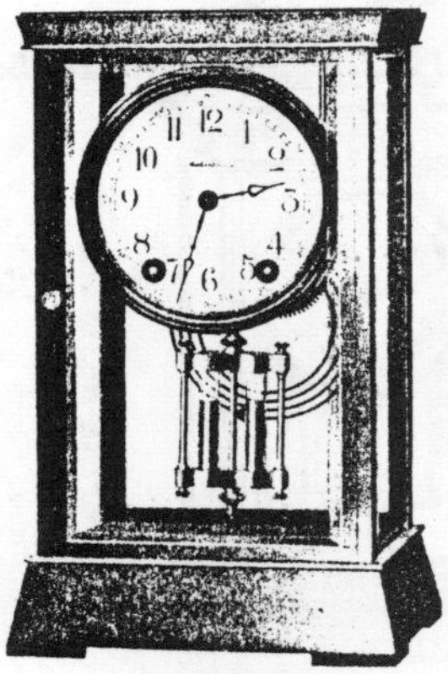

$225.00

**EMPIRE No. 24**
Rich Gold Finish
Hight, 10 3/4 inches
List, $33.00

**EMPIRE No. 24**
BRAZILIAN ONYX TOP AND BASE

$500.00

15 DAY - Half Hour Strike
4, inch Dial. Hight, 12 3/4
inches. Base, 8½ inches.
Price, $58.00.

**EMPIRE No. 25**
Eight Day. Syrian Bronze Finish.
Height, 18½ inches. Base, 10 inches.
List, $54.50

$700.00

**EMPIRE No. 26**
Rich Gold Finish.
Height, 19½ inches. Base, 10 inches.
Price of Clock only..........List, $54.50
Price of Clock with Lion.....List, 58.50

$700.00

*Only clocks in ORIGINAL MINT condition and in good running order will command top prices. A replaced or damaged dial, missing parts or chipped beveled glass will reduce prices accordingly.*

**EMPIRE No. 25**

**EMPIRE No. 26**

# 1910 SETH THOMAS CRYSTAL REGULATOR CLOCKS

Set **$950.00**
Clock only: **$675.00**

No. 27 EMPIRE SET

Rich Gold Finish and Highly Burnished. Eight Day, Half Hour Strike, Cathedral Bell. 4 inch Decorated Porcelain Dial. Height of Clock, 21½ inches; Base, 11 inches. Height of Candelabra, 23 inches; Base, 6¼ inches
List Price of Clock, $65.00 List Price of Set, $100.00
List Price of Candelabra, per pair, $35.00

Set **$1000.00**
Clock only: **$750.00**

EMPEROR SET

Bronze Art Nouveau Finish. Fifteen Day, Fine Polished Movement, Half Hour Strike, Cathedral Bell. 3 inch Decorated Porcelain Dial. Height of Clock, 16¾ inches; Base, 14½ inches. Height of Candelabra, 17 inches; Base, 5¼ inches.
List Price of Clock, $70.00
List Price of Set, $100.00
List Price of Candelabra, per pair, $30.00

**$650.00**

EMPIRE No. 28

Bronze Top and Base, Rich Gold Body, Columns and Trimmings. Height, 16½ inches. Base, 10 inches.
List, $61.00

**$650.00**

EMPIRE No. 29

Bronze Top and Base. Rich Gold Body. Fifteen Day, Fine Movement. 3½ inch Decorated Dial. Height, 17 inches. Base, 8½ inches.
List, $54.50

**$750.00**

CUPID EMPIRE

CUPID EMPIRE

Real Bronze Finish, except Base plate and the four uprights of case holding glass, which are gold plated. Fifteen Day, Fine Movement. Cathedral Bell. 3 inch Decorated Porcelain Dial. Height, 14 inches. Base, 14½ inches.
List, $70.00

# 1910 SETH THOMAS CRYSTAL REGULATOR CLOCKS

$1000.00

EMPIRE No. 30

Fifteen Day, Round Polished, High Grade Movement. Steel Cut Pinions and Springs in Barrels. 4 inch Decorated Porcelain Dial. Cut Glass Top, Sides, Columns and Base. Height, 12¾ inches. Base, 8 inches.
List, $82.00

EMPIRE No. 31

Figure and Base Bronze, Barbedienne Finish, Body and Trimmings Rich Gold, Burnished. Height, 25 inches. Base, 14½ inches.
With Cut Glass Columns..............List, $60.00
With Solid Gold Flower Decorations on Plain Glass Columns...............List, 62.00

$800.00

EMPIRE No. 31

*Most white metal and brass clock cases are gilded or silver plated. Great care must be taken when cleaning these surfaces. The thin layer of gilding or silver plating will be easily removed by cleaning solutions and strong rubbing. If you believe cleaning is necessary, hire the services of a professional and avoid the expense of re-gilding or replating.*

$1000.00

EMPIRE No. 33

Fifteen Day. Same as No. 30 above, except has Solid Gold Flower Decorations on Plain Glass Top, Sides, Columns and Base.
List, $95.00

*Only clocks in ORIGINAL MINT condition and in good running order will command top prices. A replaced or damaged dial, missing parts or chipped beveled glass will reduce prices accordingly.*

$750.00

EMPIRE No. 32

$350.00

EMPIRE No. 35

Rich Gold Finish.
Height, 10 inches. Base, 6 inches.
List, $38.00

EMPIRE No. 32

Figure Bronze Art Nouveau, Base Barbedienne Finish, Body and Trimmings Rich Gold, Burnished. Height, 20 inches. Base, 14½ inches.
With Cut Glass Columns..............List, $60.00
With Solid Gold Flower Decorations on Plain Glass Columns.............List, 62.00

# 1900 - 1910 SETH THOMAS CRYSTAL REGULATOR CLOCKS

No. 40

$225.00

15 Day, Half Hour Strike.
Fine Polished Movement.
3½ inch Dial.
Hight, 8¾ inches. Base, 5¼ inches.
Price, $25.00

$500.00

**EMPIRE No. 41**

Rich, Gold Finish, Decorated Dial.
Height, 13½ inches. Base, 6 inches.
List, $47.50

$250.00

**EMPIRE No. 42**

Rich Gold Finish.
Height, 8¾ inches. Base, 5¼ inches.
List, $37.00

$300.00

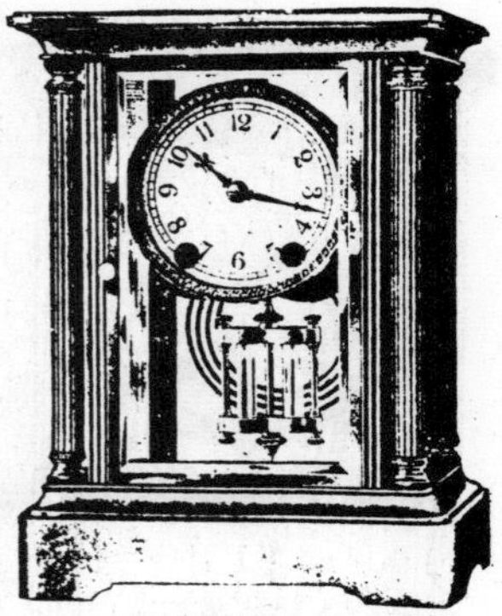

**EMPIRE No. 43**

Fifteen Day. Fine Movement. 3½ inch Dial. Polished Gold Body, Rich Gold Pillars. Height, 9¼ inches.
List, $44.00

**EMPIRE No. 45**

Same as No. 43, but has sharp square corners on Cap and Base.

$300.00

$300.00

**EMPIRE No. 48**

Polished Gold, Convex Front.
Height, 8¾ inches.
List, $43.00

$325.00

**EMPIRE No. 48 EXTRA**

Rich Gold, Convex Front.
Height, 8¾ inches.
List, $47.50

$425.00

**EMPIRE No. 44**

Polished Gold Body, Rich Gold Pillars and Top.
Height, 12¼ inches.
List, $48.00

**EMPIRE No. 46** $425.00

Same as No. 44, but with Sharp Square Corners on Base and Top.
List, $48.00

$325.00

**EMPIRE No. 47**

Rich Gold Base and Top, Polished Gold Uprights.
Height, 10 inches. Base, 6 inches.
List, $37.00

$675.00

**EMPIRE No. 49**

Rich Gold, Convex Front.
Height, 9½ inches. Base, 7 inches.
With Metal Columns.......List, $54.00
With Cut Glass Columns...List, 61.50

# 1910 SETH THOMAS CRYSTAL REGULATOR CLOCKS

**$375.00**

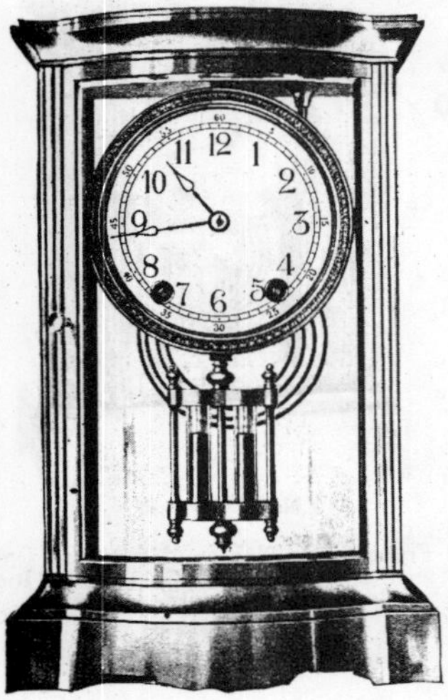

**EMPIRE No. 60**

Polished Gold, Convex Front.
Height, 11 inches.
List, $45.50

**EMPIRE No. 60 EXTRA $300.00**

Rich Gold, Convex Front.
Height, 11 inches.
List, $50.00

**$375.00**

**EMPIRE No. 63**

Rich Gold Finish.
Height, 11 inches. Base, 7½ inches.
Porcelain Dial.
List, $45.00
Metal Dial Silvered, with Raised Numerals.
List, $48.50

**$475.00**

**EMPIRE No. 64**

Rich Gold Finish, Decorated Dial.
Height, 11½ inches. Base, 9 inches.
List, $52.00

**$525.00**

**EMPIRE No. 65**

Polished Gold, Convex Front.
Height, 11 inches. Base, 9½ inches.
With Metal Columns......List, $65.00
With Cut Glass Columns.....List, 81.50
With Gold Decorated Cols. List, 86.50

**EMPIRE No. 65 EXTRA $575.00**

Rich Gold Case and Cols..List, 71.50

**$475.00**

**EMPIRE No. 67**

Rich Gold Finish.
Height, 12½ inches. Base, 9½ inches.
List, $54.00

**$300.00**

**EMPIRE No. 100**

Onyx Base and Top. Polished Gold Body. Eight Day. 4 inch Dial.
Height, 11 inches.
List, $40.00

**EMPIRE No. 101 $300.00**

Same as No. 100, but with sharp square corners on Base and Top.

**$325.00**

**EMPIRE No. 1000**

Same as No. 100, but 12½ inches high.

List, $50.00

# 1910-1915 SETH THOMAS CRYSTAL REGULATOR CLOCKS

**$250.00**

**EMPIRE No. 140**

Onyx Top and Base.
Polished Gold Body.
Height, 9 inches.
List, **$43.00**

**EMPIRE No. 142**

Same as No. 140, but with Round Corners on Base and Top.
List, **$43.00**

**$275.00**

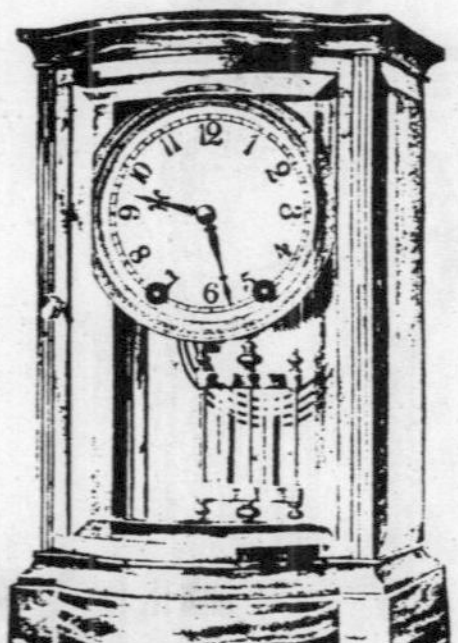

**EMPIRE No. 148**

Onyx Top and Base, Polished Gold Body, Convex Front.
Height, 9 inches.
List, **$52.00**

**$300.00**

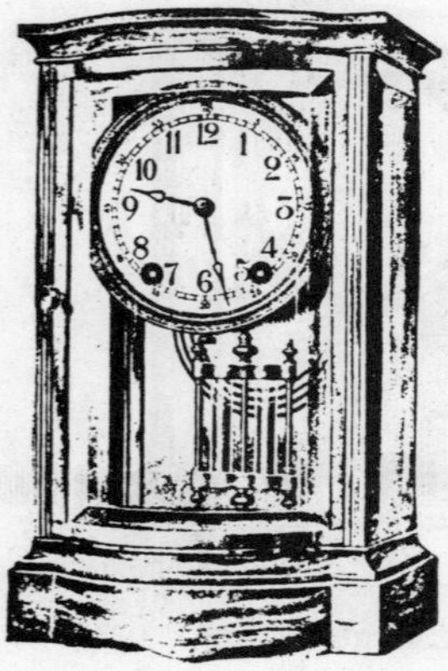

**EMPIRE No. 160**

Onyx Top and Base, Polished Gold Body, Convex Front.
Height, 11 inches.
List, **$56.00**

**$200.00**

**EMPIRE No. 200**

Polished Gold Finish. Round Corners.
Height, 9½ inches.
List, **$24.00**

**$225.00**

**EMPIRE No. 200 EXTRA**

Rich Gold Finish. Round Corners.
Height, 9½ inches.
List Price, **$40.00.**

**$200.00**

**Empire 201**

Rich gold finish. Square corners. 4-inch silvered dial. No. 12[illegible] eight-day pendulum movement with hour and half hour strike. Height 9⅝ inches. Width 7 inches.

Empire 301. Same as No. 20[illegible] in brushed brass finish.

**$200.00**

**EMPIRE No. 201 EXTRA**

Rich Gold Finish. Sharp Corners.
Height, 9½ inches.
List, **$25.00**

**$200.00**

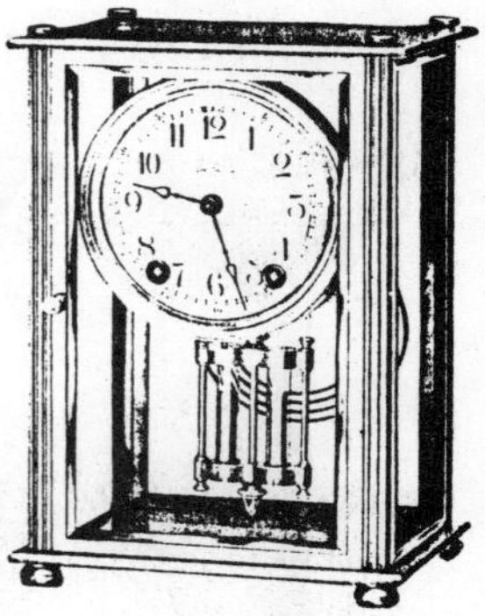

**EMPIRE No. 202**

Rich Gold Finish
Height 9½ inches
List, $25.00

**$200.00**

**EMPIRE No. 300**

Brush Brass Finish.
Height, 9½ inches.
List Price, **$40.00.**

**$200.00**

**EMPIRE No. 301**

Square Corners, Top and Base. Height, 9½ inches. List, $22.00

**$200.00**

**EMPIRE No. 302**

Brush Brass Finish.
Ground Beveled Top and Base.
Height, 10 inches.
List Price, **$40.00.**

**$200.00**

**EMPIRE No. 303**

Ground Concave Top and Base.
Height, 10 inches.
List, **$26.00**

# 1910-1915 SETH THOMAS CRYSTAL REGULATOR CLOCKS

**$185.00**

**EMPIRE No. 304**
New Style Flat Top
Height, 10 inches
List, $22.00

**$200.00**

**EMPIRE No. 305**
Ground Dome Top
Height, 10 inches
List, $25.00

**$200.00**

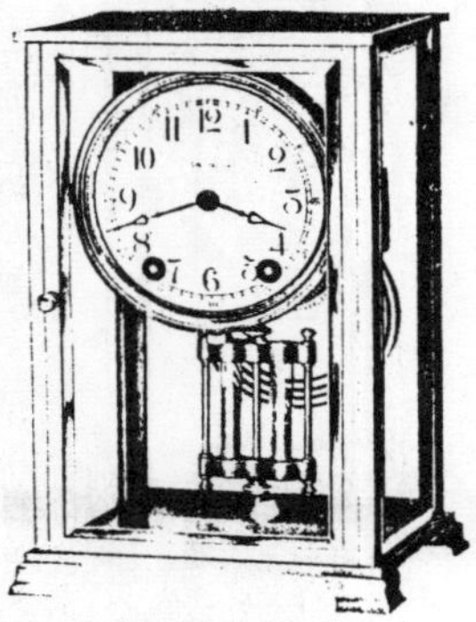

**EMPIRE No. 306**
Rich Gold Finish
Height, 9½ inches
List, $25.00

**$200.00**

**EMPIRE No. 307**
Rich Gold Finish
Height, 9,3/4 inches
List, 30.00

*Only clocks in ORIGINAL MINT condition and in good running order will command top prices. A replaced or damaged dial, missing parts or chipped beveled glass will reduce prices accordingly.*

**No. 65 EMPIRE AND LION SET**

Convex Front Case. Columns and Caps and Lion, Rich Gold, Highly Burnished. Base of Candlesticks and Base and Top of Clock Polished Gold. Also made in Verde Antique, Base, Top and Lion, with Columns Rich Gold and balance of parts Polished Gold. Eight Day, Fine Movement, Half Hour Strike, Cathedral Bell. 4 inch Decorated Porcelain Dial. Height of Clock, 14 inches; Base, 9½ inches. Height of Candlesticks, 11½ inches; Base, 3¾ inches.

List Price of Clock, $68.50 **$650.00**

List Price of Set, $87.50 **$750.00**

List Price of Candlesticks, per pair, $19.00

*Most white metal and brass cases are gilded or silver plated. Great care must be taken when cleaning these surfaces. The thin layer of gilding or silver plating will be easily removed by cleaning solutions and strong rubbing. If you believe cleaning is necessary, hire the services of a professional and avoid the expense of re-gilding or re-plating.*

# 1922 SETH THOMAS ELECTRIC CLOCKS

$1,600.00

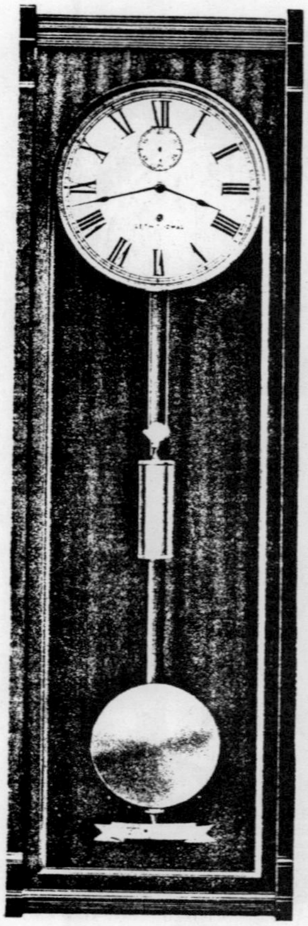

'Master Clock "K"
HAND-WOUND

$850.00

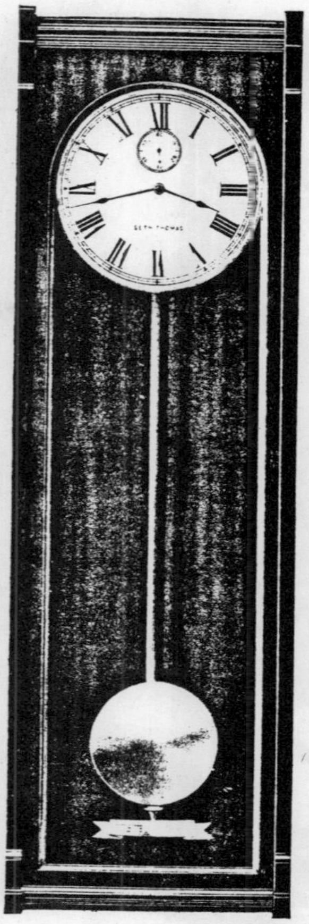

Master Clock "K-1"
SELF-WINDING

DIMENSIONS

| | |
|---|---|
| Height | 56″ |
| Width | $16\frac{1}{2}$″ |
| Depth | 8″ |
| Dial | 12″ |

Roman or Arabic Numerals
May Be Fitted with Mercury Pendulum

$1,700.00

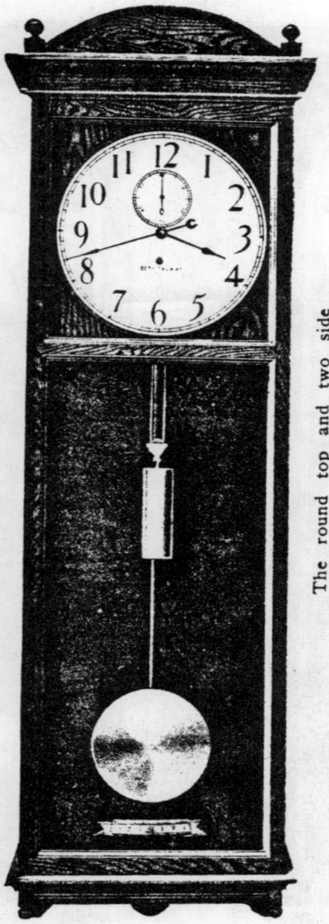

The round top and two side pieces can be removed if a plain top case is desired.

Master Clock "L"
HAND-WOUND

$1,050.00

Master Clock "L-1"
SELF-WINDING

DIMENSIONS

| | |
|---|---|
| Height | 62″ |
| Width | 17″ |
| Depth | $8\frac{1}{2}$″ |
| Dial | 14″ |

Roman or Arabic Numerals Oak or Mahogany Case
May Be Fitted with Mercury Pendulum

*To receive top prices, clocks must be in original mint condition and in good running order — A replacement dial or missing parts will reduce prices accordingly.*

## Secondary Clock Type "A"

WOOD CASE

| | DIAL | DIAMETER | DEPTH |
|---|---|---|---|
| **$75.00** | 6″ | $9\frac{3}{8}$″ | $3\frac{13}{16}$″ |
| | 8″ | $11\frac{1}{4}$″ | $3\frac{7}{8}$″ |
| | 10″ | $13\frac{1}{2}$″ | $3\frac{7}{8}$″ |
| | 12″ | $16\frac{1}{8}$″ | 4″ |
| | 14″ | 18″ | 4″ |
| **$100.00** | 18″ | $21\frac{3}{4}$″ | 4″ |

Hinged Door on Front

Roman or Arabic Numerals Oak or Mahogany Case

Special Finish to Order

# 1922 SETH THOMAS ELECTRIC CLOCKS

$1,400.00

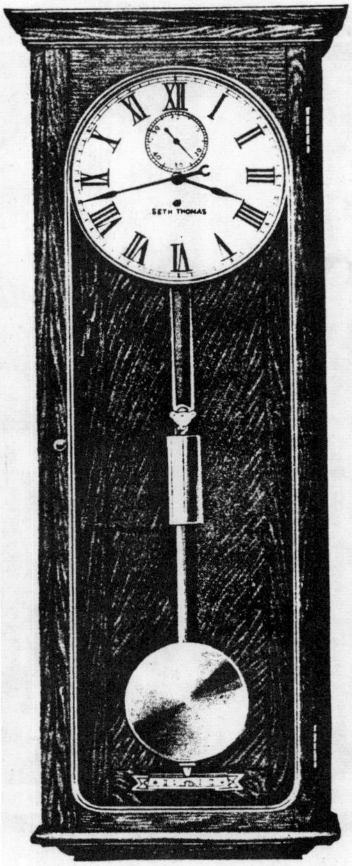

Master Clock "M"
HAND-WOUND

$700.00

Master Clock "M-1"
SELF-WINDING

DIMENSIONS

| | |
|---|---|
| Height | 56″ |
| Width | 20″ |
| Depth | 8½″ |
| Dial | 14″ |

Roman or Arabic Numerals — Oak or Mahogany Case
May Be Fitted with Mercury Pendulum

$1,800.00

Master Clock M
Finish, Mahogany and Old Oak.
18-inch Dial.
Height, 68 inches.
Width, 23 inches.
Depth, 8 inches.

## Secondary Clock Type "B"

WOOD CASE

| | DIAL | SQUARE | DEPTH |
|---|---|---|---|
| $75.00 | 6″ | 9⅛″ | 3¾″ |
| | 8″ | 11½″ | 3⅞″ |
| | 10″ | 14″ | 3⅞″ |
| | 12″ | 15¾″ | 3⅞″ |
| | 14″ | 18″ | 3⅞″ |
| $100.00 | 18″ | 22″ | 3⅞″ |

Hinged Door on Front
Roman or Arabic Numerals — Oak or Mahogany Case
Special Finish to Order

# 1922 SETH THOMAS ELECTRIC CLOCKS

**$150.00**

ROMAN.

We are prepared to furnish specially designed dials for interiors of banks, hotels, arcades, etc. On this page we illustrate types of dials suitable for such purposes and are prepared to furnish any type of dial to special drawing.

**$85.00**

Marble Dial Secondary Clock, Type "D"

| | |
|---|---|
| 12" Minute Circle | 16" Slab |
| 16" Minute Circle | 20" Slab |
| 22" Minute Circle | 26" Slab |

**$150.00**

It is well in making designs for special dials to keep them as plain as possible in order to make the time easily read, and to refrain from designing hands too elaborate or heavy. It is an established fact that the plainer a dial is, the more readily it serves its purpose.

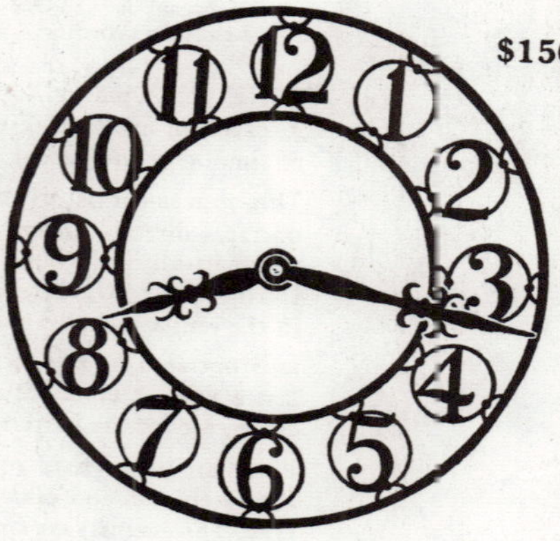

ARAB.

*To receive top prices, clocks must be in original mint condition and in good running order — A replacement dial or missing parts will reduce prices accordingly.*

Flush Secondary Clock Type "F"

| DIAL | DIAMETER | OFFSET | |
|---|---|---|---|
| 6" | 9" | 1 13/16" | **$75.00** |
| 8" | 11⅛" | 2" | |
| 10" | 13⅜" | 2" | |
| 12" | 16" | 2" | |
| 14" | 17⅞" | 2" | |
| 18" | 21⅝" | 2" | **$100.00** |

Secondary Clock Type "E"
WOOD CASE

| DIAL | DIAMETER | DEPTH | |
|---|---|---|---|
| 6" | 9⅜" | 3⅞" | **$75.00** |
| 8" | 11¼" | 3⅞" | |
| 10" | 13½" | 3⅞" | |
| 12" | 16⅛" | 3⅞" | |
| 14" | 18" | 3⅞" | |
| 18" | 21¾" | 3⅞" | **$100.00** |

Friction Door on Back

Roman or Arabic Numerals — Oak or Mahogany Case

Special Finish to Order

# 1922 SETH THOMAS ELECTRIC CLOCKS

**$175.00**

SELF WINDING

**SELF WINDING No. 1**

Mahogany and Old Oak. Fourteen-inch Dial.
Length 18 inches. Width 18 inches.

THE Self Winding Clock Movement is an excellent example of the Seth Thomas idea of simplicity in efficient time-keeping.

This movement is driven by a weight which operates directly in the center wheel. It is automatically re-wound every four minutes. During the re-winding a maintaining spring in the center wheel provides driving power.

It is operated by two dry cell batteries which last a year or more. Space is provided in the clock case for the batteries.

Plates are heavy hard rolled brass, wheels are accurately cut and pinions are of the lantern type. The magnets are enamel covered wire and contacts are pure silver.

**$150.00 *to* $250.00**

**Marble Dial Secondary Clock Type "C"**

| | |
|---|---|
| 11″ Minute Circle | 20″ Slab |
| 20″ Minute Circle | 24″ Slab |

**The above are regular stock sizes, but other sizes will be furnished on order. White Italian marble is ordinarily used, but we will furnish any other kind desired. These clocks are made in both surface and flush type and either painted or raised metal numerals will be furnished.**

**$50.00**

**No. 2 COUNCIL . . . . . $12.50**

With self-starting electric time movement

Up-to-the-minute in appearance as well as in accuracy. Genuine walnut case with sides sloping outward in graceful lines. 7¼-inch white dial with brown Arabic numerals of modern character. Dark brown hour and minute hands—red seconds hand. Has reset knob at bottom of case. 9¾-inches square. Packed singly. Shipping weight 7 pounds.

# 1904 - 1905 SETH THOMAS HALL CLOCKS

### HALL CLOCK No. 24

STRIKE.
MADE IN MAHOGANY OR OLD OAK.

93½ inches High, 24½ inches Wide,
15 inches Deep.

8 Day, Weight, striking hours and half hours on Cathedral Bell. Fitted with Movement No. 72 A. Solid, Silvered Dial.

Price, $140.00

**$3000.00**

### HALL CLOCK No. 25

CHIME.

Same as Hall Clock No. 24, but fitted with Movement No. 74 C.
Chiming Westminster at the quarters on four Cup Bells, and striking the hours on a Cathedral Bell.

Price, $185.00

**$4200.00**

### HALL CLOCK No. 26

CHIME AND MOON.

Same as Hall Clock No. 25, but fitted with Movement 74 B and Silvered Moon Dial showing Moon Changes.

Price, $195.00

**$5000.00**

### HALL CLOCK No. 22

CHIME AND MOON.
MADE IN MAHOGANY, OLD OAK AND OAK NATURAL.

98 inches High, 22 inches Wide,
15 inches Deep.

8 Day, Weight, Chiming Westminster at the quarters on four Cup Bells, and striking the hours on a Cathedral Bell. Shows Moon Changes. Fitted with Movement No. 74 B. Silvered, Moon Dial.

Price, $225.00

**$4750.00**

### HALL CLOCK No. 2272 C

STRIKE AND MOON.

Same as Hall Clock No. 22, but does not chime; instead, strikes hours and half hours on Cathedral Bell. Fitted with Movement No. 72 C.

Price, $175.00

**$3750.00**

**$2,250.00**

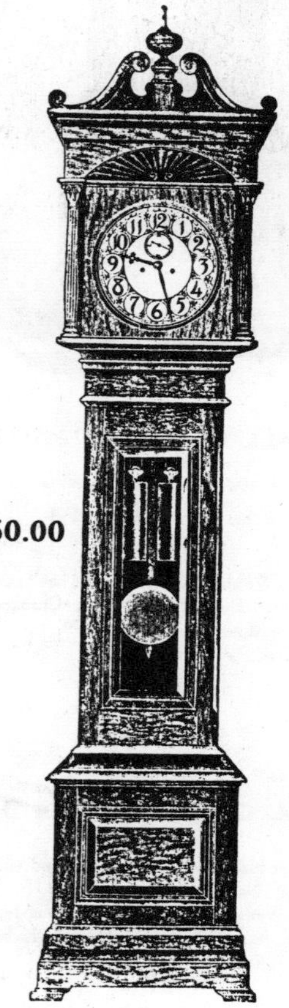

### HALL CLOCK No. 23

STRIKE.
MADE IN MAHOGANY AND GOLDEN OAK.

97 inches High, 21 inches Wide,
12½ inches Deep.

8 Day, Weight, striking hours and half hours on Cathedral Bell. Round 12-inch Brass Silvered Dial. Fitted with Movement No. 72 A.

Price, $110.00

## SETH THOMAS HALL CLOCK MOVEMENTS

### No. 72 A

8 Day, Weight, strikes hours and half hours on a 6 inch Cathedral Bell. Has Damaskeened Plates, Lantern Pinions, Graham Escapement and Maintaining power.

### No. 72 B and No. 72 C

Same as No. 72 A, but arranged for Moon Phases. "B" and "C" differ only as to arrangement for size of dial used.

### No. 74 C

8 Day, Weight, Quarter Strike, Westminster Chimes. The Chimes are struck on four Cup Bells; the hours on a 6 inch Cathedral Bell. Has Damaskeened Plates, Lantern Pinions, Graham Escapement and maintaining Power.

### No. 74 B and No. 74 D

Same as No. 74 C, but arranged for Moon Phases. "B" and "D" differ only as to arrangement for size of dials used.

### No. 84

8 Day, Weight, Quarter Strike, Westminster Chimes. The Chimes are struck on four Cathedral Bells; the hours on a large Cathedral Bell. Can be made silent or chime at will. Has Damaskeened Plates, Cut Steel Pinions, tempered, finely finished throughout, Graham Escapement and Maintaining Power. The rod is of soft steel.

This movement is equal in every detail to the finest of English makes and out-classes those of German manufacture.

*Except for those fabulous clocks for which you wouldn't mind cutting a hole in the ceiling, Grandfather or Standing Clocks under eight feet are the most desirable and practical Some manufacturers produced certain of these clocks in two styles — one to stand on the floor, and the other to hang on a wall. Of the clocks designed with two styles, the hanging style is considered to be the most valuable.*

# 1904 - 1905 SETH THOMAS HALL CLOCKS

**HALL CLOCK No. 27**

STRIKE AND MOON.
MADE IN MAHOGANY OR GOLDEN OAK.

100 inches High, 27 inches Wide,
18 inches Deep.

8 Day, Weight, striking hours and half hours on Cathedral Bell. Shows Moon Changes. Fitted with Movement No. 72 B.
Dial, Pierced Gilt.

Price, $260.00

**$3500.00**

**HALL CLOCK No. 28**

STRIKE AND MOON.
MADE IN MAHOGANY OR GOLDEN OAK.

98 inches High, 23 inches Wide,
14 inches Deep.

8 Day, Weight, striking hours and half hours on Cathedral Bell. Shows Moon Changes Fitted with Movement No. 72 B.
Dial, Pierced Gilt.

Price, $200.00

**$3200.00**

**HALL CLOCK No. 34**

STRIKE AND MOON.
MADE IN MAHOGANY OR GOLDEN OAK.

94 inches High, 24½ inches Wide,
16½ inches Deep.

8 Day, Weight, striking hours and half hours on Cathedral Bell. Shows Moon Changes. Fitted with Movement No. 72 B.
Dial, Pierced Gilt.

Price, $160.00

**$3,750.00**

**HALL CLOCK No. 33**

STRIKE AND MOON.
MADE IN MAHOGANY OR GOLDEN OAK.

97 inches High, 22½ inches Wide,
15 inches Deep.

8 Day, Weight, striking hours and half hours on Cathedral Bell. Shows Moon Changes. Fitted with Movement No. 72 B.
Dial, "Painted."

Price, $130.00

**$2,500.00**

**HALL CLOCK No. 2774 D**

CHIME AND MOON.

Same as Hall Clock No. 27, but fitted with Movement No. 74 D.
Chiming Westminster at the quarters on four Cup Bells, and striking the hours on a Cathedral Bell.

Price, $310.00

**$5500.00**

**HALL CLOCK No. 2874 D**

CHIME AND MOON.

Same as Hall Clock No. 28, but fitted with Movement No. 74 D.
Chiming Westminster at the quarters on four Cup Bells, and striking the hours on a Cathedral Bell.

Price, $250.00

**$5000.00**

**HALL CLOCK No. 3434 D**

CHIME AND MOON.

Same as Hall Clock No. 34, but fitted with Movement No. 74 D. Chiming Westminster at the quarters on four Cup Bells, and striking the hours on a Cathedral Bell.

Price, $210.00

**$4750.00**

**HALL CLOCK No. 3374 D**

CHIME AND MOON.

Same as Hall Clock No. 33, but fitted with Movement No. 74 D.
Chiming Westminster at the quarters on four Cup Bells, and striking the hours on a Cathedral Bell.

Price, $180.00

**$4000.00**

**HALL CLOCK No. 2784**

CATHEDRAL CHIME AND MOON.

Same as Hall Clock No. 27, but fitted with Fine Movement No. 84.
Chiming Westminster at the quarters on four Cathedral Bells, and striking the hours on a large Cathedral Bell. The Pierced Gilt Dial is of extra quality and design.

Price, $450.00

**$6,750.00**

*Except for those fabulous clocks for which you wouldn't mind cutting a hole in the ceiling, Grandfather or Standing Clocks under eight feet are the most desirable and practical. Some manufacturers produced certain of these clocks in two styles — one to stand on the floor, and the other to hang on a wall. Of the clocks designed with two styles, the hanging style is considered to be the most valuable.*

# 1904 - 1905 SETH THOMAS HALL CLOCKS

**$3250.00**

## HALL CLOCK No. 37

STRIKE AND MOON.
MADE IN MAHOGANY OR GOLDEN OAK.

100 inches High, 28½ inches Wide, 18 inches Deep.

8 Day, Weight, striking hours and half hours on Cathedral Bell. Shows Moon Changes. Fitted with Movement No. 72 B. Dial, Pierced Gilt.

Price, $200.00

## HALL CLOCK No. 3774 D

CHIME AND MOON.

Same as Hall Clock No. 37, but fitted with Movement No. 74 D. Chiming Westminster at the quarters on four Cup Bells, and striking the hours on a Cathedral Bell.

Price, $250.00

**$5,500.00**

## HALL CLOCK No. 3784

CATHEDRAL CHIME AND MOON.

Same as Hall Clock No. 37, but fitted with Fine Movement No. 84.
Chiming Westminster at the quarters on four Cathedral Bells, and striking the hours on a large Cathedral Bell. The Pierced Gilt Dial is of extra quality and design.

Price, $390.00

**$6,000.00**

**$6,500.00**

## HALL CLOCK No. 38

CATHEDRAL CHIME AND MOON.
MADE IN MAHOGANY AND GOLDEN OAK.

103 inches High, 29 inches Wide, 19½ inches Deep.

8 Day, Weight, Chiming Westminster at the quarters on four Cathedral Bells, and striking the hours on a large Cathedral Bell. Shows Moon Changes. Fitted with Fine Movement No. 84. The Pierced Gilt Dial is of extra quality and design.

Price, $480.00

**$4000.00**

## HALL CLOCK No. 35

STRIKE AND MOON.
MADE IN MAHOGANY OR GOLDEN OAK.

98 inches High, 27½ inches Wide, 17½ inches Deep.

8 Day, Weight, striking hours and half hours on Cathedral Bell. Shows Moon Changes. Fitted with Movement No. 72 B. Dial, Pierced Gilt.

Price, $180.00

**$4000.00**

## HALL CLOCK No. 3574 D

CHIME AND MOON.

Same case as Hall Clock No. 35, but fitted with Movement No. 74 D.
Chiming Westminster at the quarters on four Cup Bells, and striking the hours on a Cathedral Bell.

Price, $230.00

**$5000.00**

**$4500.00**

## HALL CLOCK No. 36

STRIKE AND MOON.
MADE IN MAHOGANY AND GOLDEN OAK.

103½ inches High, 28½ inches Wide, 19½ inches Deep.

8 Day, Weight striking hours and half hours on Cathedral Bell. Shows Moon Changes. Fitted with Movement No. 72 B. Dial, Pierced Gilt.

Price, $275.00

## HALL CLOCK No. 3684 D

CHIME AND MOON.

Same as Hall Clock No. 36, but fitted with Movement No. 74 D. Chiming Westminster at the quarters on four Cup Bells, and striking the hours on a Cathedral Bell.

Price, $325.00

**$5500.00**

## HALL CLOCK No. 3684

CATHEDRAL CHIME AND MOON.

Same as Hall Clock No. 36, but fitted with Fine Movement No. 84.
Chiming Westminster at the quarters on four Cathedral Bells, and striking the hours on a large Cathedral Bell. The Pierced Gilt Dial is of extra quality and design.

Price, $465.00

**$7500.00**

# 1904-1920 SETH THOMAS HALL CLOCKS

**$1,400.00**

**STRIKING HALL CLOCK No. 15**

Mahogany and Golden Oak.
88 in. High. 19 in. Wide. 11 in. Deep.
Eight-Day, Spring.
Movement.

Striking hours and half-hours on a Tubular Bell. Painted Metal Dial with 10-inch Minute Circle.

List, $125.00

Exact reproduction of case made in our factory about 1813.

**$2,000.00**

**$6,500.00**

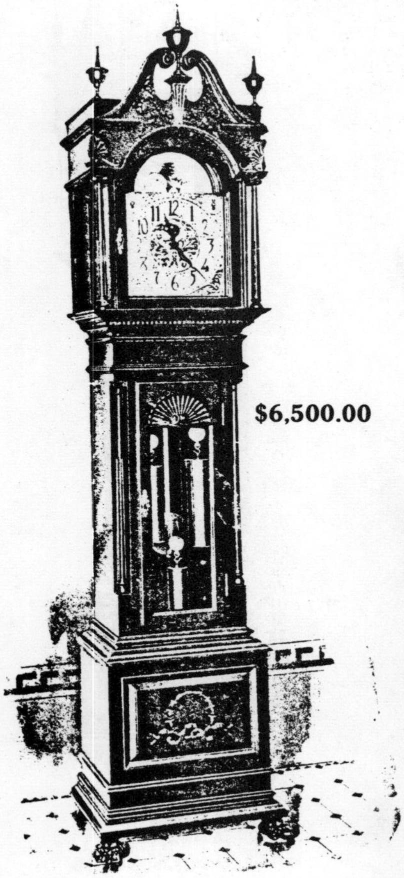

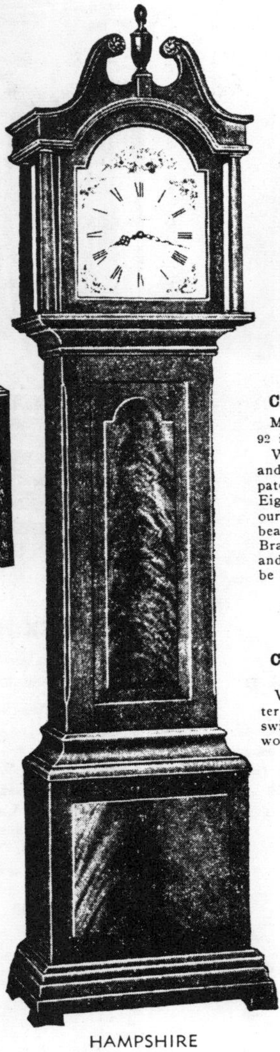

**HALL CLOCK No. 39**

CATHEDRAL CHIME AND MOON.
MADE IN MAHOGANY OR GOLDEN OAK.

105 inches High, 25 inches Wide.
16 inches Deep.

8 Day, Weight, Chiming Westminster at the quarters on four Cathedral Bells, and striking the hours on a large Cathedral Bell. Shows Moon Changes. Fitted with Fine Movement No. 84. The Pierced Gilt Dial is of extra quality and design.

Price, $445.00

**HALL CLOCK No. 3974 D**

Same case as Hall Clock No. 39. Fitted with Movement No. 74 D.
Chiming Westminster at quarters on four Cup Bells, and striking hours on a Cathedral Bell. Dial, Pierced Gilt.

Price, $305.00

**$5,750.00**

**CHIMING HALL CLOCK No. 40**

Mahogany, Flemish and Golden Oak.
92 in. High. 19 in. Wide. 11 in. Deep.

Westminster Chime every 15 minutes, and striking the hours on five "Sonora" patented deep-toned Resonated Bells. Eight-Day Spring Movement as used in our Mantel Chimes, but with a seconds beat rod and 3¾-inch brass-covered ball. Brass-etched Dial with black numerals and silvered background. Metal tip can be removed if desired. Wood door.

List, $120.00

Beveled plate glass in lower door.

List, $125.00

**CHIMING HALL CLOCK No. 240**

Same Case as No. 40.

Whittington and Westminster Quarter Hour Chime on 8 Bells. Can be switched to play either chime. With wood door.

List, $150.00

With beveled plate glass door.

List, $155.00

**$2500.00**

**$1,000.00**

**$1,750.00**

**No. 4E HALL CLOCK** . . . . . . . . . $85.00
With self-starting electric Westminster Chime movement **$850.00**

**No. 4W HALL CLOCK** . . . . . . . . . $85.00
Same with 8-day pendulum chime movement No. 113 **$1,500.00**

Mahogany cabinet with distinctive design on front panel. Ivory enameled dial with 8½-inch minute circle, black Roman numerals and colored flower decorations in the corners. Quarter hour chime. Height 72-inches. Width 13-inches. Thickness 8⅜-inches. Packed singly. Shipping weight 100 pounds.

HAMPSHIRE

**HALL CLOCK No. 3E** . . . . . . . . . $125.00
With self-starting electric Westminster Chime movement.

**HALL CLOCK No. 3W** . . . . . . . . $125.00
Same with 8-day pendulum chime movement No. 113 and etched silver plated dial.

Mahogany with walnut front panels. No. 3E has flowered antique type dial in ivory tint with 8-inch minute circle and Roman numerals. No. 3W has silvered metal dial etched in black with Arabic numerals on raised circular section. Quarter hour chime. Height 73 inches. Width 17¾ inches. Thickness 9½ inches. Packed singly. Approximate shipping weight 100 pounds.

# 1886 - 1901 SETH THOMAS HALL REGULATORS

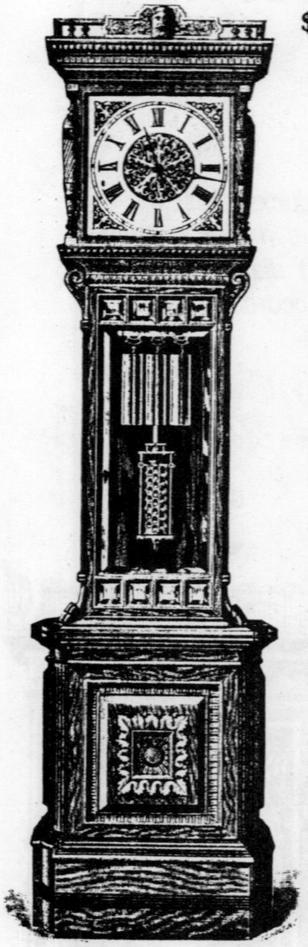

**$5,000.00**

REGULATOR No. 20.
MADE IN WALNUT, OAK AND MAHOGANY.

Strikes quarters upon Cast Metal Bells and repeats the hour upon Cathedral Bell.

8 Day, Weight, Quarter Strike, Repeating. 14 inch Ornamental Metal Dial. Hight, 100 inches. Finely finished Movement, Beating Seconds, Cut Steel Pinions, etc. Jewelled Pallets. Mercurial Pendulum.

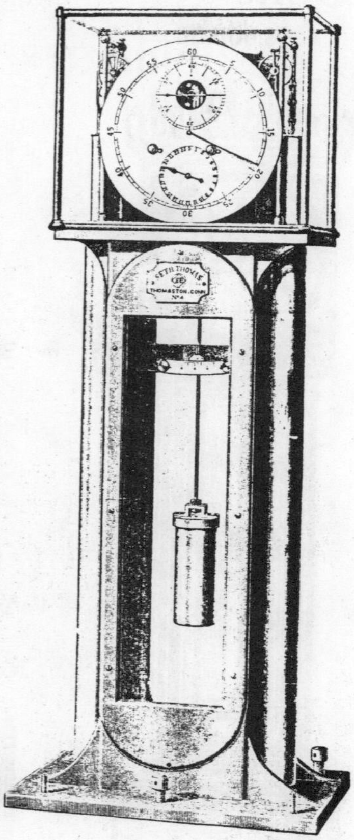

**PRECISION CLOCK**

With Dead Beat Escapement, Thirty-two Day. With Gravity Escapement, Eight Day. Height, 62 inches. 14 inch Dial.

With Dead Beat Escapement.
Price, $1000.00

**$10,000.00**

With Gravity Escapement.
Price, $1250.00

**$12,000.00**

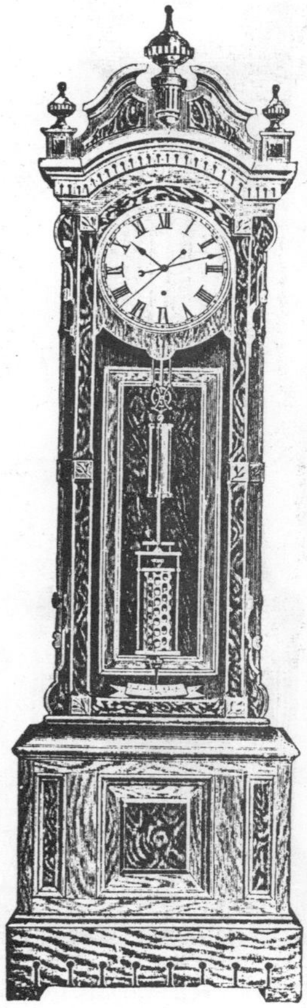

**FINE REGULATOR No. 14**

Sweep Seconds.

Walnut Case, Mercury Pendulum. Eight Day, Weight, Time. Beats Seconds. Height, 100 inches. 14 inch Silvered Dial.

Price, $455.00

**$7,500.00**

REGULATOR No. 21.
VENEERED, MAHOGANY CASE.

8 Day, Weight, Quarter Strike.
12 inch Metal Dial, Silvered. Hight, 94 inches.

Good strong Movement, Lantern Pinions. Beats Seconds.
Chimes the Quarters on four Cup Bells. For striking the hours, we fit a Cathedral Bell or a large Cup Bell, as may be preferred.

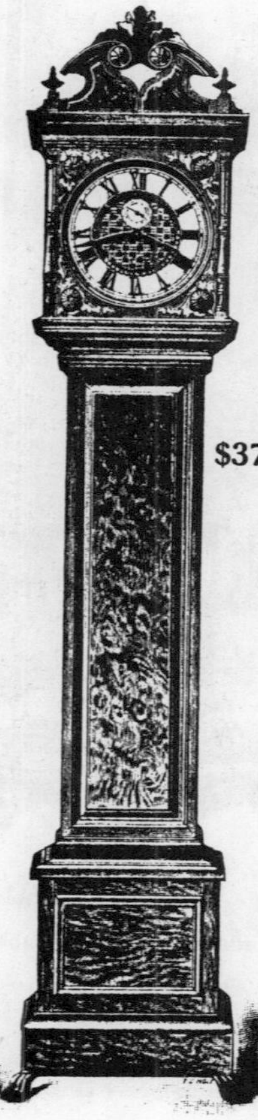

**$3750.00**

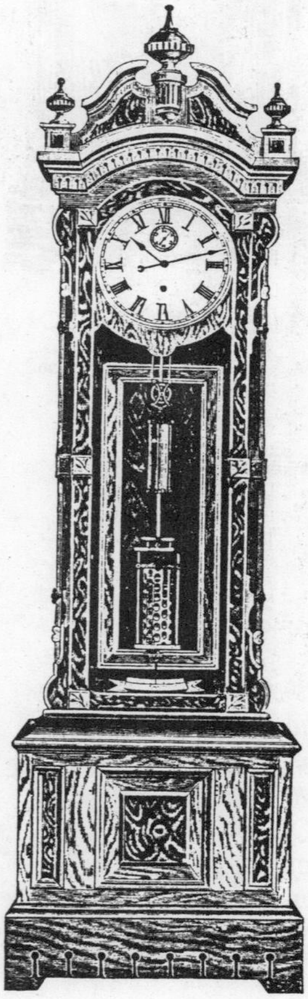

**FINE REGULATOR No. 15**

Independent Seconds.
Walnut Case, Mercury Pendulum. Eight Day, Weight, Time. Beats Seconds. Height, 100 inches. 14 inch Silvered Dial.

Price, $455.00

**$6,500.00**

# 1880 SETH THOMAS HALL REGULATOR CLOCKS

**$5,000.00**

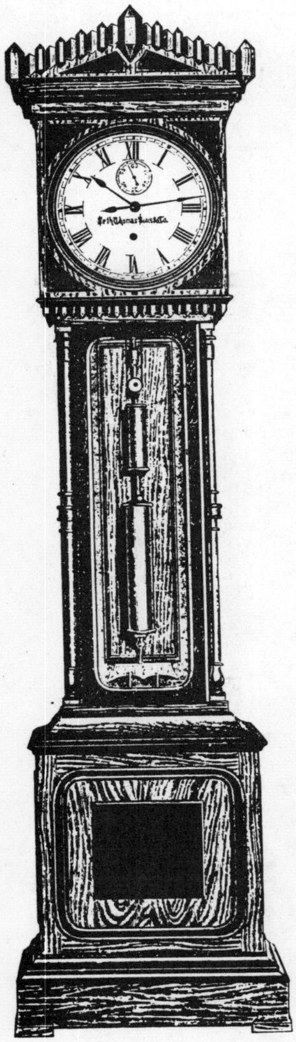

FINE REGULATOR.

No. 13, with Independent Second.

BEATS SECONDS.

8 Day, Weight, Time.

Hight, 96 inches.

**$6,000.00**

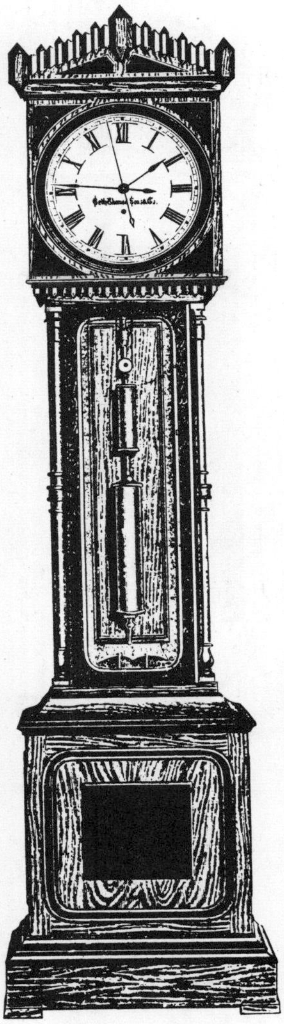

FINE REGULATOR.

No. 11, with Sweep Second.

BEATS SECONDS.

8 Day, Weight, Time.

Hight, 96 inches.

*Except for those fabulous clocks for which you wouldn't mind cutting a hole in the ceiling, Grandfather or Standing Clocks under eight feet are the most desirable and practical. Some manufacturers produced certain of these clocks in two styles — one to stand on the floor, and the other to hang on a wall. Of the clocks designed with two styles, the hanging style is considered to be the most valuable.*

# 1884-1890 SETH THOMAS HANGING CLOCKS

*To receive top prices, clocks must be in original mint condition and in good running order — A replacement dial or missing parts will reduce prices accordingly.*

No. 17 as shown

**$1800.00**

No. 17 with Double Time

***Very Rare*** **$5,000.00**

REGULATOR No. 17.

MADE IN WALNUT AND OAK.

same as Regulator No. 6, Double Time

Made to order to indicate Double Time,

8 Day, Weight, Time. Beats Seconds.
Hight, 68 inches. 14 inch Dial.

MOVEMENT SAME AS IN No. 16 REGULATOR, BUT INDEPENDENT SECONDS.

**$1,500.00**

**REGULATOR No. 9.**

Hight, 48 inches.

10 inch Dial.

WALNUT OR CHERRY.

8 Day, Weight, Time.

# 1874-1910 SETH THOMAS HANGING CLOCKS

$650.00

REGULATOR No. 30.
Walnut. Spring.
Height: 37 inches.
8 Day, Time or Strike.
Seth Thomas Movement.

REGULATOR No. 8.

$2,750.00

WALNUT, CHERRY, OAK OR OLD OAK.

8 Day, Weight, Time. 14 inch Dial. Hight, 56 inches. Polished movement. Cut steel pinions. Graham pallets.

Wood rod. Brass covered ball. Retaining power.
68 Beats to the Minute.

$3,200.00

REGULATOR No. 7

Cherry or Walnut Hand Carved Cabinet Case. 8 Day, Weight, Time. 12 inch Dial. Height, 45 inches. Cut steel pinion movement, with polished plates. Wood rod, brass-covered zinc ball. Graham dead-beat escapement, maintaining power. Seventy-two beats to the minute.
List, $60.00

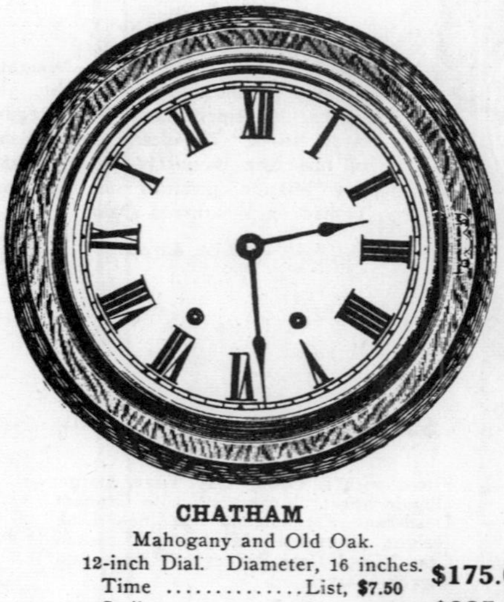

CHATHAM

Mahogany and Old Oak.
12-inch Dial. Diameter, 16 inches.
Time ..............List, $7.50 — $175.00
Strike ..............List, 8.50 — $225.00

$150.00

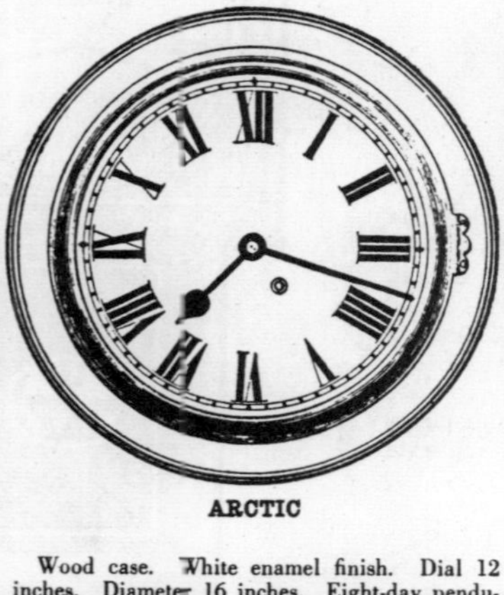

ARCTIC

Wood case. White enamel finish. Dial 12 inches. Diameter 16 inches. Eight-day pendulum time movement No. 41 or eight-day pendulum strike movement No. 89-A.

OFFICE No. 4

$300.00
$350.00

Mahogany and Old Oak. A new large pattern with plain numerals and hands at moderate price.
14-inch Dial. Diameter, 18½ inches.
Time ............List, $10.00
Strike ............List, 11.00

# 1875-1910 SETH THOMAS HANGING CLOCKS

$4500.00

REGULATOR No. 16.

8 Day, Weight, Time. Beats Seconds.

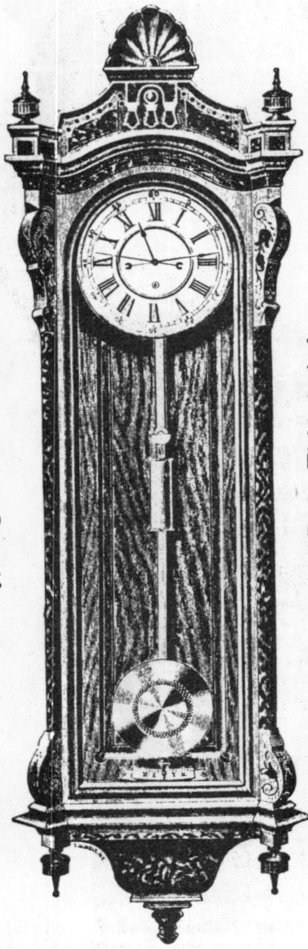

12 inch Dial. Hight, 75 inches.

$5,750.00

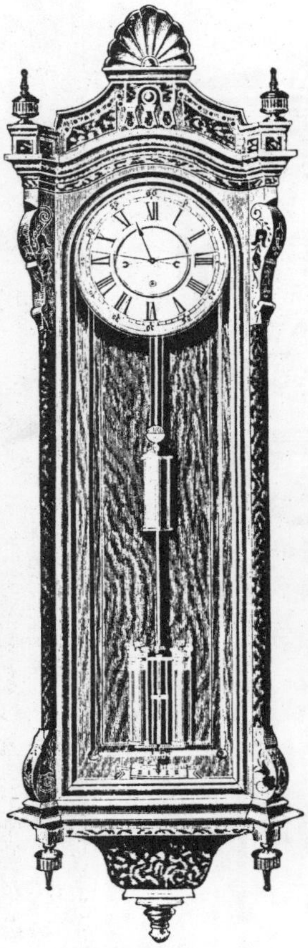

**REGULATOR No. 19**

(Sweep Seconds)
Mahogany and Old Oak. Eight Day, Weight, Time. Beats Seconds. 12 inch Dial. Height, 75 inches. Large cut steel pinion movement, with polished plates, wood rod, Mercury pendulum, Graham dead-beat escapement, maintaining power. An accurate time-piece for Railroads, Public Buildings, etc.

List, $225.00

$4500.00

REGULATOR No. 16

**REGULATOR No. 16.**
Walnut, Oak or Old Oak
8 - Day, Weight, Time
Beat Seconds, Hight 75
inches. 12 inch Dial

*Note: This is the original earlier model of No. 16. The later model No. 16 is shown at the far left from which a change in the top and bottom can be seen.*

$950.00

**REGULATOR No. 25**

Flemish Oak. Eight Day, Weight Time. 12 inch Dial. Height, 32 inches. Movement same as in Regulator No. 2. The design and finish of this case is suitable to go with any Mission furniture in Flemish, Fumed or Weathered Oak.

List, $27.00

**REGULATOR No. 20**

$1800.00

**REGULATOR No. 20**

Cherry, Mahogany Finish and Old Oak. Eight Day, Weight, Time. Beats Seconds. 14 inch Dial. Height, 62 inches. Large movement, wood rod, brass-covered zinc ball, Graham dead-beat escapement, and maintaining power.
The round top and two side-pieces can be removed if a plain top case is desired.

List, $55.00

**LITCHFIELD**

$500.00

Fitted with a Thirty Day Time, Large Double Spring Movement, with Graham Dead-beat Escapement. 12 inch Dial. Height, 31 inches. Mahogany Finished Case. Gold Leaf Border on Upper and Lower Glasses.

List, $21.50

# 1875-1913 SETH THOMAS HANGING CLOCKS

**$1,200.00**

1875

**REGULATOR. No. 1.**

8 Day, Weight, Time.

Hight, 34 inches.

12 inch Dial.

**$1,100.00**

1875

**REGULATOR No. 2**

12 inch dial.

Height, 34 inches.

EIGHT DAY, TIME.

**$2,000.00**

1875

## REGULATOR—No. 1.—Extra.

| | | | |
|---|---|---|---|
| 8 Day. | Time. | Weight. | 14 Inch Dial. |
| " | Strike. | " | " " |

Hight, 44 inches.

**$950.00**

## DOUBLE DIAL CLOCK

MADE IN MAHOGANY OR GOLDEN OAK.

1904

15 Day, Pendulum, Time.

Two 18 inch Dials, placed back to back, so that time can be read from two sides. Case arranged to hang or stand.

WHEN ORDERING SPECIFY WHICH WAY CLOCK IS TO BE USED.

Hight, 29 inches. Price, $125.00 Base, 37 inches.

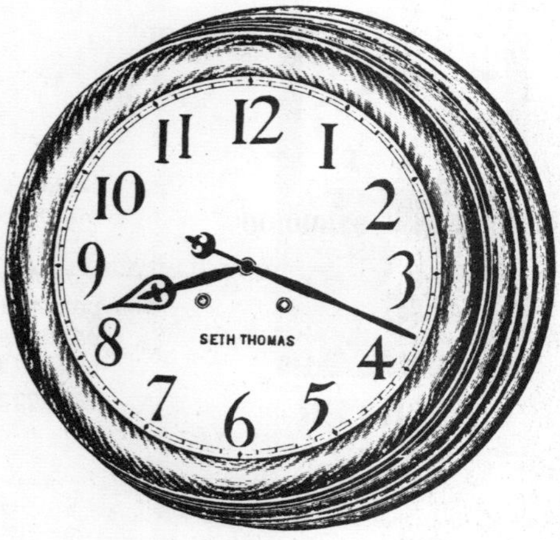


1913

**GALLERY**

Mahogany, Old Oak and Gold Leaf Finish. 18-inch Dial. Back measurement, 25½ inches. 24-inch Dial. Back measurement, 32 inches.

| | | |
|---|---|---|
| 18-inch Mahogany or Old Oak..List, | $35.00 | **$800 *to*** |
| 18-inch Gilt ....................List, | 54.00 | |
| 24-inch Mahogany or Old Oak..List, | 50.00 | |
| 24-inch Gilt ....................List, | 65.00 | **$1,200.00** |

# 1880 - 1939 SETH THOMAS HANGING CLOCKS

**$7,000.00**

Note: Number 10 shows both sweep and independent seconds display. This may be an error in the original catalogue.

**$6,000.00**

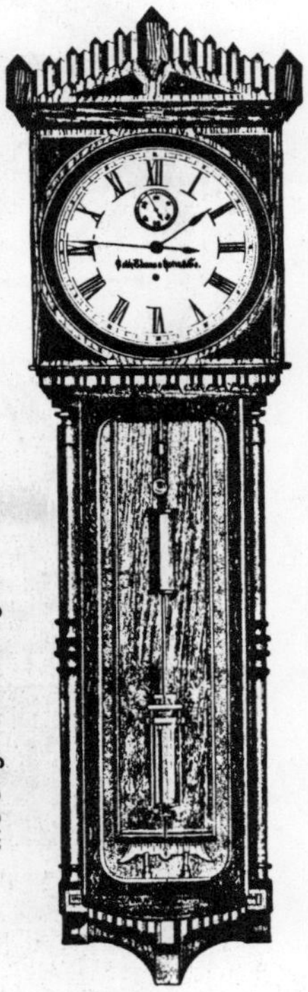

No. 10 No. 12

FINE REGULATOR.

MERCURY PENDULUM.

**8 Day, Weight, Time.**

**Hight, 72 inches.**

14 inch Dial.

BEATS SECONDS.

**$7,000.00**

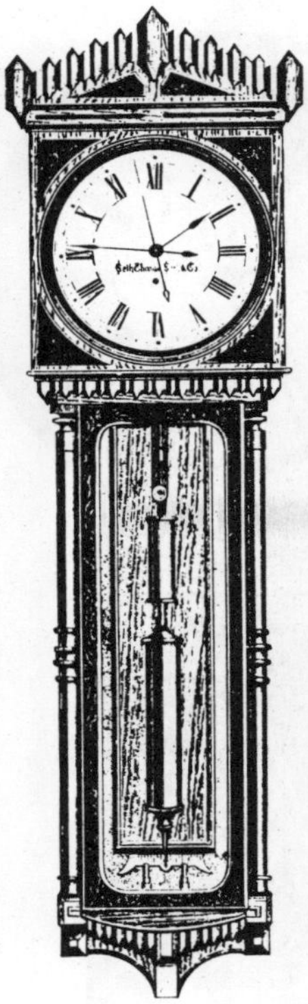

**No. 10**

Sweep Seconds.

8 Day, Weight, Time.
Hight, 72 inches. 14 inch Dial.

*Note: This is the original earlier model of No. 10. The 1884 model No. 10 is shown at the left and the later 1910 model No. 10 is shown at the right from which a change in the pendulum and case can be seen.*

**$7,000.00**

**$6,000.00**

**REGULATOR No. 10**

Sweep Second.

**REGULATOR No. 12**

Independent Second.

Walnut or Mahogany Case, Mercury Pendulum. Eight Day, Weight, Time. Beats Seconds. Height, 72 inches. 14 inch Silvered Dial.

Price, **$410.00**

1939

**$100.00**

## OBSERVER

8-day Pendulum Time Movement

An up-to-the-minute clock for schools, offices, factories and stores. The Observer has sturdily constructed oak case and is smartly designed for greatest adaptability. Legible black numerals and hands appear against white dial. SPECIFICATIONS: Height 26½ inches. Width 13⅜ inches. Depth 4¾ inches. Dial 11⅛ inches square. Packed singly. Approximate shipping weight 20 pounds.

*Except for those fabulous clocks for which you wouldn't mind cutting a hole in the ceiling, Grandfather or Standing Clocks under eight feet are the most desirable and practical. Some manufacturers produced certain of these clocks in two styles — one to stand on the floor, and the other to hang on a wall. Of the clocks designed with two styles, the hanging style is considered to be the most valuable.*

**$425.00**

Ca.1890

**Style B.**

BURGLAR ALARM INSTRUMENT.

Has constant ring, silent indicator test, battery test, cut off and **Fifteen day Seth Thomas Clock.**

This instrument has servant's call attachment and disconnects automatically any portion of the house at a given time, and can be set for any desired hour.

THE MOST PERFECT BURGLAR ALARM MADE.

# 1881 SETH THOMAS HANGING CLOCKS

$2,500.00

MARCY.

QUARTER STRIKE. MADE IN WALNUT, OAK AND CHERRY.

Dial, 8½ inches. Hight, 46 inches.

8 Day, Spring.

Striking quarter hours on two Cup Bells and hours on a Cathedral Bell.

$550.00

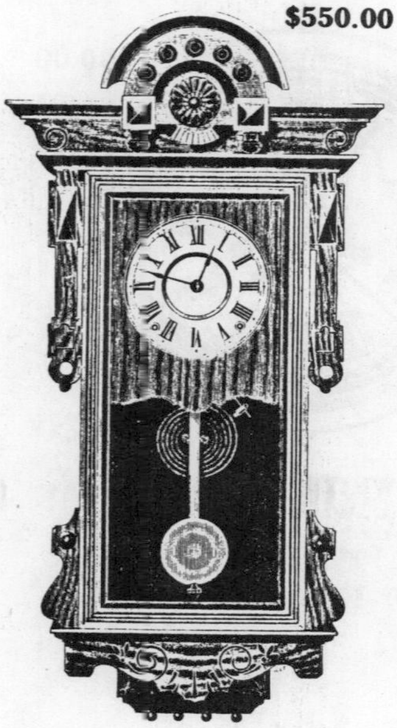

PANAMA.

MADE IN WALNUT, OAK AND CHERRY.

8 Day, Spring, Strike.

Cathedral Bell.

6 inch Dial.

Hight, 30 inches.

$1000.00

SUEZ.

8 Day, Spring, Strike, Cathedral Bell.

Hight, 44 inches.

A Superior Movement. Wooden Rod.

8½ inch Dial.

MADE IN WALNUT AND EBONY.

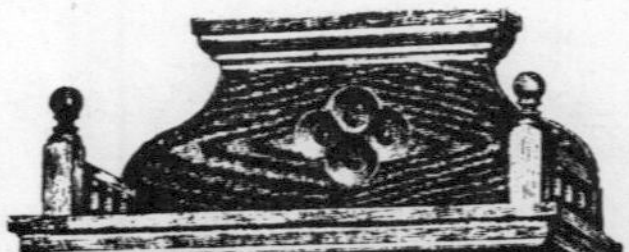

$1,275.00

8 inch Dial.

Brass Weights. Wooden Rod.

8 Day, Weight, Strike.

Cathedral Bell.

FLORA.

MADE IN WALNUT, EBONY, OAK AND CHERRY.

CASE HAND CARVED.

Hight, 38 inches.

NEW AND SUPERIOR MOVEMENT.

QUEEN ANNE.

$800.00

8½ inch Dial.

Hight, 36 inches.

We make this Clock also in Ebony finish, with a very superior movement, and a Cathedral Bell. Price, $20.00.

Price, Silent, $13.50; Striking, $15.00.

8 Day, Spring, Time.

MADE IN WALNUT, OAK AND CHERRY.

8 Day, Spring, Strike, Cathedral Bell.

MADE IN WALNUT, EBONY, OAK AND CHERRY.

$750.00

QUEEN ANNE.

8 Day, Spring. Hight, 36 inches. 8½ inch Dial.

# 1886-1887 SETH THOMAS HANGING LEVER CLOCKS

## FRENCH LEVER.

Oak or Black Walnut.

Jeweled Balance.

| | | |
|---|---|---|
| 8 Day. | 8 Inch. | Time. |
| " | " | Strike. |
| 8 Day. | 10 Inch. | Time. |
| " | " | Strike. |

**$100.00 *to* $150.00**

**$90.00**

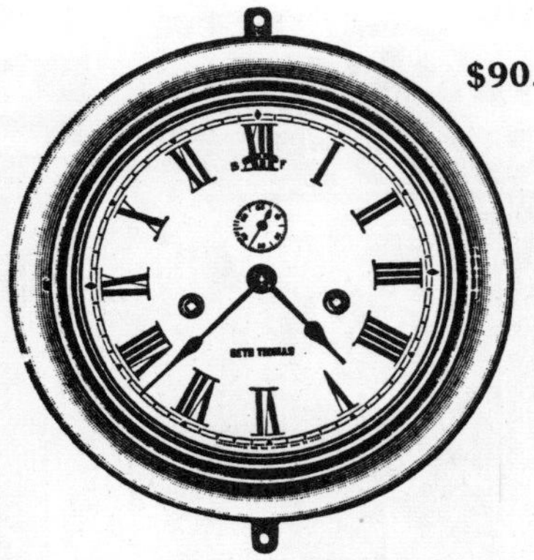

**WHITE LEVER**

Wood Case, White Adamantine Finish. Can be cleaned with a damp cloth any number of times without injury to the finish.

Eight-Day, 6-inch, Time, List **$10.00**

**$275.00**

FINE LEVER No. 1.

8 Day, Time.

12 inch Painted Steel Dial.

Diameter of Clock, 16¾ inches.

Cut Steel Pinion, Polished Jeweled Movement, Chronometer Balance.

## LEVER.

ROSEWOOD AND WALNUT VENEER, POLISHED.

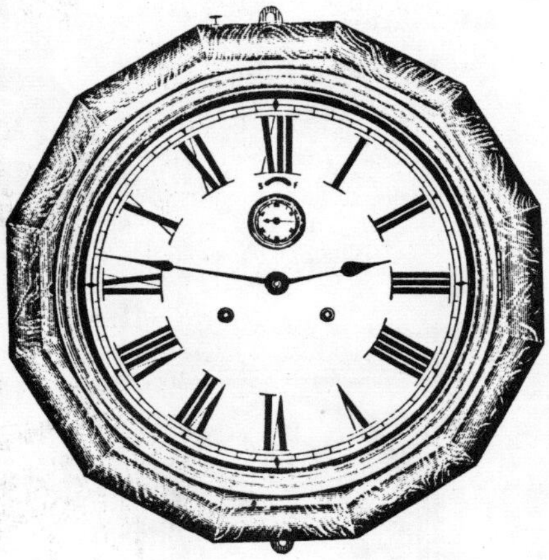

| | |
|---|---|
| 1 Day, 10 inch, Time. | **$110.00** |
| 1 Day, 10 inch, Strike. | **$125.00** |
| 8 Day, 10 and 12 inch, Time. | **$115.00** |
| 8 Day, 10 and 12 inch, Strike. | **$135.00** |

## LEVER.

ROSEWOOD, WALNUT AND MAHOGANY VENEER, POLISHED.

| | |
|---|---|
| 1 Day, 4 inch, Time. | **$85.00** |
| 1 Day, 6 and 8 inch, Time. | **$85.00** |
| 1 Day, 6 and 8 inch, Time, Alarm. | **$110.00** |
| 1 Day, 6 and 8 inch, Strike. | **$110.00** |
| 1 Day, 6 and 8 inch, Strike, Alarm. | **$150.00** |
| 8 Day, 6 and 8 inch, Time. | **$150.00** |
| 8 Day, 8 inch, Strike. | **$175.00** |

PORCELAIN DIAL.

## CHRONOMETER LEVER.

| | |
|---|---|
| 1 Day, No. 5003, 3½ inch Dial. | **$150.00** |
| 1 Day, No. 5004, 4½ inch Dial. | **$175.00** |
| 1 Day, No. 5006, 4½ inch Dial. | **$175.00** |

**The Movements are made in the best manner, with Steel Pinions and Chronometer Balance, Jeweled, and are fitted up with finely tempered Hair Springs.**

**They have a very simple and perfect regulator, which provides for the most accurate adjustment.**

**They are put up in two sizes of cast brass cases, with porcelain dials. No. 5003 is 5¼ inches, Nos. 5004 and 5006 are 7 inches in diameter.**

**Nos. 5003 and 5004 open at the front with a door. The front of 5006 is screwed on, and the winding is at the side of case. No. 5006 is dust-proof.**

**These Clocks are warranted accurate time-keepers.**

# 1886-1910 SETH THOMAS HANGING CLOCKS

$145.00

## LEVER CLOCKS

$125.00

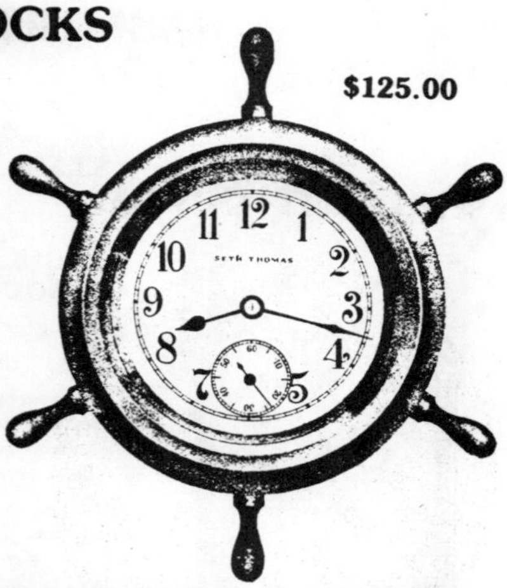

**SWITCH-BOARD LEVER**

Metal Case, Brass, Bronze or Nickel Plated. 6 inch Etched Metal, Silvered Dial, Hinged Bezel. Eight Day, Time, Double Spring Movement, with Balance Pivots running in Jewels. A superior movement and close-fitting case. Width, front and back, 8 inches. List, $25.00

**ENGINE LEVER**

Metal Case, Brass or Nickel Plated. 6 inch Dial, Hinged Bezel.

1 Day, Time........................$6.00
8 Day, Time........................ 9.00

$75.00

$125.00

**MOTOR BOAT No. 1—ONE DAY MOVEMENT**

No. 1—Flat, the top spoke used to wind and set....List, $14.00
No. 2—Offset, with spokes; with reliable lock; clock has same slanting position as Cut No. 11 " 18.00

BANNER LEVER.

**METAL, BRASS OR NICKEL PLATED.**

1 Day, 4 inch, Time.

1 Day, 6 and 8 inch, Time.

1 Day, 6 and 8 inch, Strike.

8 Day, 6 and 8 inch, Time.

8 Day, 8 inch, Strike.

$140.00
$150.00
$165.00
$175.00
$200.00

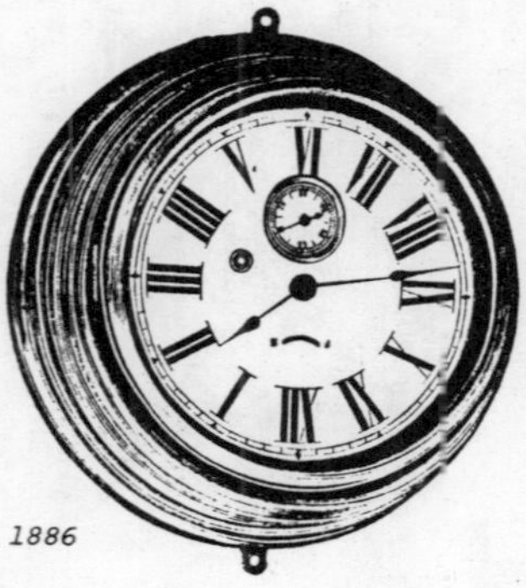

*1886*

$185.00

**CHRONOMETER LEVER**

Eight Day, Time. No. 5020

Cabinet Case, Mahogany or Golden Oak. Detached Eleven Jewel Escapement, Compensating Balance Wheel, Breguet Hairspring, Cut Steel Pinions. This Escapement is same as used in our high grade Railroad Watch Movements. 12 inch Painted Metal Dial. Dial can be had without seconds if desired. Diameter of Clock, 16¾ inches.

List, $54.50

**CHRONOMETER LEVERS**

Eight Day, Time.

For Yachts, Steamers and All Places Where Accurate Time Is Necessary and a Pendulum Clock Cannot Be Used.

MOVEMENT

Detached Eleven Jeweled Escapment, Compensating Balance Wheel, Breguet Hairspring, Cut Steel Pinions. Etched Metal Silvered Dials. This escapement is same as used in our high grade Railroad Watch Movements.

CASES

Cast Brass Cases, Polished or Nickel Plated. The jointed bezels have solid cast hinges. Crystals of heavy plate-glass, fastened with a spring and cemented. The screw bezel cases are weather-proof.

$175.00

**NAVY LEVER**

Cast Metal Case, Brass or Nickel Plated. One Day, Time. 6 inch Silvered Dial. Movement winds through side of Case, balance pivots running in Jewels.

A High Grade, Easy Running, Double Spring Movement.

Should a spring break, the other will run clock until repair can be made. **These Clocks are warranted accurate timekeepers.**

SCREW BEZEL CASES—Waterproof.

| | | | | |
|---|---|---|---|---|
| No. 4008—At front, 6½ inches; | at back, 6¾ inches. | 5 inch Dial | List, | $45.00 |
| No. 4009—At front, 7½ inches; | at back, 7¾ inches. | 6 inch Dial | List, | 50.00 |
| No. 4010—At front, 10½ inches; | at back, 10¾ inches. | 8 inch Dial | List, | 60.00 |
| No. 4012—At front, 14½ inches; | at back. 14 inches. | 12 inch Dial | List, | 80.00 |

JOINTED BEZEL WITH LOCK.

| | | | | |
|---|---|---|---|---|
| No. 5009—At front, 7½ inches; | at back, 7¾ inches. | 6 inch Dial | List, | $61.50 |
| No. 5010—At front, 10¼ inches; | at back, 10½ inches. | 8 inch Dial | List, | 73.50 |
| No. 5011—At front, 12 inches; | at back, 12 inches. | 10 inch Dial | List, | 85.00 |
| No. 5012—At front, 14½ inches; | at back, 14 inches. | 12 inch Dial | List, | 100.00 |

$200.00

to

$550.00

# 1886-1904 SETH THOMAS HANGING CLOCKS

**$3,500.00** *Rare*

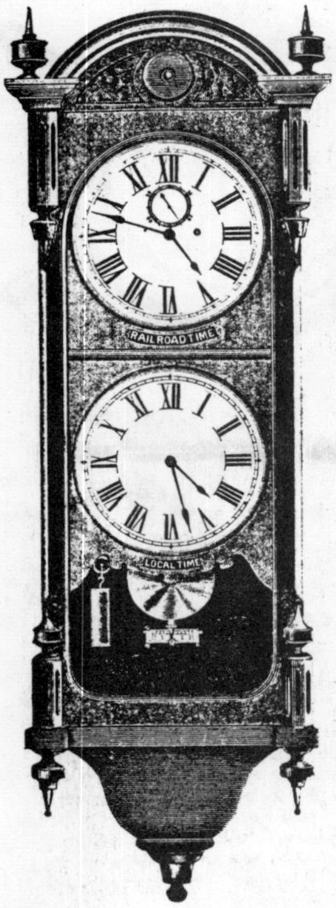

**REGULATOR No.6**
**DOUBLE TIME**

8 Day, Weight, Time. Hight, 49 inches.
10 inch Dial.

Dials are run by one movement, and being once set, will maintain relative difference. Moving hands on upper dial will change lower dial to same extent.

Serves to indicate "Railroad Time," "Local Time," "Eastern Time," "Central Time," or any other similar requirement.

Price, $40.00

**$3,000.00**

**JUPITER**

OAK, OLD OAK OR MAHOGANY.

Hight, 59 inches.

8 Day, Strike, large Cup Bell.

12 inch Dial.

Dial shows the Moon's Phases.

Clock is wound by pulling down the center chains.

Price, $45.00

**$1,600.00**

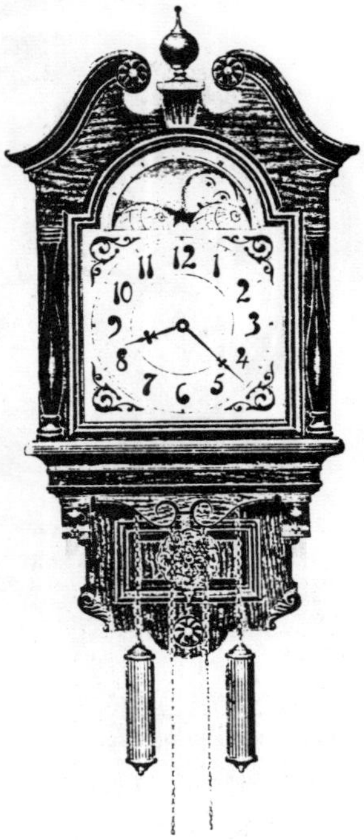

**LUNAR**

OAK, OLD OAK OR MAHOGANY.

8 Day, Strike, large Cup Bell.

12 inch Dial.

Hight, 41 inches. Width, 20 inches.

Dial shows the Moon's Phases. Clock is wound by pulling down center chains.

Price, $28.00

REGULATOR No. 6.

WALNUT CASE.

Hight, 49 inches, 10 inch Dial.

Movement same as in Regulator No. 2.

**$1,500.00**

8 Day, Weight, Silent.

Price, $25.00.

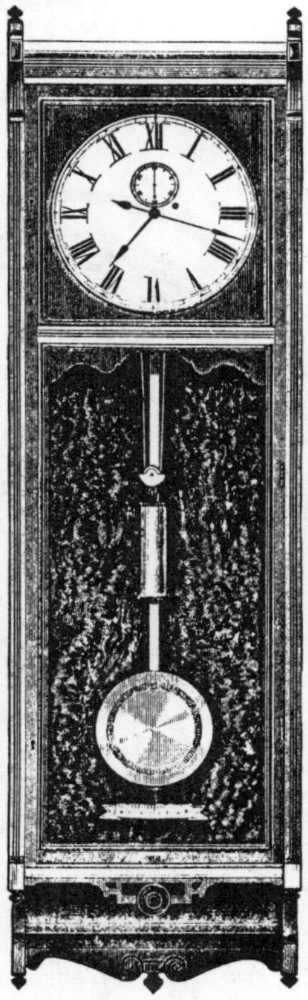

Movement same as in No. 16 Regulator, but independent seconds.

*Rare*
**$2,000.00**

WALNUT, CHERRY, OAK OR OLD OAK.

8 Day, Weight, Time. Beats Seconds.

Hight, 68 inches.

**REGULATOR No. 31,**
**with 18 inch Dial.**

**$850.00**

**UMBRIA.**

OLD OAK.

15 Day, Time. 8 Day, Strike.

Hight, 40½ inches. 10 inch Dial.

UMBRIA

# 1886-1910 SETH THOMAS HANGING CLOCKS

$3,750.00

**REGULATOR No. 60.**

Old oak or mahogany; 8-day. weight, time; beats seconds; height, 58½ inches; 14 inch dial. Movement has cut pinions of solid steel, finished plates, Graham escapement; brass covered zinc ball, wood rod....... $108 36

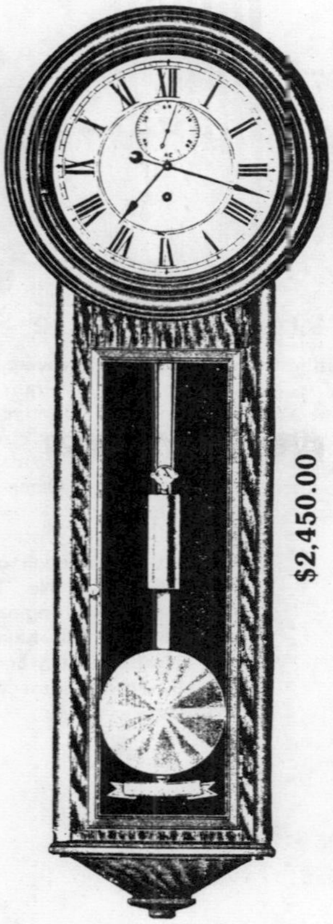

$2,450.00

**REGULATOR No. 62**

MAHOGANY OR GOLDEN OAK.

8 Day, Weight, Time, Beats Seconds.
Hight, 60 inches.
14 inch Painted Steel Dial.
Large lantern pinion movement, Graham pallets, retaining power, brass covered zinc ball, wood rod, brass weight.

Price, $56.00

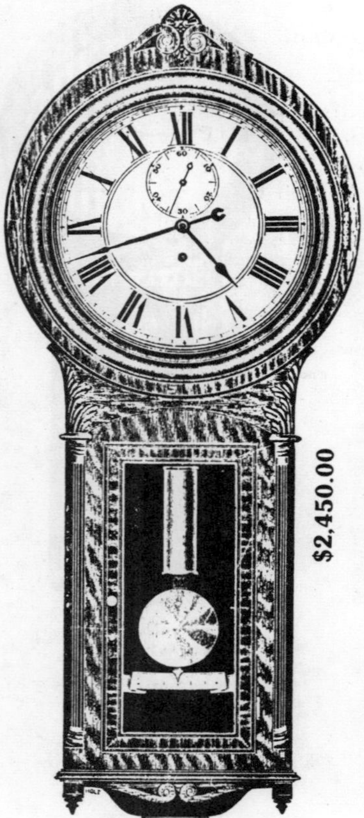

$2,450.00

**REGULATOR No. 33.**

Made in oak or mahogany.
Height: 48½ inches; 14 inch Dial.
LIST: $52.92.

$1,800.00

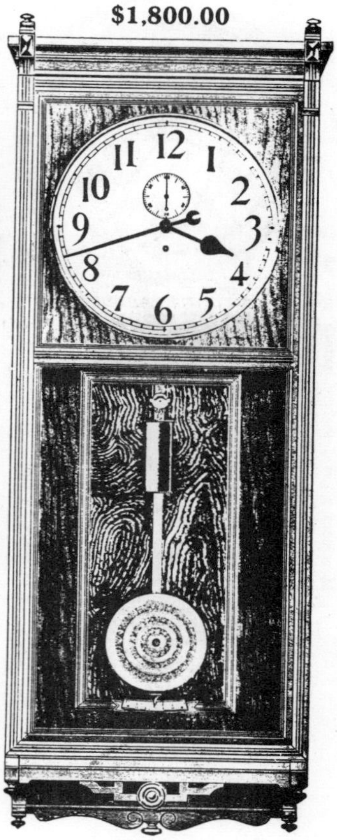

**REGULATOR No. 31**

Mahogany and Old Oak. Eight Day. Weight, Time. Beats Seconds. 18 inch Dial. Height, 68 inches. Roman or Arabic. Large cut steel pinion movement, with polished plates, wood rod, brass-covered zinc ball, Graham dead-beat escapement, and maintaining power. Suitable for a very large room.

List, $100.00

$800.00

**18-INCH LOBBY**

Mahogany and Old Oak. 18 inch Dial. Height, 38 inches. Width, 25 inches.
List, $35.00

$400.00

**ARCADE**

Fitted with a Thirty Day Time, Large Double Spring Movement, with Graham Dead-beat Escapement. 18 inch Dial. Mahogany Finish and Old Oak Cases. Front measure over all, 23½ inches. Back is an oblong box arrangement for movement, and measures 19 inches high, 10 inches wide, and 4 inches deep. The entire depth from front to back is 5½ inches.

List, $21.50

$600.00

**14 INCH LOBBY**

Mahogany and Old Oak. 14 inch Dial. Height, 30½ inches. Width, 20 inches.
List, $27.00

# 1886-1915 SETH THOMAS HANGING CLOCKS

Movement 2½x1½ inches, lantern pinions, Graham pallets, brass covered zinc ball and wood rod. Eighty beats to the minute. Movement has retaining power.

Oak $875.00
Walnut, Cherry or Mahogany $1000.00

**REGULATOR No. 2.**

Walnut and Oak Veneer, Polished.

Height, 24 inches; 12-inch Dial.

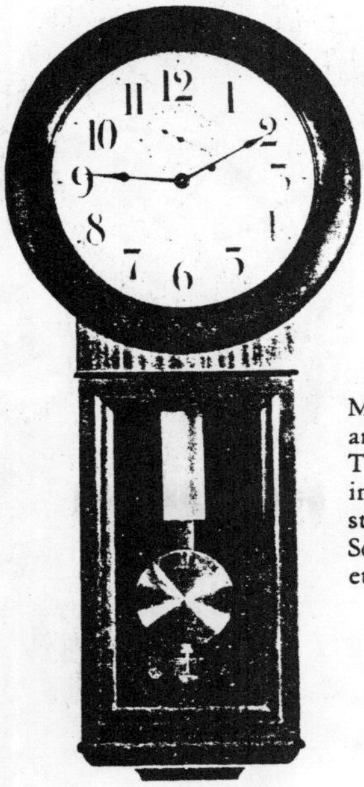

$775.00

Regulator No. 2

Mahogany and quartered oak. Veneered and polished. Eight-day, weight, time movement No. 77-A. Twelve-inch dial. Height 36½ inches. (80 beats per minute.)

*Another Regulator No. 2 ... note slight variation at bottom of case.*

$775.00

## REGULATOR No. 2

Mahogany and Old Oak. Veneered and polished. Eight-day, Weight, Time Movement No. 77-B. Twelve-inch Dial. Height 36½ inches. A standard small regulator for Railroads, Schools, Fire Departments, Offices, etc. (80 beats per minute.)

**This is a modification of earlier Regulator No. 2 model. See the original model on the left—hand page showing a fancy bottom which the later model did not have.**

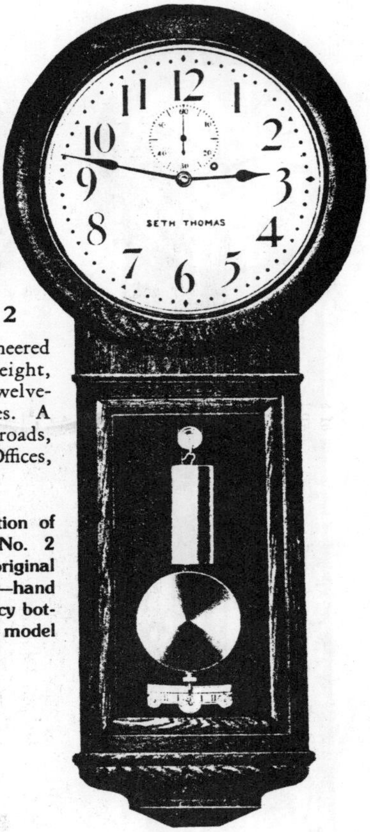


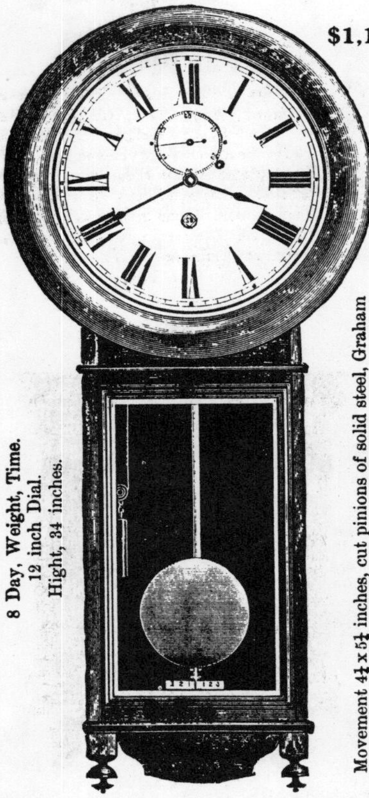

$1,100.00

8 Day, Weight, Time.
12 inch Dial.
Hight, 34 inches.

Movement 4¼ x 5¼ inches, cut pinions of solid steel, Graham pallets, brass covered zinc ball, and wood rod. Eighty beats to the minute. Movement has retaining power.

REGULATOR No. 1.

ROSEWOOD AND WALNUT VENEER, POLISHED.

REGULATOR No. 1.

ROSEWOOD AND WALNUT VENEER, POLISHED.

$1,200.00

*Another regulator No. 1...note slight variation at bottom of case.*

8 Day, Weight, Time.
12 inch Dial.
Hight, 34 inches.

Movement 4¼ x 5¼ inches, cut pinions of solid steel, Graham pallets, brass covered zinc ball, and wood rod. Eighty beats to the minute. Movement has retaining power.

Rare $2500.00

**REGULATOR No. 1**

Mahogany and Old Oak. 12-inch Dial. Lower glass painted in a Maroon with Gold Gilt Lines. Same excellent movement as in No. 2. Fine Cabinet Case. Desirable for homes as well as for office use. Has maintaining power.

Height, 34 inches.

# 1886-1923 SETH THOMAS HANGING CLOCKS

*Oak* **$1,750.00**

REGULATOR No. 3.

ROSEWOOD OR WALNUT VENEER, POLISHED.

8-Day, Weight, Time, Walnut, Oak or Old Oak Veneer .................... $ 41.34
8-Day, Weight, Strike, Rosewood or Walnut .................... 44.00 *Rare* **$2,000.00**
Time Movement 4¾ x 5¼ inches, Polished, Cut Pinions of Solid Steel, Graham Pallets, Brass Covered Zinc Ball, Wood Rod. Seventy-two beats to the minute.
Strike Movement has Lantern Pinions, Plates not Polished. Seventy-six beats to the minute. Both Movements have Retaining Power.

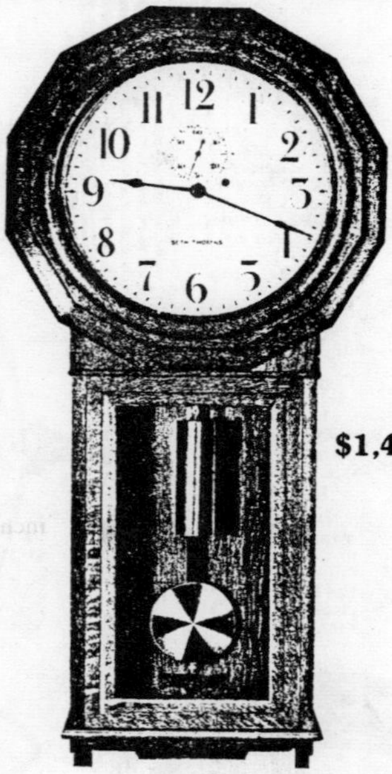

**$1,400.00**

Regulator No. 3

Quartered oak veneered case. Eight-day, weight, time movement No. 62. 14-inch dial. Height 41 inches. (72 beats per minute.)

*This is a modification of earlier Regulator No. 3 model. See the original model on the left hand page showing a fancy bottom which the later model did not have.*

*Oak* **$1,600.00**

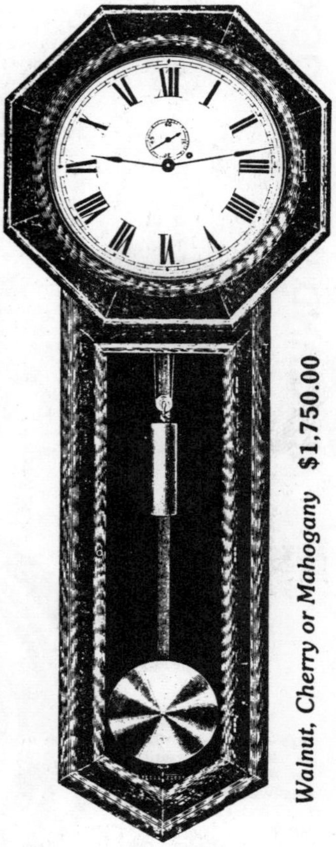

*Walnut, Cherry or Mahogany* **$1,750.00**

REGULATOR No. 18

Mahogany and Old Oak Veneer, Polished. Eight Day, Weight, Time. 14 inch Dial. Height, 54 inches. Beats Seconds. Movement has polished plates and is secured to an iron bracket, and pendulum is suspended from same. Wood rod, brass-covered zinc ball, Graham dead-beat escapement, maintaining power.

List, $40.00

REGULATOR No. 4.

**$3,000.00**

WALNUT CASE.

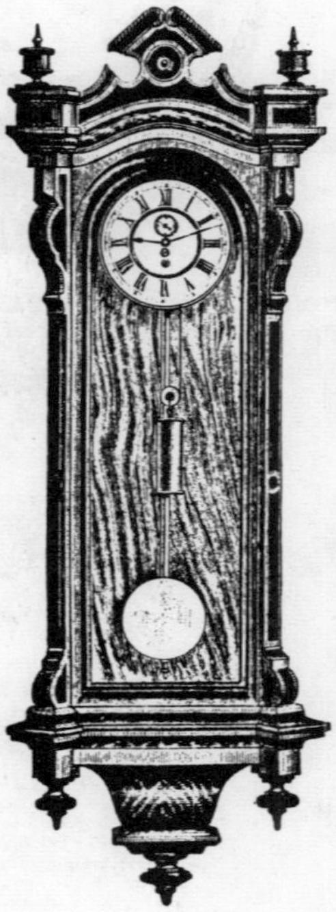

Movement 3¼ x 4¼ inches, finished same as movement in Regulator No. 3, and hung on brass bracket. It has a fine enamel sunk centre dial attached to movement, and mountings are gilded. Movement has retaining power.

Hight, 47 inches. 7 inch Porcelain Dial.

8 Day, Weight, Silent. Price, $45.00.

REGULATOR No. 5.

Solid Steel Pinions, hardened.

**$3,000.00**

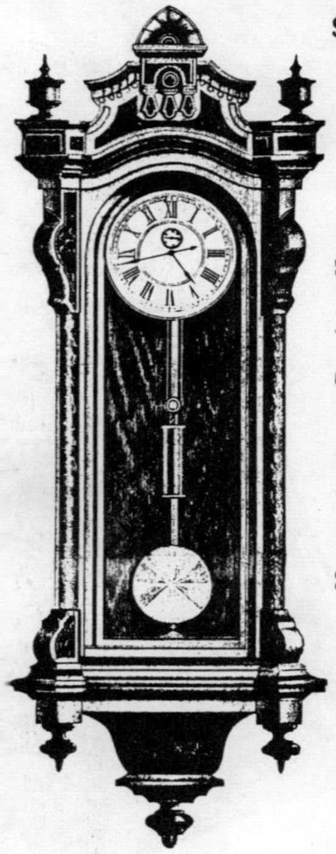

Movement same as Regulator No. 4.

8 Day, Weight, Time.

7½ inch Porcelain Dial.

Hight, 50 inches.

OFFICE No. 5

Mahogany and Old Oak.

12-inch Dial. Height, 23¼ inches. **$350.00**

Time ............List, $10.80

Strike ............List, 11.80 **$400.00**

# 1910 SETH THOMAS HANGING CLOCKS

## AUTO AND YACHT CLOCKS

### EIGHT DAY WHEEL CLOCK

For Yachts, Motor Boats, Yachtsmen's Dens, Prizes, etc.

MOVEMENT

High Grade, Lever, Time Movement. Seven Jewels, Breguet Hairspring, Cut Compensation Balance Wheel, with Timing Screws and Banking Screws. 2¾ inch Etched Metal Silvered Dial with large Seconds Circle. Movement is covered with a brass shell for additional protection and to permit of safe handling when out of case. This movement has been built specially to stand the hard usage of motor boats and automobiles.

CASE

Cast brass case, screw bezel, having a thick plate glass crystal held in place with a spring and also cemented, making clock waterproof. Clock is fastened in position from the *inside* of case, and secured by a reliable lock, as shown in Automobile Case on next page. Patent applied for, making clock burglar and waterproof. There are no visible screws that hold clock in position. The cut represents exact size.

List, $40.00

A 15 jewel movement can be had in this case for $10.00 list extra, the highest grade clock of this kind ever produced.

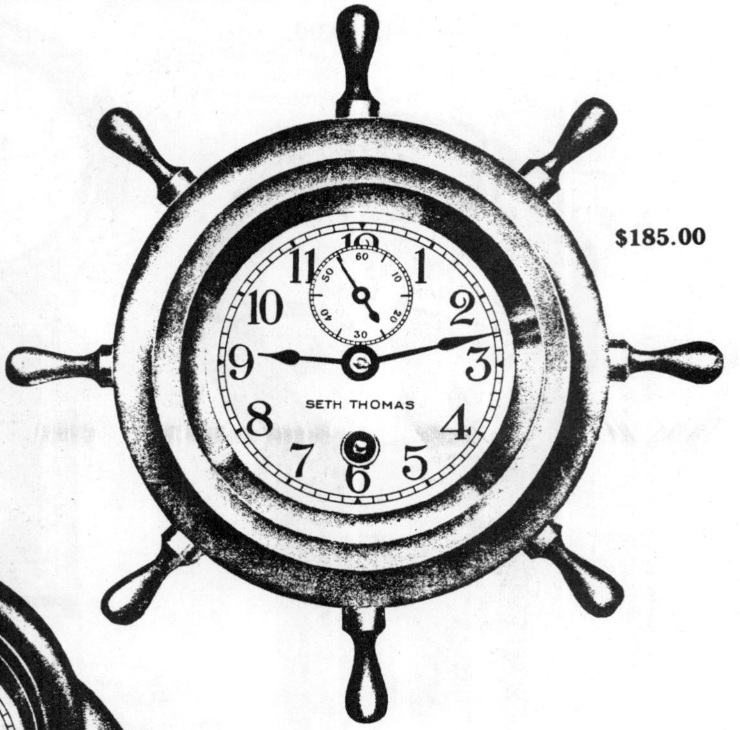


$185.00

### EIGHT DAY AUTOMOBILE CLOCKS

$175.00

MOVEMENT

High Grade, Seven Jewels, Breguet Hairspring, Cut Compensation Balance Wheel, with Timing Screws and Banking Screws, finely finished throughout. Made in two sizes, 2¾ and 3¾ inch Etched Metal Silvered Dials, with Large Seconds Circles. Movement is covered with a brass shell for additional protection and to permit of safe handling when out of case.

No. 23

CASES

Brass Cases with Screw Bezels, having a thick Plate Glass Crystal held in place with a spring; also cemented, making clocks waterproof.

Clocks are fastened in position from the inside of case and secured by a reliable lock shown in Cut No. 23. Patent applied for, making clocks burglar and waterproof. There are no visible screws that hold clock in position.

$135.00

2¾ inch DIALS

| | |
|---|---|
| No. 20—Plain Case with Lock | List, $30.00 |
| No. 21—Plain Case with Lock and Adjustable Bracket | " 33.00 |
| No. 22—Cast Case, Flat, with Flange Front and Back, no Lock | " 30.00 |
| No. 23—Cast Case, Offset, with Flange Front and Back, and Lock | " 33.00 |

3¾ inch DIALS

| | |
|---|---|
| No. 30—Plain Case with Lock | List, $33.00 |
| No. 31—Plain Case with Lock and Adjustable Bracket | " 36.00 |
| No. 32—Cast Case, Flat, with Flange Front and Back, and Lock | " 36.00 |
| No. 33—Cast Case, Offset, with Lock, but no Flange Front or Back | " 36.00 |

### AUTOMOBILE No. 11—ONE DAY MOVEMENT

| | |
|---|---|
| No. 10—Flat, same shape as No. 1, but without spokes | List, $12.00 |
| No. 11—Offset, with reliable lock; slanting position | " 16.00 |

MOVEMENT

A High Grade Seven Jewel Watch Movement, Breguet Hairspring, Cut Compensation Balance Wheel, with Timing Screws, Safety Pinion and Banking Screws. Stem wind and stem (pendant) set. This movement has been used for years successfully by the best carriage-makers as a dashboard watch.

CASES

Cast brass cases, screw bezel, having a thick crystal.

Nos. 1 and 10 are fastened in position by three screws through the edge of back flange and are waterproof.

Nos. 2 and 11 are fastened in position by screws from inside of case and secured by a lock as shown in Cut No. 11. Patent applied for, making Nos. 2 and 11 burglar and waterproof.

The plain numerals and hands make these clocks very desirable.

$135.00

No. 20 and No. 30

No. 21 and No. 31

Nos. 21 and 31 can be adjusted to various angles, and are so arranged that bracket cannot be removed without first unlocking case, taking out movement, making clock and bracket burglar proof.

A 15 Jewel Movement can be had in any of these cases for $10.00 list extra, the highest grade Automobile Clocks ever produced.

# 1910 SETH THOMAS HANGING CLOCKS

## OFFICE—No. 1.

12 Inch Dial.

8 Day. Spring.
Hight, 25 inches.

Time. $300.00

Strike. $350.00

$375.00

**RIO**

Mahogany Veneer, Polished. Eight Day, Spring, Strike. 12 inch Dial. Height, 25¼ inches. List, $10.50

$475.00

$500.00

$375.00

**OFFICE No. 3**

Mahogany Veneer, Polished. Eight Day, Spring, Strike. 10 inch Dial. Height, 21½ inches. List, $9.50

$375.00

**OFFICE No. 2**

Mahogany Veneer, Polished. Eight Day, Spring. 12 inch Dial. Height, 26 inches.
Time ....................List, $8.00
Strike ....................List, 9.60

**8-INCH DROP OCTAGON**

Mahogany Finish and Fumed Oak. Eight Day Spring. Height, 17 inches.
Time ....................List, $5.40
Strike ....................List, 7.00

$225.00

$250.00

*Another Globe...note slight variation at bottom of case.*

**GLOBE**

New Model. Mahogany and Old Oak Veneer, Polished. 12-inch Dial. Height, 32 inches.
Time ....List, $11.60
Strike ...List, 12.60

$425.00

$450.00

**WORLD**

Mahogany and Old Oak Veneer. Fifteen Day, Time. Has a large double spring movement with Graham dead-beat escapement. An excellent timepiece.
Eight Day, Spring, Strike, Cathedral Bell. Height, 32 inches. 12 inch Dial.
Time ..........List, $14.00
Strike ..........List, 16.00

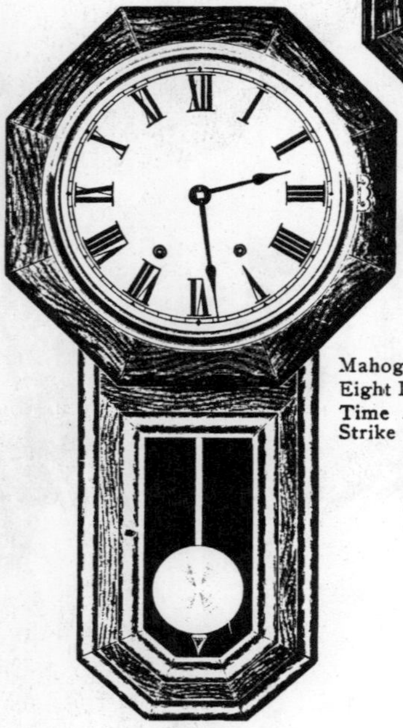

**GLOBE**

Mahogany and Old Oak Veneer, Polished. Eight Day, Spring, Wire Bell. 12 inch Dial. Height, 31 inches.
Time ............List, $11.00
Strike ...........List, 12.60

$425.00

$450.00

# 1910 SETH THOMAS HANGING CLOCKS

**10-INCH DROP OCTAGON GILT**

Mahogany Veneer. Polished with Gilt Lines, and Old Oak without Gilt Lines. Eight Day, Spring. Height, 21½ inches.

Time ....................List, $7.40
Strike ...................List, 9.00

**$250.00**

**$275.00**

SIGNET. V.V.
Height, 23 inches.
10-inch Dial
Walnut Case.

8 Day, Spring, Time **$400.00**
8 Day, Spring, Strike. **$450.00**

**12-INCH DROP OCTAGON**

Mahogany Veneer and Old Oak, Polished. Eight Day, Spring.
Height, 23½ inches.

Time ...................List, $7.80
Strike ...................List, 9.40

**$250.00**

**$275.00**

**BRIGHTON**

Made in Mahogany only. Eight Day, Spring, Strike. 12 inch Dial.
Height, 22¼ inches.
List, $9.00

**$275.00**

*Clock Only* **$400.00**

*Complete Set* **$525.00**

**LUSITANIA HANGING SET**

Rich Gold, Highly Burnished and Bronze Art Nouveau Finishes. Fitted with a High Grade Eight Day Lever Time Movement. Eleven Jewels. Compensating Balance. An accurate time piece. 4¼ inch Porcelain Dial. French Sash and Convex Beveled Glass.
Length of Clock, 20 inches; Width, 9½ inches. Length of Sconces, 13 inches; Width, 11½ inches.

List Price of Clock, $50.00 List Price of Set, $62.50 List Price of Sconces, per pair, $12.50

# 1910 SETH THOMAS HANGING CLOCKS

REGULATOR No. 17

**$1800.00**

**REGULATOR No. 17**

Cherry or Old Oak. Eight Day. Weight, Time. Beats Seconds. 12 inch Dial. Height, 68 inches. Large cut steel pinion movement, polished plates, wood rod, brass-covered zinc ball, Graham escapement, and maintaining power.

List, $85.00

**REGULATOR No. 11**

Mahogany and Golden Oak. Eight Day, Weight, Time. Beats Seconds. 12 inch Dial. Height, 56 inches. Large, cut steel pinion movement, with polished plates. Wood rod, brass-covered zinc ball. Graham dead-beat escapement and maintaining power.

List, $75.00

**$1700.00**

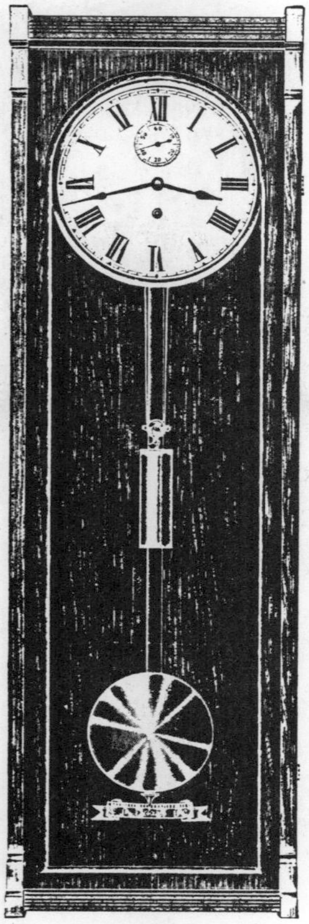

REGULATOR No. 11

**$2000.00**

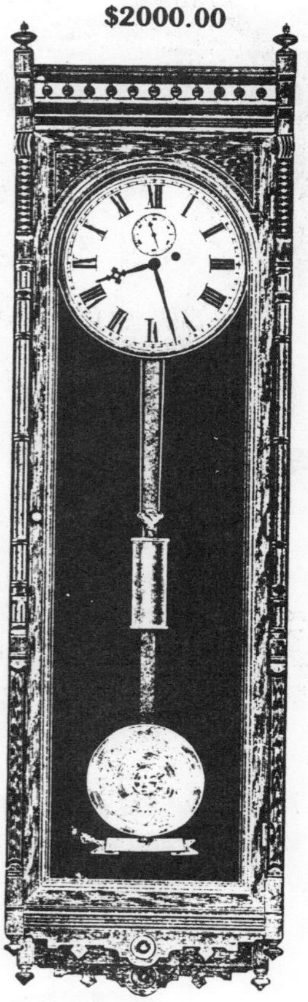

REGULATOR No. 32

Made in Old Oak only. Eight Day, Weight, Time. Beats Seconds. 12 inch Dial. Height, 68 inches. Large cut steel pinion movement, with polished plates, wood rod, brass-covered zinc ball, Graham dead-beat escapement, and maintaining power.

List, $100.00

**$450.00**

**30 DAY OFFICE**

Large Double Spring Time Movement, with Graham Dead-beat Escapement. 12 inch Dial. Height, 42 inches. Fine Hand Carved Cabinet, Old Oak Case.

List, $21.50

**$1350.00**

REGULATOR No. 30

Cherry or Old Oak. Eight Day, Weight, Time. 12 inch Dial. Height, 49 inches. Movement same as in Regulator No. 2.

List, $47.50

OFFICE No. 6

*To receive top prices, clocks must be in original mint condition and in good running order — A replacement dial or missing parts will reduce prices accordingly.*

**$350.00**

**OFFICE No. 6**

Mahogany Finish and Old Oak.
Eight Day, Spring, Time only.
12 inch Dial. Height, 36 inches.
Time ....................List, $8.00
Top can be taken off if a plainer case is preferred.

# 1913 SETH THOMAS HANGING CLOCKS

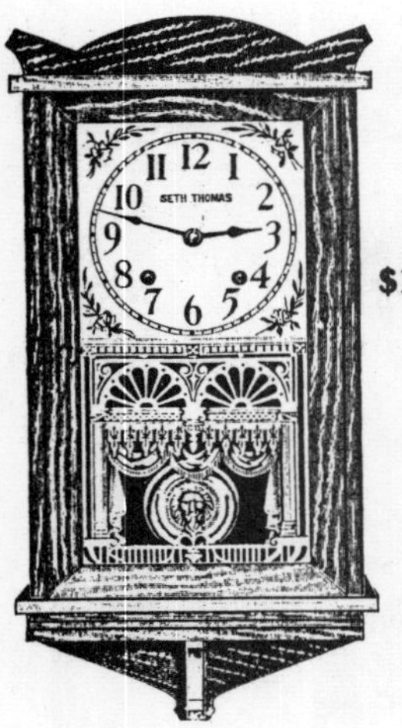


$125.00

$125.00

$125.00

$225.00

Double Dial Clock
*(Wall Bracket)*

Mahogany or quartered oak. Fitted with 30-day time large double spring movement No. 86. Dial 14 inches. Overall measurement with wall bracket 30 inches. Bracket designed to extend outward from wall or post.

**HANGING ASSORTMENT**

| | | |
|---|---|---|
| No. 33. | 6 Clocks, strike | List, $24.00 |
| No. 34. | 3 Clocks, strike<br>3 Clocks, strike and alarm | List, 25.80 |

**COMBINATION OF FOUR SHELF AND TWO HANGING CLOCKS**

| | | |
|---|---|---|
| No. 35. | 3 Clocks, assorted strike<br>3 Clocks, assorted strike and alarm | List, $25.80 |

*To receive top prices, clocks must be in original mint condition and in good running order — A replacement dial or missing parts will reduce prices accordingly.*

$125.00

**OFFICE No. 1**

Mahogany and Old Oak.
A new plain pattern.
12-inch Dial. Height, 16 inches.
Time ..........List, $7.00
Strike ..........List, 8.00

Office No. 11

Mahogany finish and quartered oak. Large double spring 30-day time movement No. 86 with Graham dead beat escapement. Twelve-inch dial. Height 15 inches. Width 15 inches.

$150.00

**OFFICE No. 10**

Thirty-Day Time.
Mahogany and Old Oak.
12-inch Dial.
Length, 22 inches.
Width, 16½ inches.
List Price, $25.00.

$185.00

$225.00

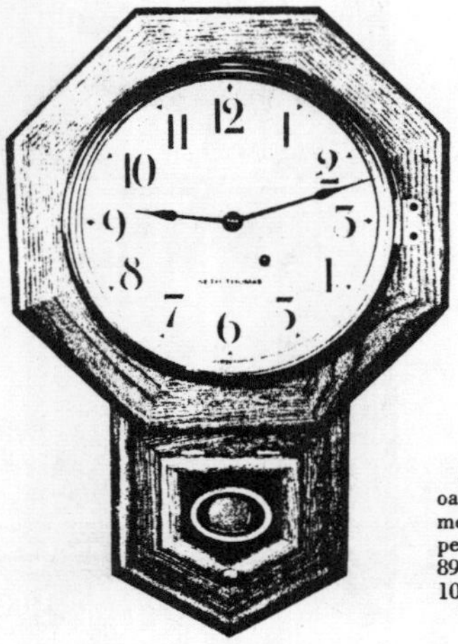

**10-inch Drop Octagon**

Mahogany finish and quartered oak. Eight-day pendulum time movement No. 41 or eight-day pendulum strike movement No. 89-C.
10-inch dial. Height 21½ inches. Width 14 inches.

$125.00

**FLAKE**

Eight-Day, Pendulum. 12-inch Dial. Wood Case, White Adamantine Finish. Can be cleaned with a damp cloth any number of times without injury to the finish. Diameter, 16 inches.

Time ..............List, $10.00
Strike ..............List, 11.00
Time and Calendar..List, 10.50
Strike and Calendar.List, 11.50

# 1913-1923 SETH THOMAS HANGING CLOCKS

## MARBLE DIAL CLOCKS

All Fifteen Day, Pendulum Time.

Large Double Spring Movements, with Graham Dead-beat Escapement. Excellent Timepieces.

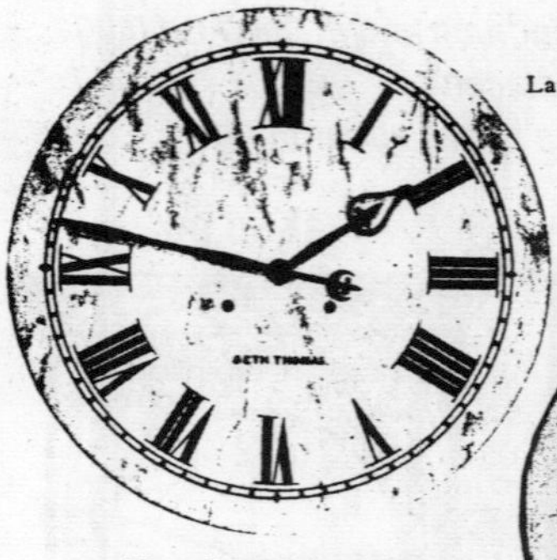

**Nos. 0 and 1 and 2**
No. 0, 12-inch Circle.
Total Diameter, 14 inches.
List, $60.00 **$425.00**
No. 1, 17-inch Circle.
Total Diameter, 20 inches.
List, $80.00 **$375.00**
No. 2, 23-inch Circle.
Total Diameter, 27½ inches.
List, $100.00 **$325.00**

**$600.00**

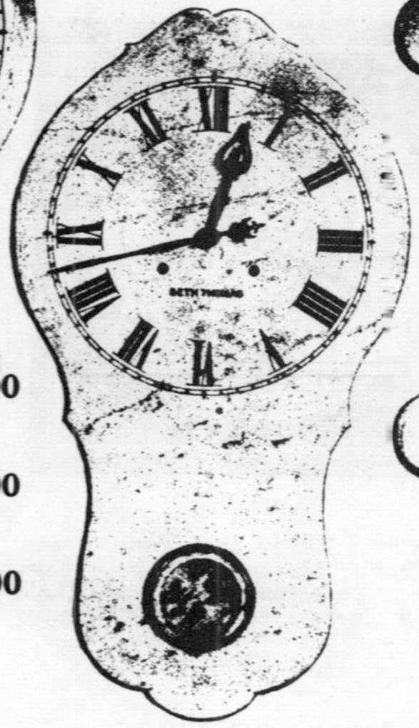

**No. 7**
No. 7, 13-inch Circle.
Height, 29½ inches. Width, 16 inches.
List, $100.00

**Nos. 3 and 4**
No. 3, 17-inch Circle.
Height, 27½ inches. Width, 23 inches
List, $90.00 **$500.00**
No. 4, 23-inch Circle.
Height, 36 inches. Width, 30 inches.
List, $110.00 **$475.00**

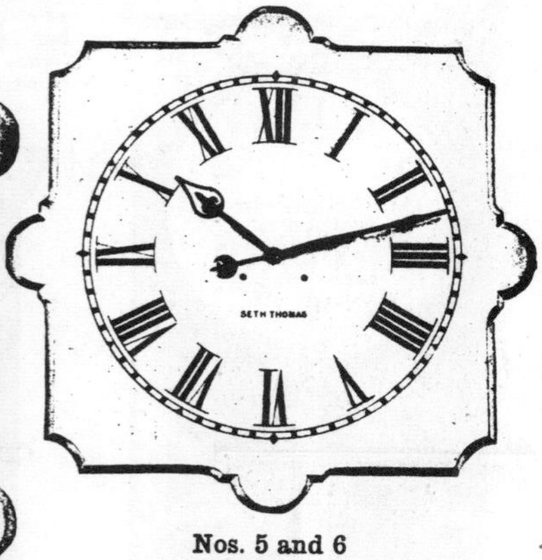

**Nos. 5 and 6**
No. 5, 17-inch Circle.
Height, 24 inches. Width, 24 inches.
List, $100.00 **$450.00**
No. 6, 23-inch Circle.
Height, 32 inches. Width 32 inches.
List, $120.00 **$400.00**

*Large clocks are generally more expensive than small models, except in the case of marble clocks where small sizes are the most valuable. They weigh less and are easier to hang in the average home.*

*Only clocks in ORIGINAL MINT condition and in good running order will command top prices. A replaced or damaged dial, missing parts or chipped beveled glass will reduce prices accordingly.*

**$2,750.00**

**REGULATOR No. 63**

Mahogany or Golden Oak. Eight Day, Weight, Time. Beats Seconds. Height, 76 inches. 14 inch Dial. Movement has polished plates, cut steel pinions, Graham escapement, brass covered zinc ball, wood rod.
Price, $126.00

**CHATEAU**
Inlaid Mahogany Case.
4-inch Convex Porcelain or Silvered Dial.
15-Day Pendulum Strike Movement No. 51.
Height, 8 inches. Width, 7½ inches.
List Price, $40.00.

**$140.00**

**$175.00**

**BRADFORD**
Mahogany and Old Oak.
Silvered Mat.
8-inch Dial, length 16½ inches.
Time. List Price, $11.00.
Strike. List Price, $13.00.

**$155.00**

**ABBEY-BRACKET**

## ABBEY-BRACKET

"Broken Arch" Old English Cabinet.
4-inch Convex Porcelain or Silvered Dial.
15-Day Pendulum Strike Movement No. 51.
Height of Clock and Bracket, 15 inches.
List Price, $50.00.
Brackets for Chateau, Amherst and Whitby.
List Price, $10.00.

# 1913-1923 SETH THOMAS HANGING CLOCKS

*To receive top prices, clocks must be in original mint condition and in good running order — A replacement dial or missing parts will reduce prices accordingly.*

**$850.00**

REGULATOR No. 4

Mahogany and Old Oak. Eight-day, Weight, Time Movement No. 77-B. Twelve-inch Dial. Height 42 inches. (72 beats per minute.)

**$1,100.00**

**Regulator No. 70**

Mahogany and quartered oak. Eight-day, weight, time movement No. 68, fitted with beat adjuster. Fourteen-inch dial. Height 56 inches. Width 20 inches. (60 beats per minute.)

*1913*

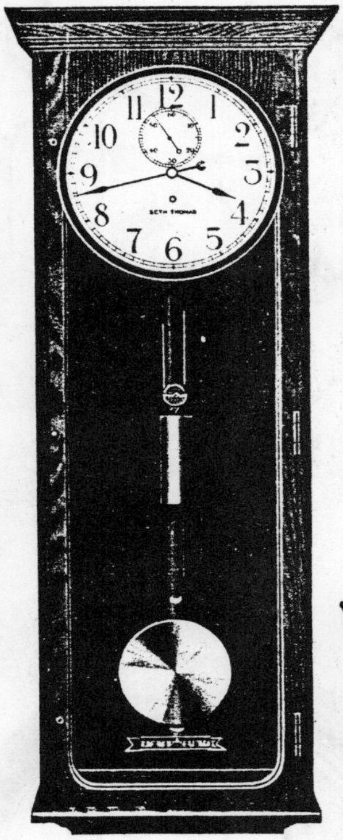

**$200.00**

**HUDSON**

Fitted with a Thirty Day Time, Large Double Spring Movement, with Graham Dead-Beat Escapement. 14 inch Dial. Case, 18 inches square. Mahogany inlaid.

List, $21.50

**$650.00**

**Regulator No. 4**

Mahogany and quartered oak veneered and polished. Eight-day weight time movement No. 77-A. Twelve-inch dial. Length 38½ inches. Width 16½ inches. Depth 6⅞ inches. 80 beats per minute.

**$1,200.00**

REGULATOR No. 80

Mahogany and Old Oak. Eight-day, Weight, Time Movement No. 68, fitted with beat adjuster. Fourteen-inch Dial. Height 55 inches. Width 18 inches. (60 beats per minute.)

**$350.00**

Office No. 16

Mahogany finish and quartered oak. Large double spring 30-day time movement No. 86 with Graham dead beat escapement. Twelve-inch dial. Length 31 inches. Width 15 inches.

**$200.00**

**8-Inch Drop**

Mahogany finish and quartered oak. Dial 8 inches. Height 18 inches. Width 11 inches. Eight-day pendulum movement No. 41 or eight-day pendulum strike movement No. 89-B.

**$350.00**

**OFFICE No. 9**

Thirty-Day Time.
Mahogany and Old Oak. 12-inch Dial. Length, 32½ inches. Width, 15 inches.
List Price, $35.00.

# 1928 SETH THOMAS HANGING CLOCKS

*Only clocks in ORIGINAL MINT condition and in good running order will command top prices. A replaced or damaged dial, missing parts or chipped beveled glass will will reduce prices accordingly.*

## *Decorative Hanging Clocks*

**$100.00**

DIANA

Antique gold on a blue base. 2½ inch dial. Height 14¾ inches. 8-day, four-jeweled time movement No. 103-A.

**$150.00**

LYRE

Antique gold. 3 inch dial. Height 21½ inches. 8-day, four-jeweled time movement No. 103-A.

**$135.00**

DuBARRY

Antique gold. 3 inch dial. Height 19 inches. 8-day, four-jeweled time movement No. 103-A.

**$110.00**

DANTE

Antique gold with red base. 3 inch dial. Height 16 inches. 8-day, four-jeweled time movement No. 103-A.

**$90.00**

DUCHESS

Old ivory finish with antique gold border. 3 inch dial. Height 9½ inches. 8-day, four-jeweled time movement No. 103-A.

**$125.00**

DAUPHIN

Antique gold. 3 inch dial. Height 15½ inches. 8-day, four-jeweled time movement No. 103-A.

# 1936-1939 SETH THOMAS HANGING CLOCKS

**$175.00**

Office No. 1

Mahogany finish and quartered oak. Dial 12 inches. Height 15 inches. Width 15 inches. Eight-day pendulum time movement No. 123-A or eight-day pendulum strike movement No. 89-A. Oak not made in strike.

**$250.00**

Hudson

Mahogany and quartered oak. Large double spring 30-day time movement No. 86 with Graham dead beat escapement. Fourteen-inch dial. Height 18 inches. Width 18 inches.

Arcade No. 1

Mahogany finish or quartered oak. Large double spring 30-day time movement No. 86 with Graham dead beat escapement. 18-inch dial. Diameter 24 inches.

**$350.00**

**$250.00**

Office No. 14

Mahogany finish and quartered oak. Large double spring 30-day time movement No. 86 with Graham beat escapement. Dial 14 inches. Diameter 18 inches.

*ASSEMBLY*

**$135.00**

## ASSEMBLY

**8-Day Time**

**Distinctive—modern—practical, the Assembly strikes a new note in business clock design. Case is finished in light mahogany with mahogany veneered front. Vertical grain of center inlay sets up a rich contrast with horizontal graining of sections on either side. 10½-inch legible dial with 5 minute markers in red. Eight-day time movement No. 41 with visible pendulum. Height 22¼ inches. Width 14 inches.**

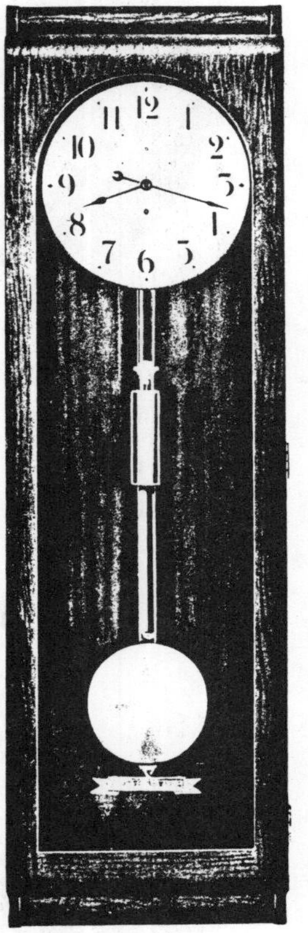

**$1,500.00** **No. 71 REGULATOR . . $150.00**
As illustrated with 8-day weight time movement No. 68 (60 beats per minute).

**$1,875.00** **No. 71 REGULATOR . . $250.00**
Mercurial pendulum.
Same except has two glass jar type mercurial pendulum and polished steel pendulum rod.

Note the up-to-date appearance of the sturdy case. Available in mahogany finish on birch or quartered oak. Features of movement include: reserve power, beat adjuster, jeweled pallets, polished cut steel pinions, polished keystone shaped plates, and it is mounted on a heavy cast iron bracket. 13-inch white enameled dial with black numerals. Height 60 inches. Width 18 inches. Packed singly. Approx. shipping weight 180 pounds (200 pounds for type with mercurial pendulum). (Specify finish in ordering.) Can also be supplied with No. 67 movement and sweep seconds hand at $10.00 extra.

## DOUBLE DIAL CLOCK

**30-Day with Ceiling Bracket $225.00**

**Mahogany or quartered oak. Fitted with substantial 30-day time double spring movement No. 86. 14-inch dial. Clock can be supplied with bracket inverted for mounting on partitions.** *Specify type in ordering.* **Ceiling bracket length overall 38 inches.**

# 1902 SETH THOMAS MANTEL CLOCKS

*Only clocks in ORIGINAL MINT condition and in good running order will command top prices. A replaced or damaged dial, missing parts or chipped beveled glass will reduce prices accordingly.*

*Most white metal and brass clock cases are gilded or silver plated. Great care must be taken when cleaning these sur- faces. The thin layer of gilding or silver plating will be easily removed by cleaning solutions and strong rubbing. If you believe cleaning is necessary, hire the services of a professional and avoid the expense of re-gilding or re-plating.*

**$300.00**

NUBIA
8 Day, Half Hour Strike, Cathedral Bell. 4¼ inch Cream Porcelain Dial French Sash, with Beveled Glass. Hight, 18 inches. Base, 9½ inches.

**$275.00**

INDIA
Cast Metal Case, Antique Brass or Barbedienne finish.
8 Day, Half Hour Strike, Cathedral Bell. Cream Porcelain 4 inch, Dial with French Sash and Beveled Glass. Hight, 15 inches. Base, 9¼ inches.

**$365.00**

LOUVRE
Marqueterie Panel. Mahogany front with polished brass inlay.
Gilt metal trimmings.
8 Day, Half Hour Strike, Cathedral Bell. 4 inch White Porcelain Dial. Hight, 16 inches. Base, 10 inches.

**$275.00**

SYRIA
8 Day, Half Hour strike, Cathedral Bell. 4¼ inch Cream Porcelain Dial French Sash, with Beveled Glass. Height, 16¼ inches. Base 9½ inches
Price, $16.50

**$275.00**

FANTASY
Bronze Art Nouveau Finish.
Height, 13 inches. Base 10 inches.
Price, $19.50

**$365.00**

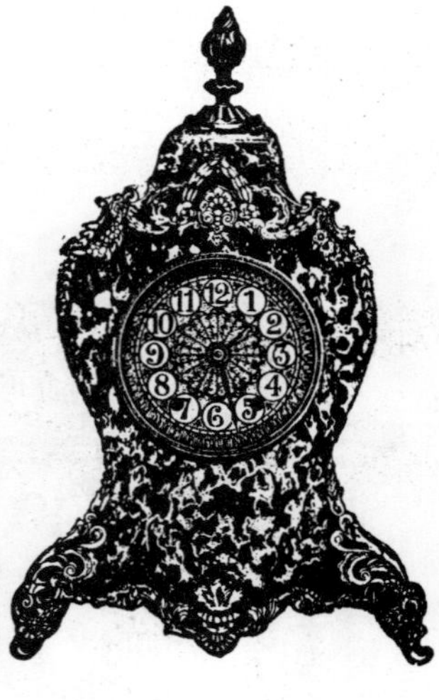

PARIS.
Tortoise Shell Adamantine. front, Gilt Metal Trimmings. Hight, 16 inches. Base 10 inches. 4 inches Gilt Dial.
8 Day, Spring, Strike, Cathedral Bell.

# 1910 SETH THOMAS MANTEL CLOCKS

**$425.00**

MARSEILLES

**MARSEILLES**

Etched Metal on Art Glass. Made in two Finishes, Verde Antique and Bronze. Eight Day, Fine Movement. Hour and Half Hour Strike, Cathedral Bell. 4 inch Porcelain Dial. French Sash and Beveled Glass. Height, 12½ inches. Base, 7½ inches.

List, $57.00

**$285.00**

DUKE

**DUKE**

Bronze and Verde Finishes. 4½ inch Dial. Five Gold Plated Top Ornaments. Height, 13 inches.

List, $18.00

*Most white metal and brass clock cases are gilded or silver plated. Great care must be taken when cleaning these surfaces. The thin layer of gilding or silver plating will be easily removed by cleaning solutions and strong rubbing. If you believe cleaning is necessary, hire the services of a professional and avoid the expense of re-gilding or replating.*

**$275.00**

**DUCHESS**

Bronze Finish. 4½ inch Dial. Height, 9¾ inches.
List, $15.00

**$275.00**

**REX**

Brass Antique and Bronze Finishes. 4½ inch Dial. Height, 12½ inches.
List, $16.00

**$300.00**

**LA REINE**

Bronze Finish. 3½ inch Decorated Dial. French Sash and Beveled Glass. Height, 13½ inches.
List, $19.00

**$285.00**

**DUCHESS WITH LION**

Bronze and Verde Finishes. 4½ inch Dial. Height, 13¾ inches.
List, $18.00

**$250.00**

**LILY**

Bronze with Gold Panel. 8 Day, Strike Hours on Cathedral Bell and Half Hours on Cup Bell. 4 inch Dial. Height, 11½ inches. Base, 8½ inches.

List, $10.50

These two patterns are also made in Rich Gold with Porcelain Dials, French Sash and Beveled Glass. List, $12.50

**$250.00**

**MEDITATION**

Bronze with Gold Panel. 8 Day. Movt. as in Lily. 4 inch Gold Dial with Porcelain Numerals. No Sash or Glass. Winding arbors piped to prevent dust getting in movement. Height, 11½ inches. Base, 8½ inches. List, $10.50.

# 1910 SETH THOMAS MANTEL CLOCKS

$325.00

TOILER

Bronze and Brass Antique Finishes. 4½ inch Porcelain Dial. Height, 27 inches. Base, 9¼ inches.
List, $27.00

$325.00

WHISTLING BOY

Bronze Finish. 4 inch Porcelain Dial. Height, 21 inches. Base, 8½ inches.
List, $27.00

$325.00

VICTORY

Bronze Finish. 4 inch Porcelain Dial. Height, 22 inches. Base, 8½ inches.
List, $27.00

$325.00

FROLIC

Bronze and Verde Antique Finishes. 4½ inch Porcelain Dial. Height, 23 inches. Base, 9¼ inches.
List, $27.00

$275.00

LA NORMA

Height, 12 inches. Base, 7 inches.
Rich Gold, List, $22.70
Special Syrian Finish, List, $18.50

$275.00

VISTA

Height, 12¾ inches. Base, 8 inches.
Rich Gold, List, $22.70
Special Syrian Finish, List, $19.00

$235.00

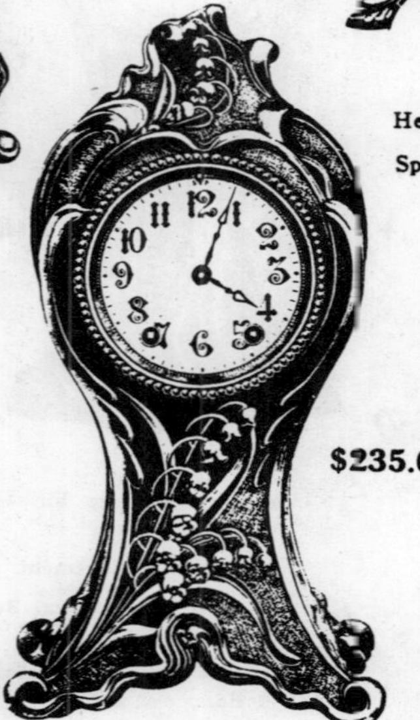

LA FLEUR

Height, 14 inches. Base, 7½ inches.
Rich Gold, List, $23.70
Special Syrian Finish, List, $19.00

$365.00

COUNT

Verde Antique and Syrian Bronze Finishes, with 4 Gold Plated Ornaments. Eight Day Movement, Striking Hours on a Cathedral Bell and Half Hours on a Cup Bell. 7 inch Gold Dial, Silver Center and Raised Black Numerals and Minute Dots. Convex Beveled Glass. Gold Plated Cast Sash and Mat. Height, 14 inches. Base, 10 inches.

List, $25.00

# 1910 SETH THOMAS MANTEL CLOCKS

## FIFTEEN DAY, ROUND POLISHED HIGH GRADE MOVEMENTS

Steel Cut Pinions and Springs in Barrels.

**$350.00**

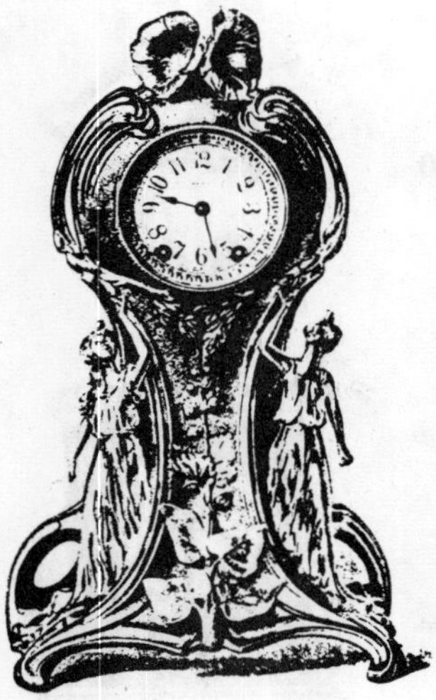

**LUCERNE**

Rich Gold and Bronze
Art Nouveau Finishes.
Fifteen Day, Fine Polished Movement,
Half Hour Strike, Cathedral Bell.
3 inch Porcelain Dial.
Height of Clock, 15½ inches;
Base, 9 inches.
List Price of Clock, $41.00

**$300.00**

**CLUNY**

Rich Gold Finish and Highly Burnished.
Fifteen Day, Fine Polished Movement,
Half Hour Strike, Cathedral Bell.
3½ inch Porcelain Dial.
Height of Clock, 15½ inches;
Base, 8 inches.
List Price of Clock, $30.00

**$375.00**

**TRISTAN**

Rich Gold Finish. Fifteen Day,
Half Hour Strike, Cathedral Bell.
3 inch Porcelain Dial.
Height of Clock, 12 inches;
Base, 8½ inches.
List Price of Clock, $45.00

**$450.00**

**IMPERIAL**

Rich Gold and Highly Burnished.
Fifteen Day, Fine Polished Movement,
Half Hour Strike, Cathedral Bell,
Convex Beveled Glass.
4½ inch Porcelain Dial.
Height of Clock, 21 inches;
Base, 12½ inches.
List Price of Clock, $81.00

**$375.00**

**ROSELLE**

Rich Gold Finish and Highly Burnished.
Fifteen Day, Fine Polished Movement,
Half Hour Strike, Cathedral Bell.
4 inch Decorated Porcelain Dial
Height of Clock, 15 inches; Base, 9 inches.
List Price of Clock, $47.00

**$300.00**

**GARNISH**

Rich Gold Finish and Highly Burnished.
Fifteen Day, Fine Polished Movement,
Half Hour Strike, Cathedral Bell.
3 inch Porcelain Dial.
Height of Clock, 12½ inches;
Base, 9 inches.
List Price of Clock, $32.50

**$285.00**

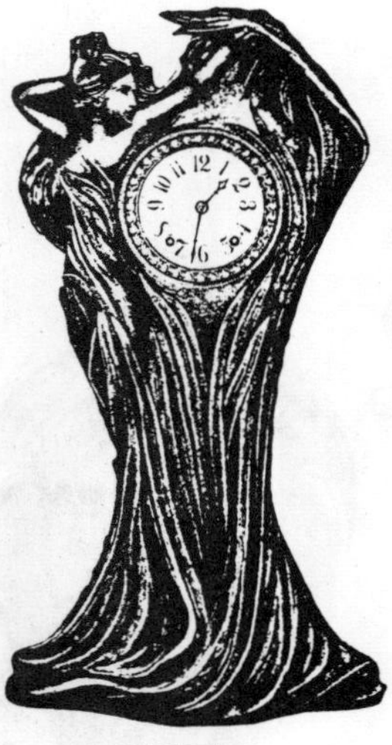

**NYMPH**

Bronze Art Nouveau Finish.
Fifteen Day,
Fine Polished Movement,
Half Hour Strike, Cathedral Bell.
3½ inch Porcelain Dial.
Height of Clock, 18 inches;
Base, 9½ inches.
List Price of Clock, $34.00

**$285.00**

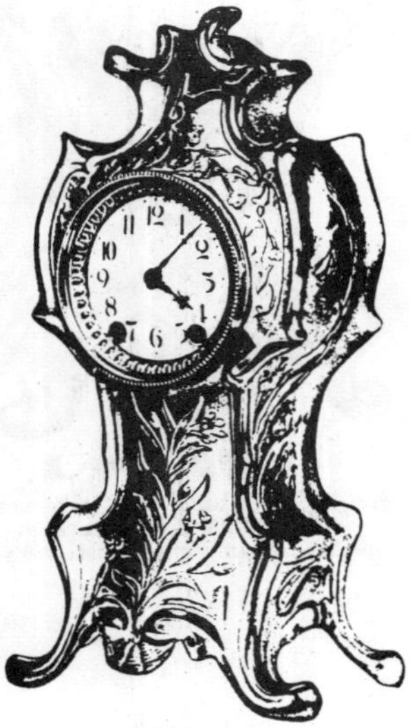

**BLOSSOM**

Rich Gold Finish. Fifteen Day,
Fine Polished Movement,
Half Hour Strike, Cathedral Bell.
3½ inch Porcelain Dial.
Height of Clock, 13 inches;
Base, 6½ inches.
List Price of Clock, $30.00

# 1910 SETH THOMAS MANTEL CLOCKS

**FIFTEEN DAY, ROUND POLISHED HIGH GRADE MOVEMENTS**

Steel Cut Pinions and Springs in Barrels

**$300.00**

**FLORA**

Rich Gold and Bronze Art Nouveau Finishes. Fifteen Day, Fine Polished Movement, Half Hour Strike, Cathedral Bell. 3 inch Porcelain Dial. Height of Clock, 14 inches; Base, 7 inches. List Price of Clock, $34.00

**$285.00**

**DOROTHY**

Bronze Art Nouveau Finish. Fifteen Day, Fine Polished Movement, Half Hour Strike, Cathedral Bell. 3½ inch Decorated Porcelain Dial. Height of Clock, 16 inches; Base, 9 inches.

List Price of Clock, $34.00

**$285.00**

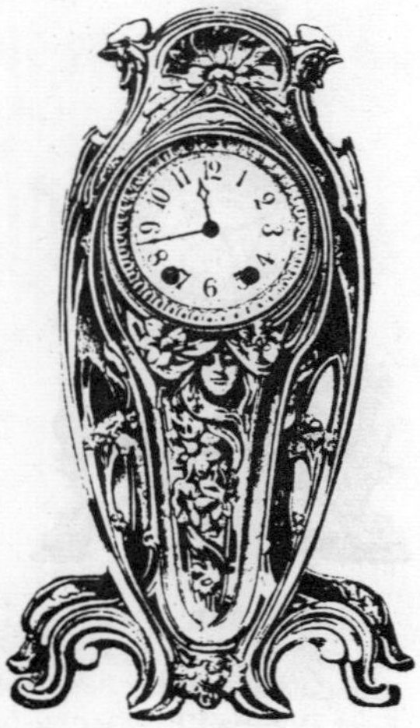

**LUCRECE**

Rich Gold and Bronze Art Nouveau Finishes. Fifteen Day, Fine Polished Movement, Half Hour Strike, Cathedral Bell. 3½ inch Porcelain Dial. Height of Clock, 16 inches; Base, 9 inches. List Price of Clock, $35.00

**$285.00**

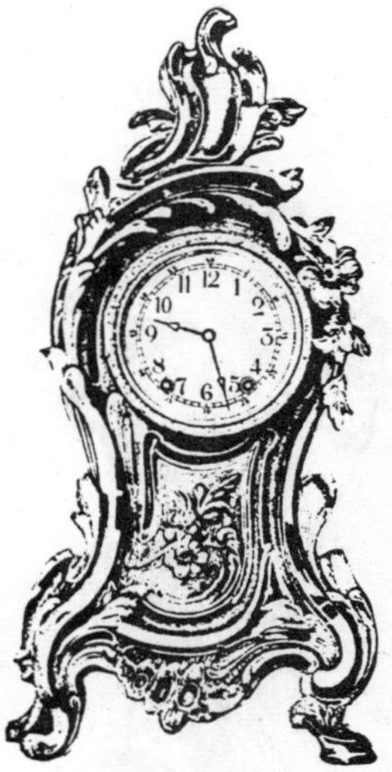

**DAUPHIN**

Rich Gold Finish and Highly Burnished. Eight Day, Half Hour Strike, Cathedral Bell. 4 inch Porcelain Dial. Height of Clock, 18½ inches; Base, 8½ inches. List Price of Clock, $35.00

**$300.00**

**ISABEL**

Bronze Art Nouveau Finish. Half Hour Strike, Cathedral Bell. Fifteen Day, Fine Polished Movement, 3½ inch Decorated Porcelain Dial, Visible Pendulum, Gold Plated, Burnished and Lacquered. Height of Clock, 18 inches; Base, 11 inches. List Price of Clock, $45.00

**$300.00**

**TANYA**

Bronze Art Nouveau Finish. 4 inch Porcelain Dial. French Sash, Beveled Glass. Height, 25 inches. Base, 10 inches. List, $54.50

**$250.00**

**ARTFUL**

Rich Gold and Bronze Art Nouveau Finishes. Fifteen Day, Fine Polished Movement, Half Hour Strike, Cathedral Bell. 3½ inch Decorated Porcelain Dial. Height of Clock, 16 inches; Base, 8 inches. List Price of Clock, $34.00

**$225.00**

**POPPY**

Bronze Art Nouveau Finish. Fifteen Day, Fine Polished Movement Half Hour Strike, 3½ inch Porcelain Dial. Cathedral Bell. Height of Clock, 19 inches; Base, 9 inches. List Price of Clock, $34.00

# 1910 SETH THOMAS MANTEL CLOCKS

**$300.00**

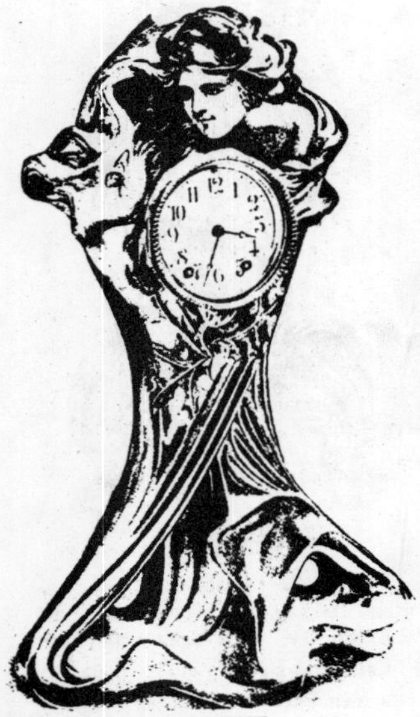

**JOSEPHINE**

Bronze Art Nouveau Finish. 4 inch Porcelain Dial. French Sash, Beveled Glass. Height, 21 inches. Base, 12½ inches. List, $51.30

**$365.00**

**ROYAL**

Rich Gold and Highly Burnished.
Eight Day, Half Hour Strike,
Cathedral Bell.
4 inch Decorated Porcelain Dial.
Height of Clock, 16 inches; Base, 10 inches.
List Price of Clock, $43.00

**$400.00**

**VALENTINE**

Fifteen Day, High Grade Movement. 4 inch Porcelain Dial. French Sash, Beveled Glass. Real Bronze Finish with Verde Ball. Height, 22½ inches. Base, 14 inches. List, $50.00

**$300.00**

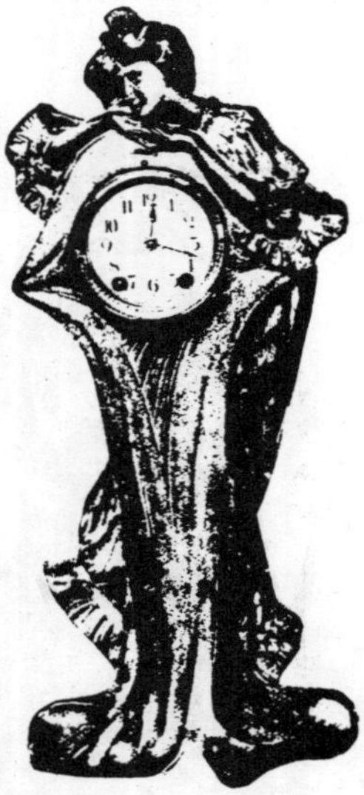

**JEANNETTE**

Bronze Art Nouveau Finish.
Fifteen Day, Fine Polished Movement, Half Hour Strike, Cathedral Bell. 3½ inch Decorated Porcelain Dial.
Height of Clock, 20 inches; Base, 8½ inches.
List Price of Clock, $38.00

**$250.00**

**ORLEANS**

Rich Gold, Highly Burnished.
Eight Day, Half Hour Strike,
Cathedral Bell.
4 inch Decorated Porcelain Dial.
Height of Clock, 14½ inches; Base, 9½ inches.
List Price of Clock, $25.00

**$350.00**

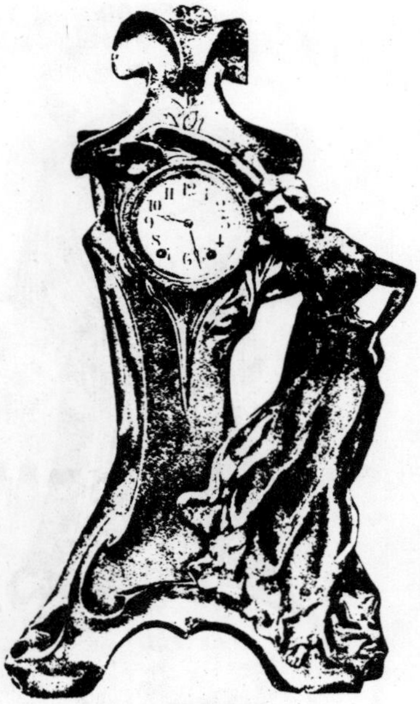

**MERCEDES**

Bronze Art Nouveau Finish. Eight Day, Fine Movement. Hour and Half Hour Strike, Cathedral Bell. 4 inch Porcelain Dial. French Sash and Beveled Glass. Height, 26 inches. Base, 15 inches.

List, $68.00

**$400.00**

**FOLLY**

Bronze Art Nouveau Finish. Fifteen Day, High Grade Movement. Hour and Half Hour Strike, Cathedral Bell. 3½ inch Porcelain Dial. French Sash and Beveled Glass. Height, 21½ inches. Base, 9 inches.

List, $54.50

**$300.00**

**THISTLE**

Rich Gold and Highly Burnished. Fifteen Day, High Grade Movement. Hour and Half Hour Strike, Cathedral Bell. 3½ inch Decorated Porcelain Dial. French Sash and Beveled Glass. Height, 14½ inches. Base, 7 inches.

List, $30.00

# 1910 SETH THOMAS MANTEL CLOCKS

$300.00

IRMA

Rich Gold and Bronze Art Nouveau Finishes 4½ inch Decorated Porcelain Dial, French Sash and Beveled Glass. Height, 18 inches. Base, 11 inches.
List, $36.00

$600.00

INSPIRATION AND INNOCENCE

Bronze Art Nouveau Finish. Eight Day, Fine Movement. Hour and Half Hour strike, Cathedral Bell. 4 inch Porcelain Dial, French Sash and Beveled Glass. Height, 21 inches. Base, 26 inches.
List, $68.00

$450.00

No. 8004

Bronze Art Nouveau Finish. Fifteen Day, High Grade Movement. Hour and Half Hour Strike, Cathedral Bell. 4 inch Porcelain Dial, French Sash and Beveled Glass. Height, 22 inches. Base, 13½ inches.
List, $68.00

*Only clocks in ORIGINAL MINT condition and in good running order will command top prices. A replaced or damaged dial, missing parts or chipped beveled glass will reduce prices accordingly.*

*Most white metal and brass clock cases are gilded or silver plated. Great care must be taken when cleaning these surfaces. The thin layer of gilding or silver plating will be easily removed by cleaning solutions and strong rubbing. If you believe cleaning is necessary, hire the services of a professional and avoid the expense of re-gilding or replating.*

$450.00

INNOCENCE

Bronze Art Nouveau Finish. Eight Day, Fine Movement, Hour and Half Hour Strike, Cathedral Bell. 4 inch Porcelain Dial, French Sash and Beveled Glass. Height, 21 inches. Base, 21 inches.
List, $50.00

$600.00

CREATION

Rich Gold Finish. Fifteen Day, High Grade Polished Movement, Half Hour Strike, Cathedral Bell. 3½ inch Porcelain Dial. Height of Clock, 28 inches; Base, 14 inches.
List Price of Clock, $80.00

$450.00

INSPIRATION

Bronze Art Nouveau Finish. Eight Day, Fine Movement, Hour and Half Hour Strike, Cathedral Bell. 4 inch Porcelain Dial, French Sash and Beveled Glass. Height, 21 inches. Base, 21 inches.
List, $50.00

# 1910 SETH THOMAS MANTEL CLOCKS

*Clock Only* **$485.00**

*Complete Set* **$600.00**

EIGHT DAY, ROUND FINE MOVEMENTS

VERSAILLES SET

LOUVRE SET

Etched Metal on Art Glass. Made in two Finishes, Verde Antique and Bronze. Eight Day, Half Hour Strike, Cathedral Bell. 4 inch Porcelain Dial. French Sash and Beveled Glass. Height of Clock, 12½ inches; Base, 8 inches. Height of Candlesticks, 10 inches.

List Price of Clock, $57.00

List Price of Set, $85.00

List Price of Candlesticks, per pair, $28.00

Etched Metal. Made in two Finishes, Verde Antique and Bronze. Eight Day, Half Hour Strike, Cathedral Bell, 4 inch Porcelain Dial. French Sash and Beveled Glass. Height of Clock, 11½ inches; Base, 7½ inches. Height of Candlesticks, 8¾ inches.

List Price of Clock, $50.00

List Price of Set, $75.00

List Price of Candlesticks, per pair, $25.00

*Clock Only* **$385.00**

*Complete Set* **$475.00**

*Only clocks in ORIGINAL MINT condition and in good running order will command top prices. A replaced or damaged dial, missing parts or chipped beveled glass will reduce prices accordingly.*

*Clock Only* **$1,000.00**

*Complete Set* **$1,350.00**

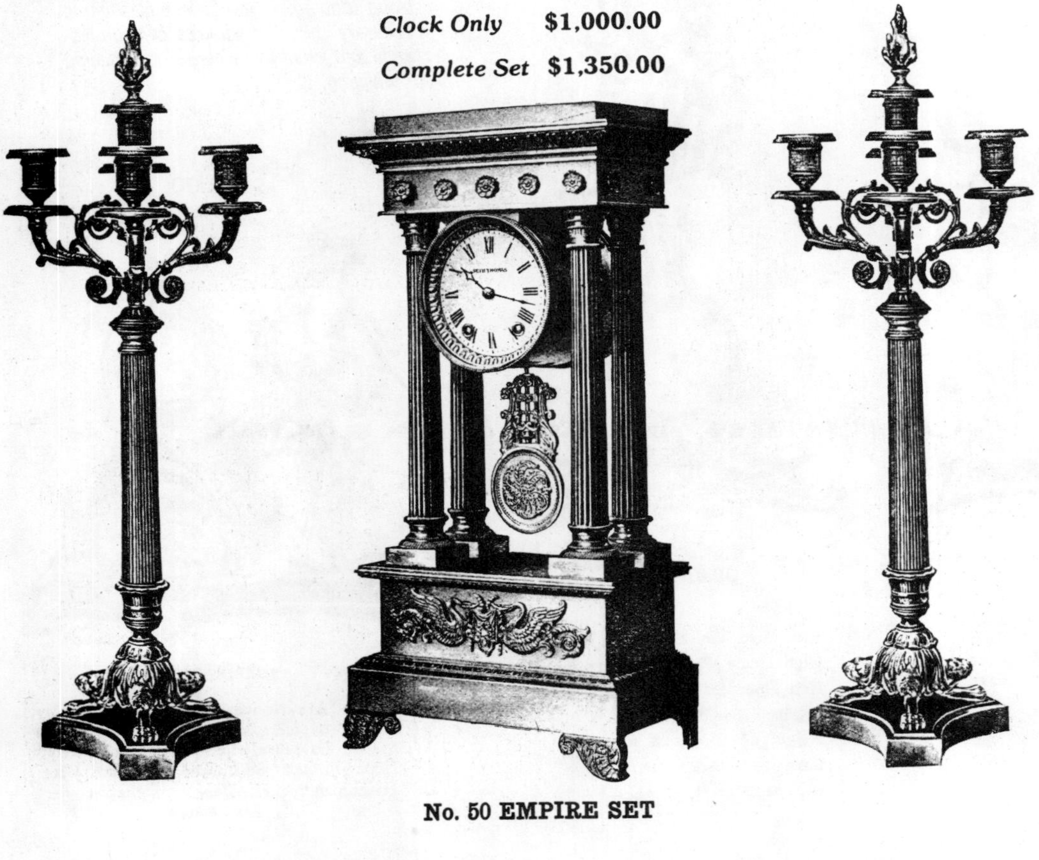

No. 50 EMPIRE SET

**No. 50 EMPIRE SET**

Rich Gold Finish. Fifteen Day, Round, Polished, High Grade Movement, Steel Cut Pinions, and Springs in Barrels. Hour and Half Hour Strike, 3½ inch Porcelain Dial. Convex Beveled Glass. Case is Gold Plated, Burnished and Lacquered and is of the highest grade and workmanship. Height, 18½ inches; Base, 9 inches.

List Price of Clock, $135.00

List Price of Set, $200.00

List Price of Candelabra, per pair, $65.00

# 1910-1915 SETH THOMAS MANTEL CLOCKS

VIVIEN SET

Bronze Art Nouveau Finish with Brazilian Onyx Inlay. Fifteen Day, Porcelain Dial with French Sash and Beveled Glass. Height of Candelabra, 28 inches; Base, 8 inches.

Fine Polished Movement, Half Hour Strike, Cathedral Bell. 4 inch Height of Clock, 27½ inches; Base, 16½ inches.

List Price of Set, $122.00 **$775.00**

List Price of Clock, $81.00 **$600.00**

List Price of Candelabra, per pair, $41.00

GAIETY SET

Real Bronze Finish with Verde Ball. Fifteen Day, Fine Polished Movement, Half Hour Strike, Cathedral Bell. 3½ inch Porcelain Dial. Height of Clock, 22½ inches; Base, 14 inches. Height of Candelabra, 22 inches; Base, 7 inches.

List Price of Set, $100.00 **$650.00**

List Price of Clock, $65.00 **$525.00**

List Price of Candelabra, per pair, $35.00

MIGNON SET

Rich Gold and Bronze Art Nouveau Finishes. Fifteen Day, Fine Polished Movement, Half Hour Strike, Cathedral Bell. 3 inch Porcelain Dial. Height of Clock, 15 inches; Base, 8 inches. Height of Candelabra, 13½ inches; Base, 4 inches.

List Price of Set, $54.00 **$400.00**

List Price of Clock, $30.00 **$300.00**

List Price of Candelabra, per pair, $24.00

**$400.00**

NEW REPOSE

**NEW REPOSE**

Eight-Day Fine Round Movement. Hour and Half-Hour Rack Strike on Cathedral Bell. 5-inch Convex Silvered Dial with Raised Bronze Numerals. Convex Beveled Glass and Brass Sash. Height, 12 inches. Base, 20 inches.

Bronze Finish Figure and Clock on Dull Gold Base..List, $35.00

Dull Gold Finish, Figure, Clock and Base ..........List, 40.00

**$450.00**

REPOSE

REPOSE

Empire Gilt Finish. 5-inch Dial. Convex Beveled Glass Height, 15 inches. Base 20 inches. List........$40.00

# 1910-1913 SETH THOMAS MANTEL CLOCKS

## REAL BRONZE "DORIC" CLOCKS

Real Bronze Cases. Convex Silvered Dials, with Raised Bronze Numerals, Bronze Sash and Mat with Convex Beveled Glass.

Fifteen-Day, High-Grade Movement with Round, Polished Plates, Cut Steel Pinions and Springs in Barrels. Hour and Half-Hour Rack Strike on Cathedral Bell.

**$250.00**

**DORIC No. 0**

Real Bronze.
Fifteen-Day Strike. 3¾-inch Dial.
Height, 8 inches.
List, **$38.50**

**$300.00**

**DORIC No. 1**

Real Bronze.
Fifteen-Day Strike. 5-inch Dial.
Height, 10¾ inches.
List, **$44.00**

**$350.00**

**DORIC No. 2**

Real Bronze.
Fifteen-Day Strike. 6-inch Dial.
Height, 13 inches.
List, **$55.00**

**$200.00**

**DORIC No. 00**

Real Bronze.
Eight Day Time in this case. 2⅝ inch Dial. Height, 5 inches.
This size only, has 7 jewel, lever, time movement and etched dial described on page 22.

List, **$35.00**

Made also in brass, brush finish. Same price.

**$400.00**

**DORIC No. 3**

Real Bronze.
Fifteen Day Strike. 8 inch Dial.
Height, 16 inches. Depth, 8½ inches.
List, **$70.00**

**DORIC No. 4**

**$385.00**

Real Bronze and Brass, Brush Finish. Fifteen Day, High Grade Movement, Hour and Half Hour Strike, Cathedral Bell. 8 inch Silvered Convex Dial, with Raised Numerals and Minute Dots. Convex Beveled Glass. Height, 12¼ inches. Depth. 8½ inches.

List, **$70.00**

## REAL BRONZE "GOTHIC" CLOCKS

**GOTHIC No. 0**

Real Bronze.
Fifteen-Day Strike. 3¾-inch Dial.
Height, 9 inches.
List, **$38.50**

**$250.00**

**GOTHIC No. 0**

*Only clocks in ORIGINAL MINT condition and in good running order will command top prices. A replaced or damaged dial, missing parts or chipped beveled glass will reduce prices accordingly.*

**$300.00**

**GOTHIC No. 1**

Real Bronze.
Fifteen-Day Strike. 5-inch Dial.
Height, 11¼ inches.
List, **$44.00**

**GOTHIC No. 1**

# 1910-1913 SETH THOMAS MANTEL CLOCKS

**$225.00**

**SOLAR**

Bronze Finish Case.
Eight Day Movement, Striking Hours on a Cathedral Bell and Half Hours on a Cup Bell. 8 inch Silvered Convex Dial, with Raised Numerals and Minute Dots. Convex Beveled Glass. Height, 10½ inches. Base, 9¼ inches.

List, $25.00

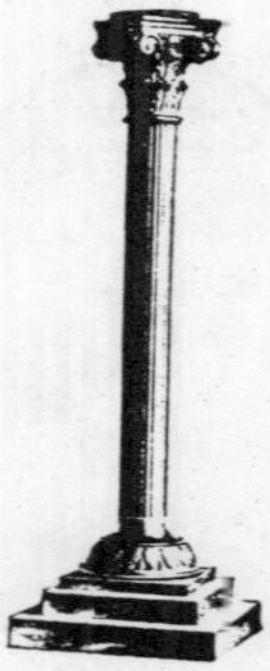

*Clock Only*

**$265.00**

*Complete Set*

**$325.00**

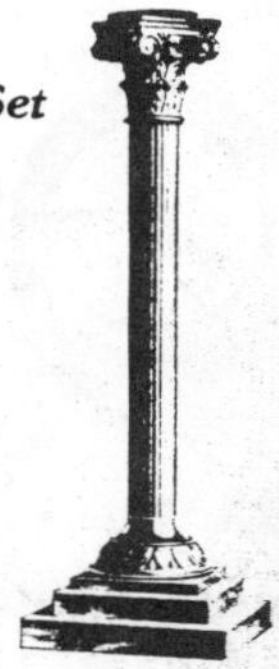

**DENTON SET**

Gold-Plated Sash and Mat. 6-inch Silvered Convex Dial with Raised Bronze Numerals. Cut Serpentine Hands. Made in following finishes:

Clock Rich Gold-Plated....List, $35.00
Candlesticks, Rich Gold to match....List per pair, 16.50
List of Gold Set, $51.50

Clock Bronze Finish Body and Gold Base and Cylinder....List, 30.00
Candlesticks, Bronze Finish and Gold to match....List per pair, 16.50
List of Bronze Set, $46.50

Height of Clock, 8½ inches. Base, 8½ inches.
Height of Candlesticks, 8½ inches. Base, 3¾ inches.
Candlesticks have ball feet same as on clock.

**$285.00**

**PARIS**

Dull Gold Plated Brass Base and Feet, Verde Antique Top, Gold Plated Sash and Mat and Beveled Glass. 15 Day, High Grade Movement. 3½ inch Porcelain Dial. Height, 8½ inches. Base, 9 inches.

List, $25.00

**$135.00**

**ELECT**

Metal Case and Dial. Verde Antique Base. Gun Metal Top and Hands. 7 inch Gold Dial, Silver Center and Raised Black Numerals and Minute Dots. Height, 10 inches. Base, 7½ inches.

List, $8.50

Bronze Base and Top, List, $8.50
Gold Plated Base and Top, List, $10.50

**$300.00**

**COLOSSAL**

Cast Metal Body - Cast Brass Base Dull Gold Plated. 8-inch Silvered convex Dial with raise Bronze numerals. Height, 10, 3/4 inches.
Syrian Bronze Finish.......$38.00
Rich Gold Plated...........$55.00

**$285.00**

**TAMPA**

Gold Plated Sash and Mat. Verde Antique Body of Clock, Eight Day, Fine Movement, Half Hour Strike, Cathedral Bell. 4½ inch Porcelain Dial. Height of Clock, 8¾ inches; Base, 13 inches.

List Price of Clock, $25.00

*Most white metal and brass clock cases are gilded or silver plated. Great care must be taken when cleaning these sur- faces. The thin layer of gilding or silver plating will be easily removed by cleaning solutions and strong rubbing. If you believe cleaning is necessary, hire the services of a professional and avoid the expense of re-gilding or re-plating.*

**$300.00**

**CELTIC**

Gold-Plated Sash and Mat. 6-inch Silvered Convex Dial with Raised Bronze Numerals. Cut Serpentine Hands.

Height of Clock, 9 inches. Base, 11 inches.

Clock, Rich Gold Plated........List, $55.00

# 1863-1875 SETH THOMAS MANTEL CLOCKS

*8 Day* **$150.00**

*30 Hour* **$115.00**

**30 HOUR CABINET.**
**8 DAY CABINET.**
**Time—No. 1.—Extra.**
**Strike—No. 1.—Extra.**
Hight, 9½ inches.

**$175.00**

8 DAY.
**TUDOR, No. 3.—Spring—Strike.**
(Strikes Half-Hours.)
Hight, 12 inches.

**$185.00**

8 DAY.
**TUDOR, No. 1.—Spring—Strike.**
Hight, 16 inches.

*8 Day* **$150.00**

*30 Hour* **$115.00**

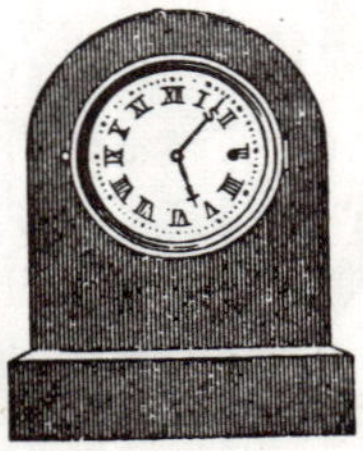

8 DAY CABINET TIME—No. 1.
(With or without Alarm.)
THIRTY HOUR—CABINET
STRIKE—No. 1.
Height of Clock 9½ inches.

*8 Day* **$150.00**

*30 Hour* **$115.00**

**CABINET.**

8 DAY. TIME.
30 HOUR. STRIKE.
Hight, 9 1-2 inches.

**$185.00**

**TUDOR—No. 1.**

8 DAY. SPRING. STRIKE.
Hight, 16 inches.

**$175.00**

**TUDOR No. 3.**

8 DAY. SPRING. STRIKE.
Strikes Half-Hours.
Hight, 12 inches.

*8 Day* **$150.00**

*30 Hour* **$115.00**

8 DAY CABINET TIME—No. 2.
(With or without Alarm.)
THIRTY HOUR—CABINET
STRIKE—No. 2.
Height of Clock 9½ inches.

*8 Day* **$225.00**

*30 Hour* **$185.00**

8 DAY OR 30 HOUR.
**GOTHIC—Spring—Strike.**
Hight, 15¾ inches.
(8 Day Strikes Half-Hours.)

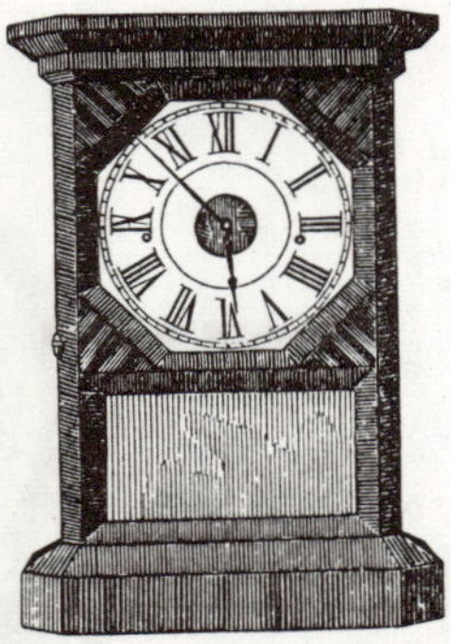

8 Day or 30 Hour.
PARLOR.
Spring - Strike.
Height of Clock
14 inches.

*8 Day* **$250.00**

*30 Hour* **$185.00**

8 Day or 30 Hour.
Octagon Top.
Spring - Strike
Height of Clock
14 inches.

*8 Day* **$225.00**

*30 Hour* **$175.00**

*8 Day* **$200.00**

*30 Hour* **$150.00**

CABINET No. 3.
8 Day or 30 Hour.
With or without
Alarm.
Height of Clock
10 inches.

# 1863-1890 SETH THOMAS MANTEL CLOCKS

**$185.00**

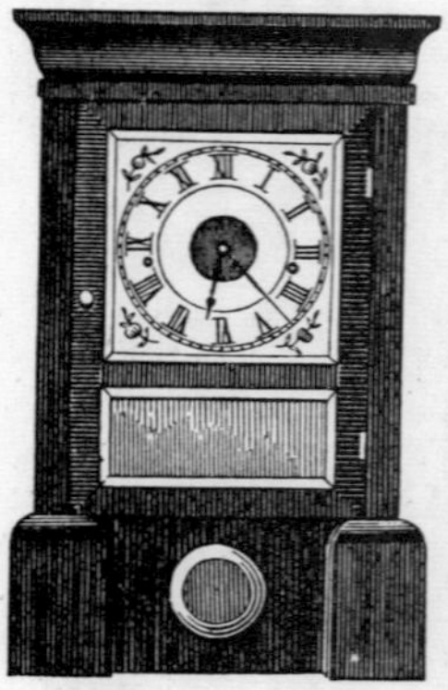

30 HOUR. O.G. TOP.
Spring - Strike.
Hight, 15 inches.

*8 Day* **$185.00**
*30 Hour* **$150.00**

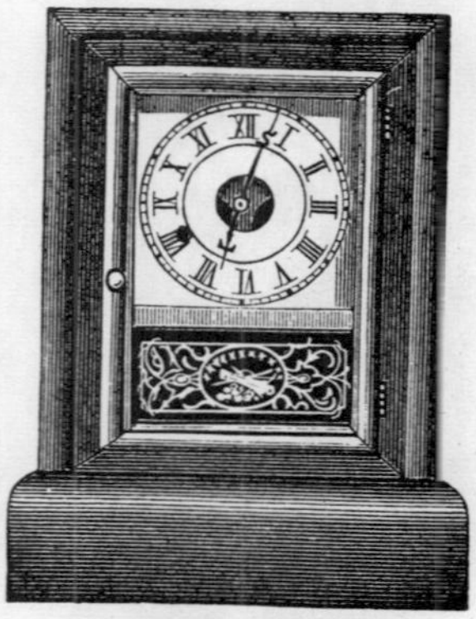

COTTAGE, Extra.
Spring - Strike.
8 Day or 30 Hour.
Hight, 15 inches.

*8 Day* **$185.00**
*30 Hour* **$150.00**

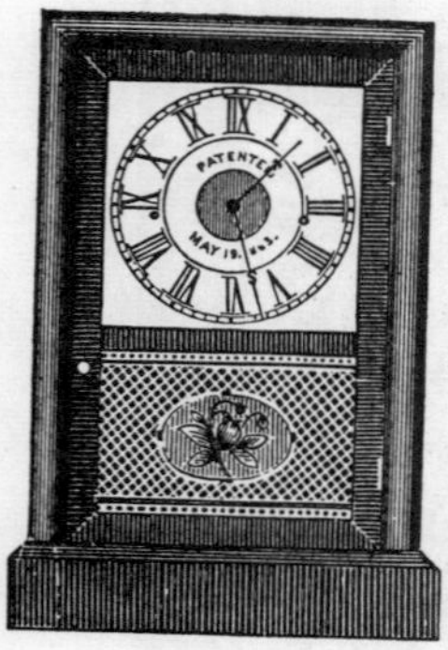

8 DAY OR 30 HOUR
Cottage - Spring
Strike.
Hight 14 inches.

*8 Day* **$175.00**
*30 Hour* **$135.00**

**COTTAGE.**

8 Day or 30 Hour. Time.
30 Hour. Small. Strike.
Hight, 9 inches.

**COTTAGE—Extra.**

Hight, 15 inches.

**COTTAGE.**

Hight, 14 1-2 inches.

8 Day or 30 Hour. Spring. Strike.

*8 Day* **$185.00**
*30 Hour* **$150.00**

**$135.00**

COTTAGE.

ROSEWOOD OR WALNUT VENEER, POLISHED.

1 Day, Time.
4 inch Dial.
Hight, 9 inches

**$135.00**

COTTAGE TIME PIECE.
With or without alarm.
Hight 7½ inches.

*8 Day* **$185.00**
*30 Hour* **$150.00**

COTTAGE.

ROSEWOOD OR WALNUT VENEER, POLISHED.

1 Day, Spring Strike.
8 Day, Spring Strike.
6 inch Dial. Hight, 14½ inches.
WITH AND WITHOUT ALARM.
Price, 1 Day, Strike, $3.75; 8 Day, Strike, $4.95
Alarms, 50 cents additional.

**OCTAGON TOP.**

**$175.00**

**$175.00**

30 Hour. Time. Hight, 9 inches.

**$175.00**

OCTAGON TOP.

OCTAGON TIME PIECE.
With or without Alarm.
Hight, 7½ inches.

# 1863-1875 SETH THOMAS MANTEL CLOCKS

## KITCHEN.

*8 Day* **$250.00**

*30 Hour* **$185.00**

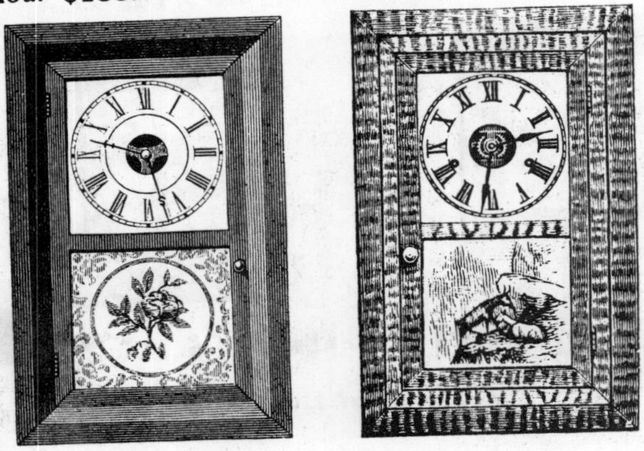

8 DAY OR 30 HOUR. SPRING. STRIKE.

Hight, 16 inches.

## ROUND BAND.

*8 Day* **$250.00**

*30 Hour* **$185.00**

8 DAY OR 30 HOUR. SPRING. STRIKE.

Hight, 17 inches.

*Only clocks in ORIGINAL MINT condition and in good running order will command top prices. A replaced or damaged dial, missing parts or chipped beveled glass will reduce prices accordingly.*

## O. G. **$285.00**

ROSEWOOD OR WALNUT VENEER, POLISHED.

8 Day, Weight, Strike.

8 inch Dial.

Hight, 29½ inches.

## O. G.

ROSEWOOD OR WALNUT VENEER, POLISHED.

1 Day, Weight, Strike.

7½ inch Dial.

Hight, 25 inches.

*8 Day* **$275.00**

*30 Hour* **$200.00**

## O. O. G. **$275.00**

8 DAY. WEIGHT. STRIKE.

Hight, 25 inches.

## LARGE O. G.

**$300.00**

8 DAY. WEIGHT. STRIKE.

Hight, 29 1-2 inches.

# 1863-1890 SETH THOMAS MANTEL CLOCKS

*8 Day* **$400.00**
*30 Hour* **$225.00**

**ROSEWOOD-**
**-Gilt Column.**

30 Hour or Small 8 Day.
Strike. Weight.
Hight, 25 inches.

**$225.00**

30 HOUR - ROSEWOOD.
GILT or GILT CAP.

Strike - Weight.
Height of clock
25 inches.

**$225.00**

30 HOUR - COLUMN.
ROSEWOOD SHELL.

Strike - Weight.
Heights of clock
25 inches.

*8 Day* **$400.00**
*30 Hour* **$225.00**

**ROSEWOOD**
**-Shell Column.**

30 Hour or Small 8 Day.
Strike. Weight.
Hight, 25 inches.

**COLUMN.**
Height, 16 inches.

1 Day, Spring, Strike.

8 Day, Spring, Strike.

6 inch Dial.

Hight, 16 inches.

*8 Day* **$250.00**
*30 Hour* **$175.00**

*To receive top prices, clocks must be in original mint condition and in good running order — A replacement dial or missing parts will reduce prices accordingly.*

*8 Day* **$400.00**
*30 Hour* **$225.00**

**ROSEWOOD**
**COLUMN.**

30 Hour or Small 8 Day.
Strike. Weight.
Hight, 25 inches.

ROSEWOOD COLUMN.
8 Day
**$450.00**

Strike - Weight.
Height of clock
32 inches.

**$450.00**

SHELL COLUMN.
Rosewood, 8 Day

Strike - Weight
Height of clock
32 inches.

**$450.00**

ROSEWOOD - 8 DAY
GILT or GILT CAP

Strike - Weight.
Height of clock
32 inches.

# 1863-1906 SETH THOMAS MANTEL CLOCKS

**$225.00**

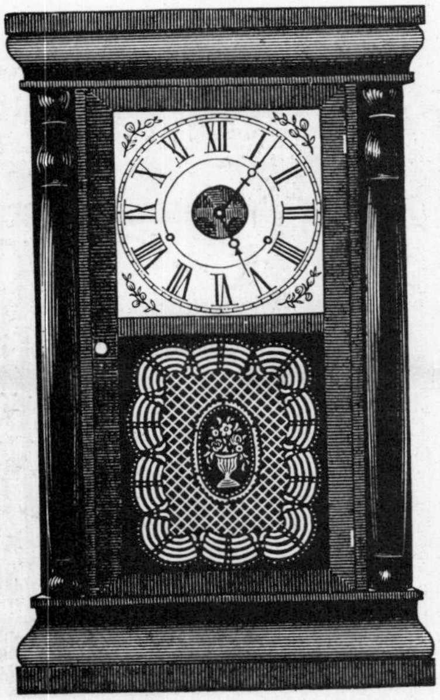

30 HOUR ROSEWOOD COLUMN
Strike - Weight - Height
of clock 25 inches.

**$450.00**

COLUMN.

ROSE AND WALNUT VENEER, POLISHED.
ROSEWOOD, SHELL OR GILT COLUMN.

8 Day, Weight, Strike.
8 inch Dial.
Hight, 32 inches.

**$225.00**

COLUMN WEIGHT.

ROSEWOOD OR WALNUT VENEER, POLISHED.

1 Day, Weight, Strike.
7¼ inch Dial.
Hight, 25 inches.

**LARGE ROSEWOOD COLUMN.**

**$450.00**

8 Day. Strike. Weight.
Hight, 32 inches.

**$500.00**

8 DAY. PARLOR.
Strike - Weight.
Height of Clock
30½ inches.

**LARGE ROSEWOOD Shell Column**

**$450.00**

8 Day. Strike. Weight.
Hight, 32 inches.

# Ca. 1815 - 1904 SETH THOMAS MANTEL CLOCKS

**$450.00**

**LARGE ROSEWOOD**
1875 **Gilt Column.**
8 DAY. STRIKE. WEIGHT.
Hight, 32 inches.

**$550.00**

**$550.00**

*Franklin & Putnam are in 1902 Seth Thomas catalog.*

**FRANKLIN.**
8-Day. Weight Strike. Cathedral Bell.
10 inch dial.
Gold Gilt Columns. Hand Carved
Scroll at Head.
Made in Dark Mahogany only.

**PUTNAM.**
8-Day. Weight Strike. Cathedral Bell.
10 inch Dial. Hand Carved Columns,
Feet and Head Ornament.
Made in dark Mahogany only.
Height, 36¼ inches. Base, 18 inches.
Price, **$18.00**

**$475.00**

1904
**YORKTOWN**
MAHOGANY CASE, HAND CARVED COLUMNS, FEET AND ORNAMENT.
8 Day, Spring, Strike. Cathedral Bell.
10 inch Dial.
Hight, 36 inches. Base, 18 inches.
Price, $24.00

*To receive top prices, clocks must be in original mint condition and in good running order — A replacement dial or missing parts will reduce prices accordingly.*

OFF-CENTER PENDULUM PILLAR & SCROLL

With door open.

*Visible escapement off-center pendulum pillar and scroll clock. Dimensions, 29½" to top of center finial x 17⅜" x 4⅜". Circa 1815.*

*Very Rare* *Woodworks* **$3,750.00**

Off-Center Pendulum
Pillar and Scroll
Dimensions, 31 x 17½ x 4½

*Rare* *Woodworks*
**$3,200.00**

Standard Pillar & Scroll
Dimensions, 31 x 17½ x 4½

*Woodworks*
**$1,500.00**

# 1913-1928 SETH THOMAS MANTEL CLOCKS

$400.00

**COLONY**

Mahogany Cabinet Case. 6-inch Dial with colored flowers in corners. Fine tablet. This is a new and convenient size. Mahogany Tips.
Height, 19 inches.
Base, 10 inches.
List, **$16.50**

Eight-Day, Hour and Half Hour Strike on Cathedral Bell.

$800.00

**CONTINENTAL**

Mahogany Case, Brass or Wood Tips. Eight Day, Spring, Half Hour Strike, Cathedral Bell. 10 inch Dial. Height, 32 inches. Base, 17½ inches.
List, $30.70

$375.00

**Concord**

Mahogany. Decorated dial. Eight-day movement No. 89 striking hour and half hour on Bell Metal Gong. Height 17 inches. Base 9½ inches.

$700.00

CAMBRIDGE

Mahogany. Hand painted dial and lower panel. Authentic reproduction of original Seth Thomas. 8-day, No. 89 movement. Hour and half hour strike on finely toned gong. Height 25 inches. Base 14 inches.

*To receive top prices, clocks must be in original mint condition and in good running order — A replacement dial or missing parts will reduce prices accordingly.*

$500.00

**KINGSBURY**
*8-Day Strike*

**KINGSBURY**
*8-Day Strike*

Case: Red Mahogany. Mount Vernon panel in full color. Polished brass top ornaments

Dial: Antique Roman with black and gold decorations

Height 17 inches
Width 10 inches
8-day pendulum movement No. 89

$500.00

**PLYMOUTH**

Eight-Day, Hour and Half-Hour Strike, Cathedral Bell.

Mahogany Cabinet Case.
Six-inch Dial. Colored Tablet.
Mahogany Tips.
Height, 24 inches. Width, 12 inches.
List Price, $32.00.

**Bedford**

Mahogany finish with burled mahogany panel. Decorated 6-inch dial. Eight-day movement No. 89. Strikes hour and half hour on cathedral gong. Height 15 inches. Base 9 inches.

$285.00

**Sharon**

Mahogany finish. 5-inch decorated dial. Eight-day movement No. 89. Hour and half hour strike on Cathedral gong. Height 14¼ inches. Base 9¼ inches.

$125.00

$185.00

**SHARP GOTHIC, V. P.**

ROSEWOOD, WALNUT OR MAHOGANY VENEER. POLISHED.

8 Day, Spring, Strike.
6 inch Dial  Hight, 19½ inches.
WITH AND WITHOUT ALARM.
Price, Strike, $5.80; Strike, Alarm, $6.30

# 1894 SETH THOMAS MANTEL CLOCKS

8 Day, Half Hour Strike, Cathedral Bell. 4 inch White Porcelain Dial.

**$200.00**

## CYPRUS.

WOOD CASE, ADAMANTINE FINISH.
DECORATION IN COLOR, ASSORTED DESIGNS.
GILT METAL TRIMMINGS.

Hight, 13 inches. Base, 9 inches.

**$200.00**

## CORSICA.

WOOD CASE, ADAMANTINE FINISH.
BRONZE ADAMANTINE INLAY.
GILT METAL TRIMMINGS.

Hight, 13½ inches. Base, 8½ inches.

**$175.00**

## MADRID.

WOOD, ADAMANTINE FINISH.
DECORATION IN COLOR. ASSORTED DESIGNS.
GILT METAL TRIMMINGS.

Hight, 11 inches. Base, 12 inches.

## MEXICO.

WOOD CASE, ADAMANTINE FINISH.
DECORATION IN COLOR, ASSORTED DESIGNS.
GILT METAL TRIMMINGS.

**$175.00**

Hight, 12 inches. Base, 13 inches.

*Only clocks in ORIGINAL MINT condition and in good running order will command top prices. A replaced or damaged dial, missing parts or chipped beveled glass will reduce prices accordingly.*

## PUEBLO.

WOOD CASE, ADAMANTINE FINISH.
BRONZE ADAMANTINE INLAY.
GILT METAL TRIMMINGS.

**$150.00**

Hight, 11 inches. Base, 12 inches.

**$140.00**

## TUSCANY.

WOOD CASE, ADAMANTINE FINISH.
DECORATION IN COLOR, ASSORTED DESIGNS.
GILT METAL TRIMMINGS.

Hight, 10¾ inches. Base, 9 inches

**$150.00**

## No. 700.

WOOD CASE.

BLACK FINISH AND ENGRAVED.

8 Day, Half Hour Strike.
Cathedral Bell.
5 inch Dial.
Hight, 13 inches; Base, 12 inches.

# 1904 SETH THOMAS MANTEL CLOCKS

$125.00

## ELBA

WOOD CASE, ADAMANTINE FINISH.
GILT METAL FEET AND ORNAMENTS.

8 Day, Half Hour Strike.
Half Hours on Cup Bell, Hours on Cathedral Bell.
5 inch Dial.

Hight, 12 inches. Base, 13 inches.
Price $7.50

## PASHA

WOOD CASE, ADAMANTINE FINISH.
ADAMANTINE ONYX COLUMNS, GILT METAL ORNAMENTS AND BASES.

$150.00

8 Day, Half Hour Strike, Cathedral Bell.
4¼ inch White Porcelain Dial, French Sash with Beveled Glass.
Hight, 11½ inches. Base, 16½ inches.
Price, $10.50

## ARNO

$125.00

ADAMANTINE ONYX COLUMNS.

8 Day, Half Hour Strike.
Half Hours on Cup Bell, Hours on Cathedral Bell.
5 inch Dial.

Hight, 11½ inches. Base, 12½ inches.
Price $6.00

## WINDSOR

ADAMANTINE ONYX COLUMNS, GILT METAL CAPS AND BASES.

$125.00

8 Day, Half Hour Strike, Cathedral Bell.
4¼ inch White Porcelain Dial,
French Sash, with Beveled Glass.
Hight, 10¾ inches. Base, 9 inches.
Price $8.25

$150.00

## TEXEL

WOOD CASE, ADAMANTINE FINISH. GILT METAL FEET AND ORNAMENTS.

8 Day, Half Hour Strike, Half Hours on Cup Bell,
Hours on Cathedral Bell. 5 inch Dial.
Hight, 11 inches. Base, 16½ inches.

Price, $8.10

## TOULON

$150.00

WOOD CASE, ADAMANTINE FINISH, GILT METAL ORNAMENTS.
ADAMANTINE ONYX COLUMNS.

8 Day, Half Hour Strike,
Cathedral Bell, Superior Movement.
4¼ inch White Porcelain Dial, French Sash with Beveled Glass.
Hight, 12 inches. Base, 11¾ inches.
Price $9.25

## DELOS

$125.00

WOOD CASE, ADAMANTINE FINISH. BRONZED METAL FEET AND ORNAMENTS.

*Only clocks in ORIGINAL MINT condition and in good running order will command top prices. A replaced or damaged dial, missing parts or chipped beveled glass will reduce prices accordingly.*

8 Day, Half Hour Strike, Half Hours on Cup Bell, Hours on Cathedral Bell. 5 inch Dial.

Hight, 11 inches. Base, 17 inches.

Price, $7.80

## RAVENNA

WOOD CASE, ADAMANTINE FINISH. GILT METAL FEET AND ORNAMENTS.

$125.00

8 Day, Half Hour Strike, Half Hours on Cup Bell, Hours on Cathedral Bell. 5 inch Dial.
Hight, 10¾ inches. Base, 17 inches.

Price, $7.15

# 1904-1910 SETH THOMAS MANTEL CLOCKS

**$135.00**

LYONS

Adamantine Onyx Columns. Gilt Metal Feet and Ornaments. Height, 12½ inches. Base, 16 inches.
List, $9.90

**$135.00**

TYRO

Green Mottled Finish only, Gilt Metal Columns, Feet and Ornaments. Height, 10¾ inches. Base, 16 inches.
List, $9.50

**$135.00**

SUCILE

Adamantine Onyx Columns and Sides. Gilt Metal Feet and Ornaments. Height, 11 inches. Base, 17½ inches.
List, $13.00

**$135.00**

LEEDS

Adamantine Onyx Columns. Gilt Metal Feet and Ornaments. Height, 11½ inches. Base, 16¼ inches.
List, $9.00

*Only clocks in ORIGINAL MINT condition and in good running order will command top prices. A replaced or damaged dial, missing parts or chipped beveled glass will reduce prices accordingly.*

**$135.00**

**MILO**

WOOD CASE, ADAMANTINE FINISH.

GILT METAL FEET AND ORNAMENTS. ADAMANTINE ONYX COLUMNS.

8 Day, Half Hour Strike, Half Hours on Cup Bell, Hours on Cathedral Bell. 5 inch Dial.

Hight, 11¾ inches. Base, 17½ inches.

Price, $9.30

**$135.00**

**NIPHON**

WOOD CASE, ADAMANTINE FINISH, GILT METAL FEET AND ORNAMENTS. ADAMANTINE ONYX COLUMNS.

8 Day, Half Hour Strike, Half Hours on Cup Bell, Hours on Cathedral Bell. 5 inch Dial.

Hight, 11 inches. Base, 17½ inches.

Price, $9.00

# 1904-1910 SETH THOMAS MANTEL CLOCKS

$90.00

**DOMINO**

5 inch Dial. Height, 10 inches. Base, 9 inches.
List, $6.50

$135.00

**ROSLYN**

Adamantine Onyx Columns. 5 inch Dial. Bronzed Metal Feet and Ornaments. Height, 11 inches. Base. 16¾ inches. List, $10.30

$115.00

**DON**

Gold Plated Metal Ornaments. 4¼ inch Dial. Height, 11¾ inches. Base, 10 inches.
List, $7.50

$135.00

**PATMOS**

WOOD CASE, ADAMANTINE FINISH. BRONZED METAL FEET AND ORNAMENTS.

8 Day, Half Hour Strike, Half Hours on Cup Bell, Hours on Cathedral Bell. 5 inch Dial.

Hight, 11 inches, Base, 16¾ inches.

Price, $7.50

$135.00

**BOSNIA**

WOOD CASE, ADAMANTINE FINISH. BRONZED FEET AND ORNAMENTS.

8 Day, Half Hour Strike, Half Hours on Cup Bell, Hours on Cathedral Bell. 5 inch Dial.

Hight, 10¼ inches. Base, 17 inches.

Price, $7.00

$115.00

**SUSSEX**

BRONZED METAL FEET.

8 Day, Half Hour Strike.
Half Hours on Cup Bell, Hours on Cathedral Bell.
5 inch Dial.
Hight, 10¾ inches, Base, 11½ inches.
Price $5.40

$135.00

**PEQUOD**

WOOD CASE, ADAMANTINE FINISH. BRONZED METAL FEET AND ORNAMENTS.

8 Day, Half Hour Strike, Half Hours on Cup Bell,
Hours on Cathedral Bell. 5 inch Dial.
Hight, 11 inches. Base, 17½ inches.
Price, $7.00

$135.00

**ADELE**

Adamantine Onyx Columns. Gilt Metal Feet and Ornaments. Height, 11½ inches. Base, 12½ inches.
List, $8.50

# 1904-1910 SETH THOMAS MANTEL CLOCKS

*Only clocks in ORIGINAL MINT condition and in good running order will command top prices. A replaced or damaged dial, missing parts or chipped beveled glass will reduce prices accordingly.*

**$125.00**

**TIBER**

Adamantine Onyx Columns. 5 inch Dial. Gilt Metal Feet and Ornaments. Height, 11½ inches. Base, 12½ inches.
List, $8.50

**$135.00**

**DALE**

Adamantine Onyx Columns. Gilt Metal Feet and Ornaments. 5 inch Dial. Height, 12½ inches. Base, 17 inches.
List, $11.30

**$135.00**

**IDEAL**

Bronzed Metal Feet and Ornaments. Height, 11½ inches. Base, 16¼ inches.
List, $9.30

**$135.00**

**PETREL**

WOOD CASE, ADAMANTINE FINISH. GILT METAL FEET AND ORNAMENTS.

8 Day, Half Hour Strike, Half hours on Cup Bell, Hours on Cathedral Bell. 5 inch Dial
Hight, 10¾ inches. Base, 16¾ inches.
Price, $7.00

**$135.00**

**VIKING**

WOOD CASE, ADAMANTINE FINISH, GILT METAL FEET AND ORNAMENTS.

8 Day, Half Hour Strike, Half Hours on Cup Bell, Hours on Cathedral Bell. 5 inch Dial.
Hight, 12⅜ inches. Base, 17¼ inches.
Price, $8.70

**$135.00**

**SPARTA**

WOOD CASE, ADAMANTINE FINISH.
GILT METAL FEET AND ORNAMENTS. ADAMANTINE ONYX COLUMNS.

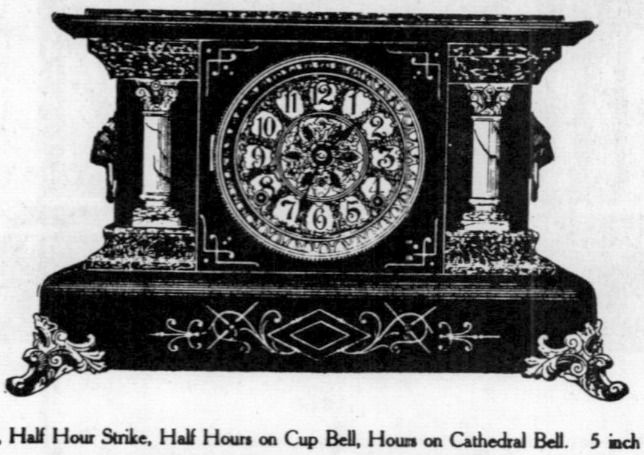

8 Day, Half Hour Strike, Half Hours on Cup Bell, Hours on Cathedral Bell. 5 inch Dia.
Hight, 10¾ inches. Base, 16 inches.
Price, $7.15

**$135.00**

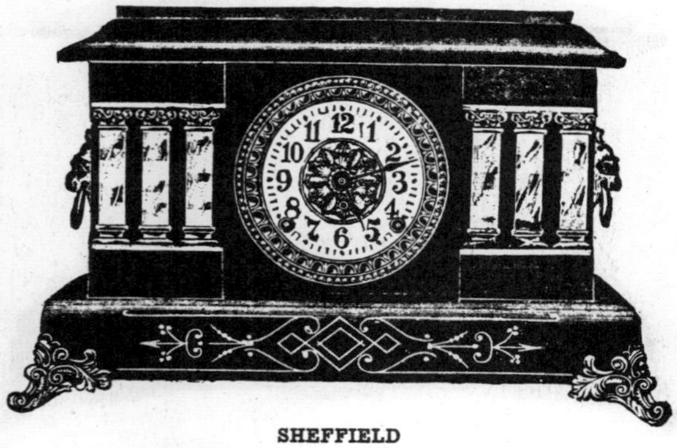

**SHEFFIELD**

Adamantine Onyx Columns. Gilt Metal Feet and Ornaments. Height, 11¼ inches. Base, 17¾ inches.
List, $11.00

# 1910 SETH THOMAS MANTEL CLOCKS

Eight Day, Half Hour Strike, Half Hour on Cup Bell, Hour on Cathedral Bell. 5 inch Dials.

**$110.00**

**PELHAM**

Made in Black, also in Mahogany Adamantine. 4¼ inch Dial. Height, 10 inches. Base, 14 inches.

List, $8.00

**$110.00**

**HASTINGS**

Gold Plated Metal Ornaments. 5 inch Dial. Made in Black, also in Mahogany Adamantine. Height, 11½ inches. Base, 14 inches.

List, $9.00

**$135.00**

**HALLIE**

Gold Plated Columns. 5 inch Dial. Gilt Metal Feet and Ornaments. Height, 11 inches. Base, 16 inches.

List, $9.50

**$135.00**

**EVERETT**

Adamantine Onyx Columns. 5 inch Dial. Gilt Metal Feet and Ornaments. Height, 11 inches. Base 16 inches.

List, $9.50

**$135.00**

**HOOD**

Adamantine Onyx Columns. 4¼ inch Cream Dial. Bronze Metal Feet and Ornaments. Height, 12 inches. Base, 17 inches.

List, $12.20

**$145.00**

**SHASTA**

Adamantine Onyx Columns and Sides. 5 inch Dial. Gilt Metal Feet and Ornaments. Height, 11 inches. Base, 17½ inches. List, $13.00

# 1910 SETH THOMAS MANTEL CLOCKS

*Only clocks in ORIGINAL MINT condition and in good running order will command top prices. A replaced or damaged dial, missing parts or chipped beveled glass will reduce prices accordingly.*

$135.00

**HULL**

Adamantine Onyx Columns. Gilt Metal Feet and Ornaments. Height, 10¾ inches. Base, 16 inches.
List, $9.50

$135.00

**KESWICK**

Black Metal Columns, Gilt Feet and Ornaments. Height, 10¾ inches. Base, 16 inches.
List, $9.50

$125.00

**WANDA**

Adamantine Onyx Columns. Gold Plated Metal Caps and Bases. Height, 11¼ inches. Base, 15½ inches.
List, $10.80

$110.00

**HARRISON**

Gold Plated Metal Columns, Feet and Ornaments. Made in Black, also in Mahogany Adamantine. Height. 12 inches. Base, 16 inches.
List, $9.50

$135.00

**MANCHESTER**

Bronzed Metal Columns, Feet and Ornaments. Height. 12½ inches. Base. 17 inches.
List. $10.70

$110.00

**ADNAW**

Gold Plated Ornaments. 5 inch Dial. Height, 11¼ inches. Base, 15½ inches.
List, $10.80

# 1910 SETH THOMAS MANTEL CLOCKS

## WOOD CASE, WHITE ADAMANTINE FINISH

Extra Gold Plated Metal Feet and Ornaments. These cases can be cleaned with a damp cloth without injury to the finish. Eight Day, Half Hour Strike. Half Hour on Cup Bell, Hour on Cathedral Bell.

$100.00

**AVON**

5 inch Dial. Height, 10 inches. Base, 9 inches.
List, $7.50

*Only clocks in ORIGINAL MINT condition and in good running order will command top prices. A replaced or damaged dial, missing parts or chipped beveled glass will reduce prices accordingly.*

$100.00

**DWIGHT**

Gold Plated Columns and Ornament. 4¼ inch Cream Dial. Height, 11¾ inches. Base, 10 inches.
List, $8.50

$125.00

**BEULAH**

Adamantine Onyx Columns. 5 inch Dial. Height, 10½ inches. Base, 14½ inches.
List. $9.00

$135.00

**BLANCHE**

Adamantine Onyx Columns. 4¼ inch Cream Dial. Height, 12 inches. Base, 17 inches.
List, $13.20

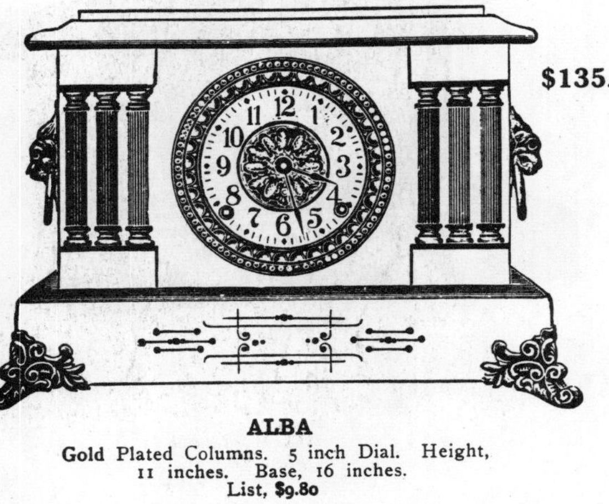

$135.00

**ALBA**

Gold Plated Columns. 5 inch Dial. Height, 11 inches. Base, 16 inches.
List, $9.80

$135.00

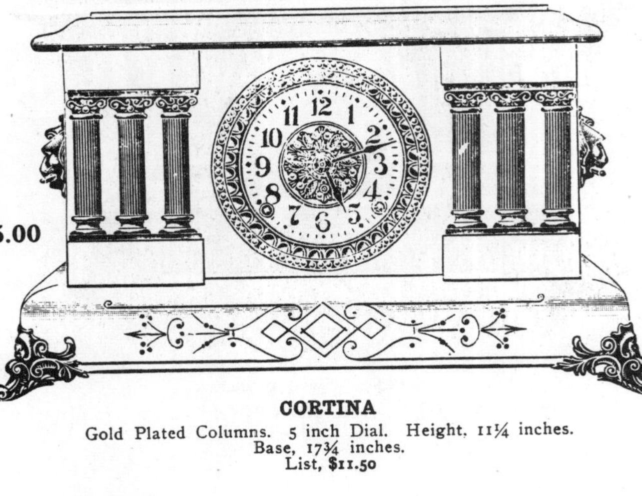

**CORTINA**

Gold Plated Columns. 5 inch Dial. Height, 11¼ inches. Base, 17¾ inches.
List, $11.50

# 1915 SETH THOMAS MANTEL CLOCKS

**WOOD CASE, MAHOGANY ADAMANTINE FINISH—EXCEPT No. 5**

Eight-Day, Hour and Half-Hour Strike on Cathedral Bell. 5-inch Dials.

**$100.00**

**BELMONT No. 2**
Mahogany Finish.
2 Oval Columns.
Base, 13 inches.
List, $8.30

**Dull gold-plated sash, mat and feet.**

**$100.00**

**BELMONT No. 3**
Mahogany Finish.
2 Flat Columns.
Base, 13 inches.
List, $8.30

*Only clocks in ORIGINAL MINT condition and in good running order will command top prices. A replaced or damaged dial, missing parts or chipped beveled glass will reduce prices accordingly.*

**$100.00**

**BELMONT No. 5**
Golden Bronze Finish.
Base, 13 inches. List, $8.30

**$100.00**

**BELMONT No. 1**
Mahogany Finish.
Base, 13 inches.
List, $8.00

**$150.00**

**BERKLEY**

Mahogany with Gold Plated Metal Trimmings. 4½ inch Dial. Height, 11½ inches. Base, 10 inches.

List, $13.00

**$125.00**

**NEW SUCILE**
Mahogany Adamantine Finish. Adamantine Onyx Columns and Sides.
Gold-Plated Metal Ornaments and Feet.
Height, 11 inches. Base, 16 inches.
List, $11.00

**$100.00**

**BELMONT No. 4**
Mahogany Finish.
Base, 13 inches. List, $8.30

# 1915 SETH THOMAS MANTEL CLOCKS

## WOOD CASE, MAHOGANY ADAMANTINE FINISH

Eight-Day, Hour and Half-Hour Strike on Cathedral Bell. 5-inch Dials.

$90.00

**DANA No. 1**
Mahogany Finish.
Height, 10 inches.
List, $7.00

$90.00

**DANA No. 2**
Mahogany Finish.
Height, 11 inches.
List, $7.30

$90.00

**DANA No. 3**
Mahogany Finish.
Height, 11 inches.
List, $7.30

$90.00

**DANA No. 4**
Mahogany Finish.
Height, 11 inches.
List, $7.30

$125.00

**BOLO**
Height, 9 inches, Base
14 inches. List, $9.50

$125.00

**ALTO**
Height 9½ inches. Base
16 inches. List $10.50

$125.00

**DING DONG STRIKE No. 1**
Mahogany Adamantine Finish.
Eight-Day. Striking the hours on two patented Resonated Bells.
Half-Hour Strike on one bell.
Height, 11¾ inches. Base, 14 inches.
List, $11.00

$125.00

**BASO**
Height 10 inches. Base
17 inches. List $11.00

# 1913-1921 SETH THOMAS MANTEL CLOCKS

## MAHOGANY CABINET CASES

**Dull Rubbed Finish.**

Eight-Day Hour and Half-Hour Strike on Cathedral Bell. 5-inch Embossed Etched Silvered Dial.

**$100.00**

**CLIFTON No. 1**
Height, 9½ inches.
List, **$10.50**

**$100.00**

**CLIFTON No. 2**
Height, 10½ inches.
List, **$11.00**

**$100.00**

**CLIFTON No. 3**
Height, 10½ inches.
List, **$11.00**

**$225.00**

**ABBEY**
Mahogany Cabinet Case.
15-Day Pendulum Strike Movement.
Four-inch Dial.
Plain Mahogany.
Height, 10 inches.
List Price, **$40.00.**

**$200.00**

**CHATEAU**
Mahogany Cabinet Case.
Fine 15-Day Pendulum Strike Movement No. 51.
List Price, **$40.00.**
Also fine 8-Day Lever 7-Jewel Strike Movement No. 105
List Price, **$50.00.**
Also with fine 8-Day 7-Jeweled Lever "Ship's Bell" Strike Movement No. 115.
List Price, **$50.00.**
Four-inch Porcelain Dial.
Height, 8 inches. Width, 7½ inches.

**$200.00**

**$225.00**

**ABBEY**
Mahogany Cabinet Case.
15-Day Pendulum Strike Movement.
Four-inch Dial.
Inlaid Marqueterie.
Height, 10 inches.
List Price, **$43.00.**

PERTH

**$180.00**

## MAHOGANY CABINET CASES

**Dull Rubbed Finish.**

Eight-Day, Hour and Half-Hour Strike on Cathedral Bell. Silvered Dials, with 4½-inch Minute Circle, Silvered and Attached.

**DUNDEE**
Height, 10¼ inches. Width, 8¾ inches.
List Price, **$32.00.**

**PERTH**
Height, 11¼ inches. Width, 8¾ inches.
List Price, **$32.00.**

**$180.00**

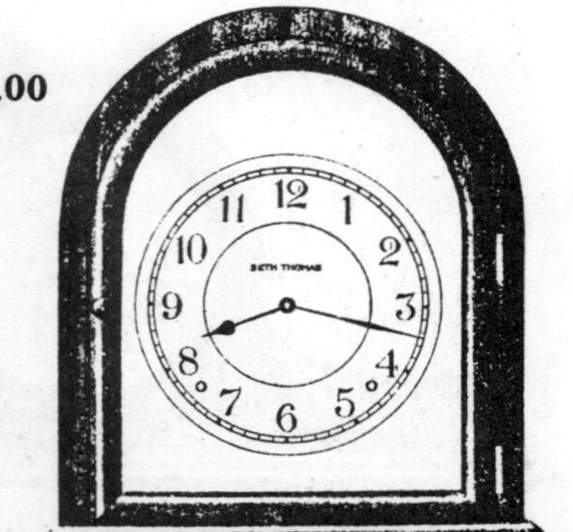

DUNDEE

# 1913-1921 SETH THOMAS MANTEL CLOCKS

$115.00

**RIDEAU**

Mahogany Cabinet Case.
Five-inch Porcelain Convex Dial, Dull Gold Cast Sash with Silvered Mat, and Beveled Convex Glass. Height, 9 inches. Length, 21½ inches.
List Price, **$65.00.**

These patterns can be had fitted with
No. 105, 7-Jewel, Lever Highest Grade Movement, Striking Hour and Half Hour.
No. 115, 7-Jewel, Lever Highest Grade Movement, Striking 8-Bell Ship's Strike.

**No. 2 TAMBOUR**
Mahogany Cabinet Case.
Four-inch Dial.
Height, 7 inches. Base, 13¼ inches.
List Price, **$30.00.**

$100.00

**No. 2 TAMBOUR**

**Barnard Green**

Jade green wood case. 2-inch porcelain dial. Height 3¼ inches. Base 6½ inches. One day lever time movement.

**Barnard Ivory**

Same with Ivory case.

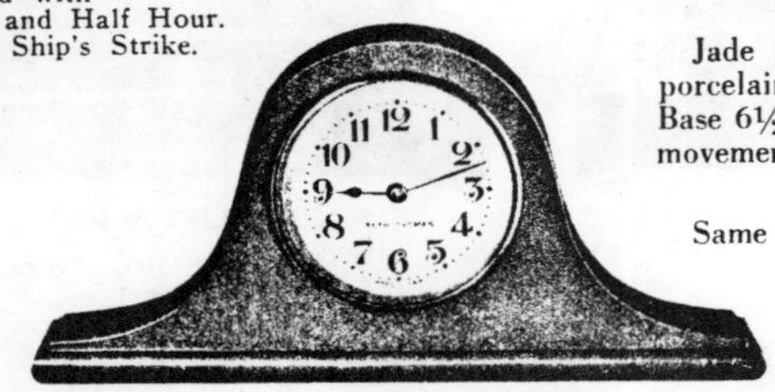

$85.00

$100.00

**TRENT**

Mahogany Cabinet Case.
Five-inch Porcelain Convex Dial, Dull Gold Cast Sash with Silvered Mat, and Beveled Convex Glass. Height, 8½ inches. Length, 19 inches.
List Price, **$60.00.**

$110.00

**TAMBOUR No. 9**
Mahogany Cabinet Case.
Height, 6½ inches. Length, 14½ inches.
Fitted with fine 8-Day 7-Jeweled Movement, No. 105, striking hours and half hours.
List Price with No. 105 Movement, **$55.00.**

$150.00

**Gloucester**

Mahogany base. Length 16½ inches. Height 6⅞ inches. Cast brass head with screw bezel and 3½-inch silvered dial. Seven-jewel time movement No. 112.

$100.00

**TAMBOUR No. 18**
Mahogany Case, 5-inch Dial, Cast Sash.
Height, 9 inches. Length, 20¼ inches.
List Price, Silvered Dial, $42.00
With Raised Bronze Numerals, $48.00

$170.00

**TYNE**
Height, 8½ inches. Width, 12 inches.
List Price, **$55.00.**

# 1906 SETH THOMAS MANTEL CLOCKS

**$165.00**

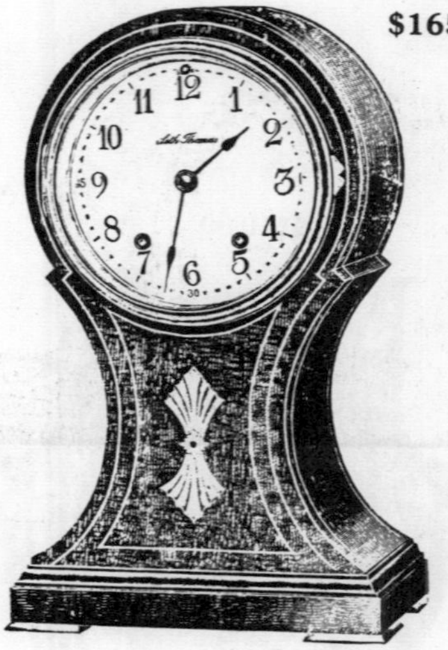

**PARMA**
Height, 12 inches. Base, 7½ inches.
List, $22.50

**$165.00**

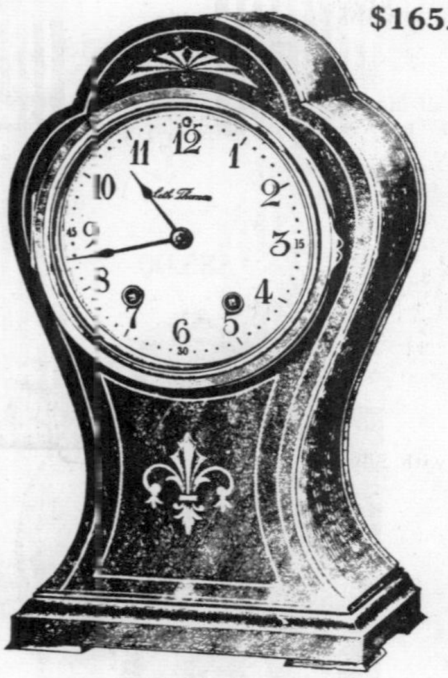

**FLORENCE**
Height, 12¼ inches. Base, 7 inches.
List, $24.50

**$175.00**

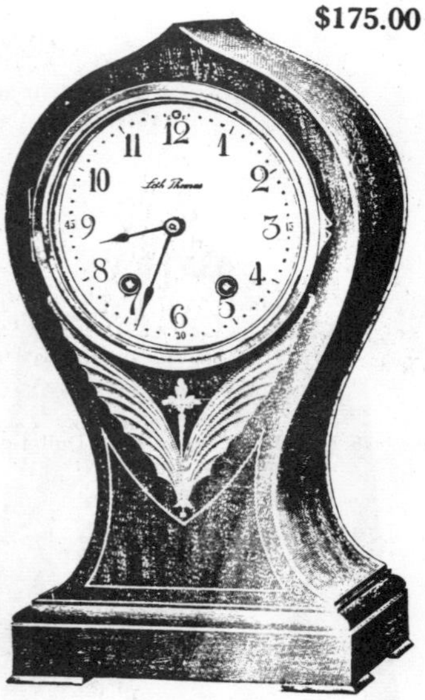

**SAVOY**
Height, 13¼ inches. Base, 8½ inches.
List, $26.50

**KENT**

**$100.00**

**KENT**
Made in Dark Old Oak Only. 5 inch Dial. Height, 14¾ inches.
List, $10.00

**$200.00** **RIVOLI**

Fine Polished Cabinet Case, Hand Carved Mahogany or Golden Oak. Fifteen Day, Half Hour Strike, Cathedral Bell. Fine Polished Movement. Gold Gilt Dial, with Raised Ornaments in Corners and 5 inch Raised Silvered Ring. Height, 12¾ inches.
Price, $39.00

**$175.00**

**PALERMO**
Also made in Plain Flemish Oak and Plain Mahogany.
Height, 14½ inches. Base, 9¼ inches.
List, $30.80

**$120.00**

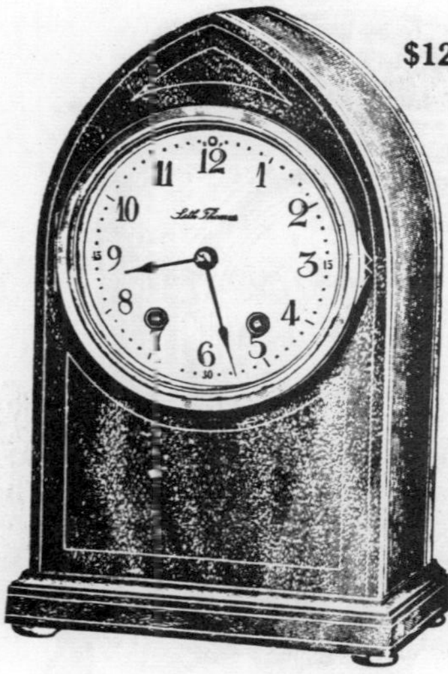

**WHITBY**
Mahogany Inlaid.
Height, 12 inches.
Base, 7¾ inches.
List, $20.50

**$175.00**

**STRATFORD**
Also made in Plain Mahogany.
Height, 15 inches. Base, 9¼ inches.
List, $30.80

# 1906 SETH THOMAS MANTEL CLOCKS

## ARTS AND CRAFTS

Eight Day, Half Hour Strike, Half Hour on Cup Bell, Hour on Cathedral Bell. 6 inch Exposed Metal Dial. The panels are veneered to the body in such a manner as to make these cases very strong and durable.

**$85.00**

**ZUNI**

Panel Dark Red Oak, Body Weathered Oak.
Height, 14½ inches. Base, 12 inches.
List, $8.80

**$85.00**

**ONAVA**

Panel Dark Olive Green Oak, Body Silver Gray Oak. Visible Pendulum.
Height, 15¾ inches. Base, 11½ inches.
List, $9.30

**$110.00**

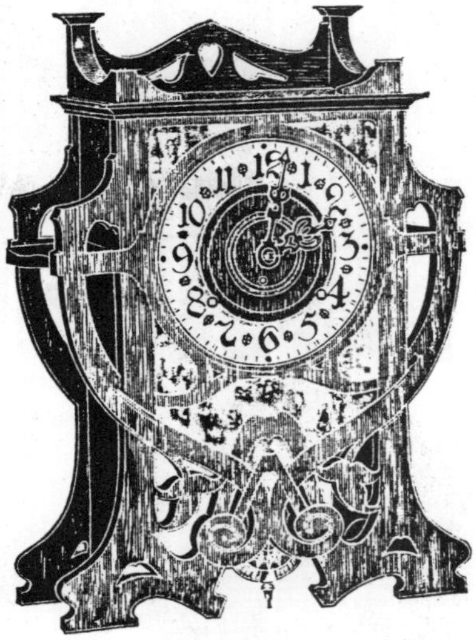

**AZTEC**

Panel Birdseye Maple, Body Dark Red Oak. Visible Pendulum.
Height, 17 inches. Base, 11¾ inches.
List, $9.80

## ENGLISH CABINETS

5 INCH DIALS

Mahogany Cabinet Cases, with Marqueterie and Brass Feet. Eight Day. Round Fine Movement. Hour and Half Hour Rack Strike, Cathedral Bell, 5 inch Convex White Porcelain Dials, Cast Sashes, with Convex Beveled Glass.

**$110.00**

**MODENA**

Height, 12 inches. Base, 8 inches.
List, $20.50

**$125.00**

**TOURAINE**

Height, 13¾ inches. Base, 9¼ inches.
List, $25.50

**$175.00**

**TORY**

Height, 13½ inches. Base, 10¼ inches.
List, $29.50

# 1910-1913 SETH THOMAS MANTEL CLOCKS

## CABINET CASES

Eight-Day, Hour and Half-Hour Strike on Cathedral Bell. 5-inch Dials. Dull Gold-Plated Sash and Mat.

PROSPECT NO. 1

Quartered Oak and Mahogany with Marqueterie. 5 inch Dial. Height, 13 inches. Base, 9 inches.

List, $10.00

**$115.00**

PROSPECT No. 00

**$100.00**

**PROSPECT No. 00**

Mahogany, Golden Oak and Flemish Oak Veneered.

Height, 10½ inches.

List, **$7.50**

**PROSPECT No. 0**

**PROSPECT No. 0**

Mahogany Veneered.

Height, 13 inches.

List, **$8.50**

**$110.00**

**PROSPECT No. 2**

Mahogany with Marqueterie. 5 inch Dial. Height, 13 inches. Base, 9 inches.

List, **$11.00** **$135.00**

**PROSPECT No. 3**

Mahogany with Marqueterie. 5 inch Dial. Height, 13 inches. Base, 9 inches.

List, **$12.00** **$135.00**

**PROSPECT No. 9**

Mahogany Veneered

Height, 13¼ inches

List, $11.00 **$125.00**

**PROSPECT No. 10**

Mahogany Veneered

Height, 13¼ inches

List, $9.50 **$115.00**

**PROSPECT No. 11**

Mahogany Veneered

with shade Marqueterie

Height, 13¼ inches

List, $12.00 **$135.00**

# 1910-1915 SETH THOMAS MANTEL CLOCKS

$145.00

**ETON**

Mahogany, plain....List, $20.00
Mahogany, inlaid....List, 21.00
Height, 10 inches.

$80.00

**DERBY**

Mahogany, plain only.
Height, 8½ inches.
List, $13.50

$150.00

**YORK**

Hand Carved Case. Height, 10 inches.
List, $26.00

$125.00

**TURIN**

Mahogany and Golden Oak.
Plain only.
Height, 8¾ inches.
List, $16.00

$125.00

**TUDOR**

Mahogany, plain....List, $18.00
Mahogany, inlaid....List, 19.00
Height, 8¾ inches.

$90.00

**WINDSOR**

Mahogany, Flemish Oak and Old Oak.
5-inch Dial. Height, 9½ inches.
List, $6.80

$145.00

**WALES**

Plain only, Mahogany and Golden Oak.
5-inch Dial. Height, 10¾ inches.
List, $18.00

$125.00

**ESSEX**

Mahogany, plain....List, $18.00
Mahogany, inlaid....List, 19.00
Height, 9½ inches.

$145.00

**AMHERST**

Mahogany, plain ..............List, $20.00
Mahogany, inlaid ..............List, 21.00
5-inch Dial. Height, 10½ inches.

$400.00

**HOTEL**

Mahogany or Old Oak. Eight Day, Spring, Strike, Cathedral Bell. 8 inch Dial. Height, 18 inches.
List, $30.00

# 1915 SETH THOMAS MANTEL CLOCKS

## SOLID MAHOGANY CASES

Eight-Day Lever Time Movements. Brass Sash and Beveled Glass.

**$60.00**

**CAPEL**

Eight-Day Time. 2½-inch Porcelain Dial
Height, 4½ inches
List, **$5.50**

**$60.00**

**CANBY**

Eight-Day Time
2½-inch Porcelain Dial
Height, 4½ inches
List, **$6.50**

**$60.00**

**CISCO**

Eight-Day Time. 2½-inch Porcelain Dial.
Height, 4½ inches.
List, **$6.00**

**$60.00**

**CANNOCK**

Eight-Day Time. 2½-inch Porcelain Dial.
Inlaid Lines on front edge of body.
Height, 4½ inches.
List, **$6.00**

**$75.00**

**CARY**

Eight-Day Time. 2½-inch Porcelain Dial.
Height, 4½ inches. Base, 9 inches.
List, **$7.00**

**BOLTON**

Eight-Day Time. 4-inch Porcelain Dial.
Height, 6¼ inches.
List, **$10.00**

**$100.00**

**$60.00**

COMO
Eight Day Time
Inlaid Lines on
front edge of body
2½ inch Porcelain
Dial. Height 5½
inches.List $5.50

**COMO**

## MAHOGANY CABINET CASES

**Dull Rubbed Finish.**

Eight-Day Hour and Half-Hour Strike on Cathedral Bell.

5-inch Convex White Porcelain Dials.

Heavy Brass Sash with Convex Beveled Glass.

**$75.00**

**HENLEY**

Height, 10 inches. Base, 9½ inches.
List, **$13.50**

**$100.00**

**NEWPORT**

Height, 9 inches. Base, 13 inches.
List, **$16.00**

**$100.00**

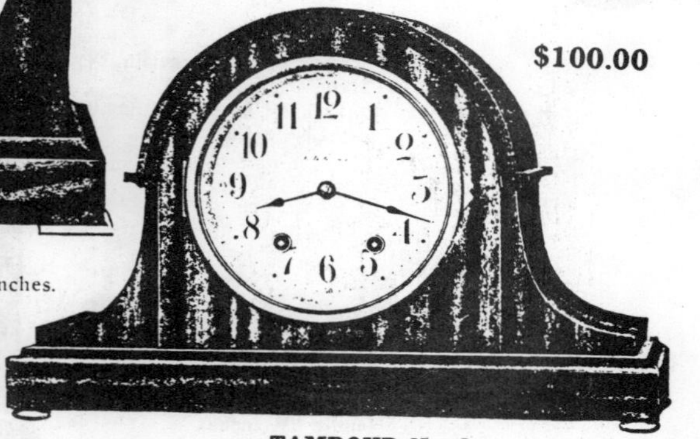

**TAMBOUR No. 3**

Height, 9 inches. Length, 13½ inches.
List, **$18.00**

# 1915 SETH THOMAS MANTEL CLOCKS

SOLID MAHOGANY CASES

**$40.00**

**BISLEY**
One-Day Time. 2-inch Porcelain Dial.
Gold-Plated Ball Feet.
Height, 3 inches.
List, **$3.00**

**$40.00**

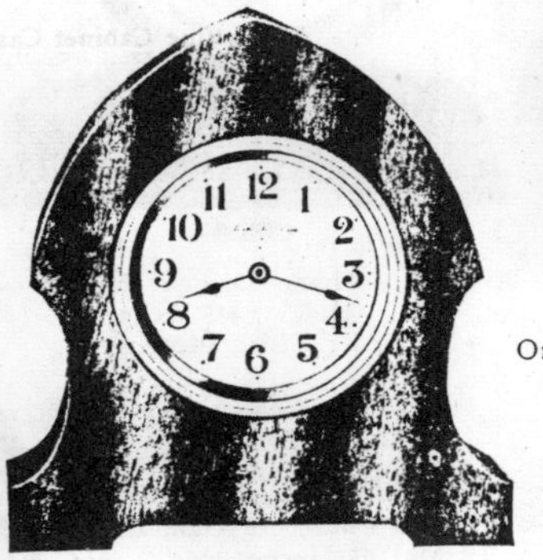

**BANGOR**
One-Day Time. 2-inch Porcelain Dial.
Height, 4½ inches.
List, **$3.50**

**$40.00**

**BARD**
One-Day Time. 2-inch Porcelain Dial.
Height, 4½ inches.
List, **$3.30**

**$50.00**

**BOULDER**
Eight-Day Time. 4-inch Porcelain Dial.
Inlaid Lines.
Height, 4¾ inches.
List, **$8.50**

**$40.00**

**BARNET**
One-Day Time. 2-inch Porcelain Dial.
Height, 4½ inches.
List, **$3.50**

**$40.00**

**BATH**
One-Day Time. 2-inch Porcelain Dial.
With offset arranged body for desk use.
Height, 4 inches.
List, **$3.50**

**$40.00**

**BARFORD**
One-Day Time. 2-inch Porcelain Dial.
Height, 4½ inches. Base, 6½ inches.
List, **$3.30**

**$80.00**

**ANSON**
Eight-Day Time. 4-inch Porcelain Dial.
Hand-Carved Base.
Height, 7½ inches. Base, 13½ inches.
List, **$11.00**

**$75.00**

**BETHEL**
Eight-Day Time. 4-inch Porcelain Dial.
Height, 6½ inches.
List, **$8.50**

**BETHEL**

**$90.00**

**BRISTOL**
Eight-Day Time. 4-inch Porcelain Dial.
Height, 7¾ inches.
List, **$9.00**

# 1910-1922 SETH THOMAS MANTEL CLOCKS

## HIGHEST GRADE CABINET CASES

Dull Rubbed Finish.

Round Fine Movements.

Hour and Half-Hour Rack Strike.

Cathedral Bell. Heavy Brass Sash.

**TAMBOUR**

Eight-Day. Mahogany, Hand-Carved Cabinet Cases, Convex Beveled Glass.
No. 1—5-inch Porcelain Dial. Spade Hands. Height, 10¼ inches. Base, 20 inches. List, $36.50 **$250.00**
No. 2—8-inch Convex Silvered Dial with Raised Bronze Numerals and Serpentine Hands. Height, 14 inches. Base, 32½ inches. List, $90.00

**$375.00**

**$125.00**

**TAMBOUR**

Fine Cabinet Case, Mahogany or Golden Oak. Eight Day, Half Hour Strike, Cathedral Bell. 4¼ inch Cream Porcelain Dial, French Sash, with Beveled Glass. Height, 9¾ inches. Base, 16 inches. List, $33.50

**$115.00**

**TAMBOUR No. 4**

Mahogany Cabinet Case.
Four-inch Porcelain Convex Dial, Dull Gold Cast Sash with Silvered Mat and Beveled Convex Glass. Height, 7½ inches. Base, 15 inches.
List Price, $40.00.

**$175.00**

**PEMBROKE**

Mahogany Cabinet Case.
Silvered Dial, with 4½-inch Minute Circle Silvered and attached.
Height, 10 inches. Length, 18 inches.
List Price, $65.00.

**ENFIELD**

Solid Mahogany Cabinet Case.
Height, 5 inches. Length, 12 inches. Depth, 2½ inches.
List Price, $32.00.

**$160.00**

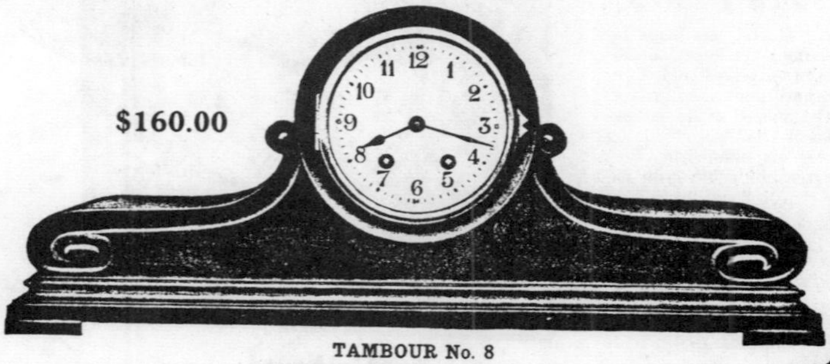

**TAMBOUR No. 8**

Mahogany Cabinet Case.
Four-inch Porcelain Convex Dial, Dull Gold Cast Sash with Silvered Mat, and Beveled Convex Glass. Height, 7½ inches. Length, 18 inches.
List Price, $55.00.

**$115.00**

**ENFIELD**

**$155.00**

**PEER**

Mahogany Cabinet Case.
Four-inch Porcelain Convex Dial, Dull Gold Cast Sash with Silvered Mat and Beveled Convex Glass. Height, 8 inches. Base, 17½ inches.
List Price, $50.00.

This pattern shown with Raised Bronze Numerals on Silvered Dial. List extra, $5.00.

These patterns can be had fitted with
No. 105, 7-Jewel, Lever Highest Grade Movement, Striking Hour and Half Hour.
No. 115, 7-Jewel, Lever Highest Grade Movement, Striking 8-Bell Ship's Strike.

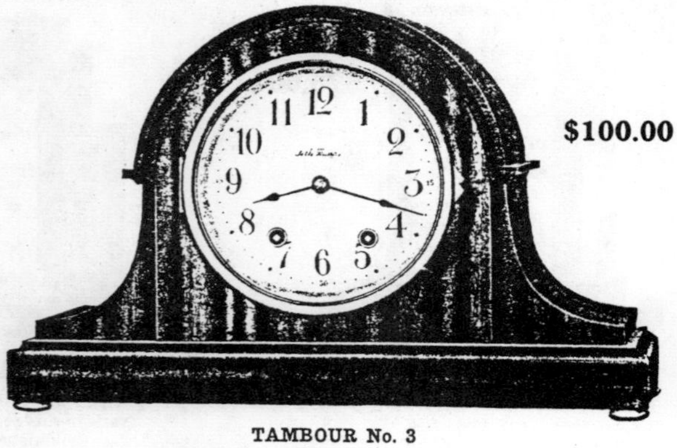

**$100.00**

**TAMBOUR No. 3**

Mahogany.
5-inch Dial.
Height, 9 inches. Length, 13½ inches.

# 1928-1929 SETH THOMAS MANTEL CLOCKS

$65.00

**Mayfair**

Gold finished case with hand painted decorations in six colors. 2½-inch silvered dial. 4-jeweled lever time movement No. 103-A. Height 5½ inches. Base 4½ inches.

$65.00

**Tivoli**

Florentine tooled leather case in blue, green or brown with 22 carat gold decorations. 2½-inch silvered dial. Height 5½ inches. Base 4½ inches. Four jeweled lever time movement No. 103-A.

$85.00

**Lido**

Florentine tooled leather case in green or brown with 22 carat gold decorations. 3-inch silvered dial. Height 7 inches Base 5¼ inches. Four-jeweled lever time movement No. 103-A

$65.00

**Milan**

Florentine tooled leather case in red or brown with 22 carat gold decorations. 3-inch silvered dial. Height 5¼ inches. Base 5¼ inches. 4-jeweled lever time movement No. 103-A.

## *Florentine Tooled Leather and Gold Decorated Clocks*

THE cases of Maywood and Mayfair are made in Italy. The flower decorations are hand painted in six colors on a background of burnished gold. Two coats of colorless lacquer are then applied to protect the artistic workmanship. The method of decorating the cases is technically known as Graffito.

Milan, Tivoli and Lido cases are finished in Florentine tooled leather in a variety of colors with 22 carat gold decorations.

*Only clocks in ORIGINAL MINT condition and in good running order will command top prices. A replaced or damaged dial, missing parts or chipped beveled glass will reduce prices accordingly.*

$85.00

**Maywood**

Gold finished case with hand painted decorations in six colors. 2½-inch silvered dial. 4-jeweled lever time movement No. 103-A. Height 6⅛ inches. Base 6¼ inches.

$75.00

**GRAMERCY**

**The gold dial blends attractively with the two-toned oak case in a size that adapts itself to many places around the home or attractively furnished office. Eight-day, No. 103-A time movement. Height 9 inches. Base 7 inches. $22.50**

$60.00

**RIVERSIDE**

**All of the beauty of modern art prevails in this desk or boudoir clock. The case is finished in natural mahogany and black. Silver dial. No. 103-A, 8-day time movement. Height 6⅛ inches. Base 7⅞ inches. $18.50**

$60.00

**SUTTON**

**The black and silver case of the Sutton finds its most favorable setting on a bookcase or with a lamp that has been created in the modernistic manner. Silver dial. No. 103-A, 8-day time movement. Height 5½ inches. Base 9 inches. $15.00**

# Ca. 1890 SETH THOMAS MUSICAL CLOCKS

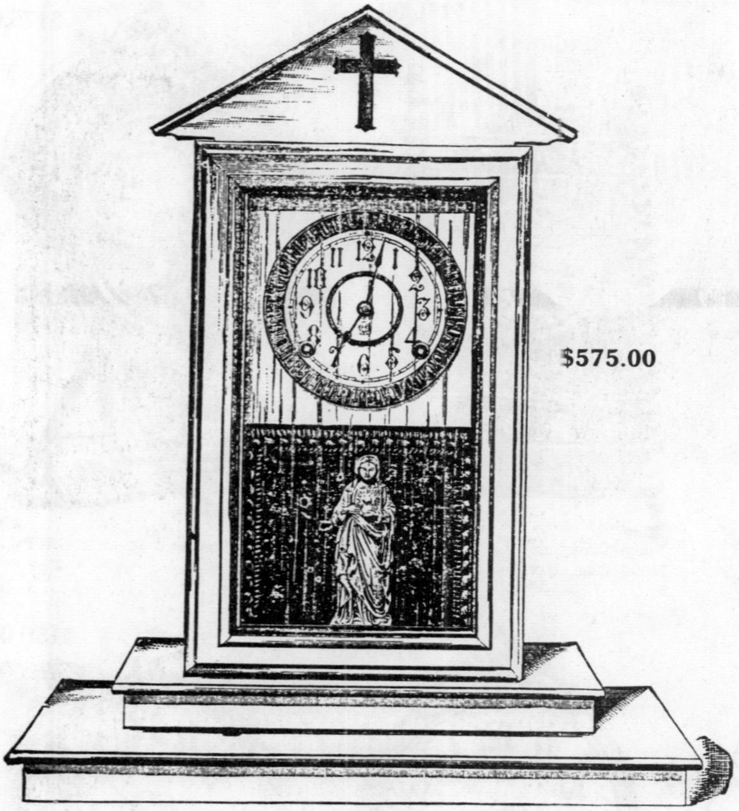

$575.00

The Devotional Musical Clock.

The Devotional Musical Clock.

This beautiful Clock contains Musical Works playing two Sacred Hymns, such as Te Deum Laudamus, Stabat Mater, (Cujus Animam,) Ave Maria, etc. It has an eight day "Seth Thomas" movement and plays and strikes every hour. The music plays eight times with one winding.
Height, 22 inches.
Width, 18 inches.
Price, $10.00

*Only clocks in ORIGINAL MINT condition and in good running order will command top prices. A replaced or damaged dial, missing parts or chipped beveled glass will reduce prices accordingly.*

$400.00

No. 526.

THE MUSICAL PARLOR CLOCK.

Striking and playing every hour; Seth Thomas eight day movement. The music plays eight to ten times with one winding. Price, playing two tunes, $12.00. four tunes, $15.00. Size, 16-1/2 x 11-1/2 inches.
Bronze Statuettes for this Clock, $2.00 extra.

No. 526.

THE MUSICAL PARLOR CLOCK.

# 1890-1910 SETH THOMAS NOVELTIES

## METAL NOVELTIES

One Day, 2 inch Porcelain Dials.

**SHETLAND.**
Height, 4½ inches.

$60.00

1-day, Time..............$ 2.80
Nickel Case, Gilt Handle.

$40.00

**PONY.**
2-inch Dial.

1-day, Time..............$2.00
Nickel Case. Smallest Clock with Second Hand.

$140.00

**HURDLE.**
Height, 7½ inches.

1-day, Time..................$ 3.46
Metal Case, Oxydized Silver or Gilt Finish.

**KITE.**
Height, 8½ inches.

$175.00

1-day, Time..............................$ 3.46
Metal Case, Oxydized Silver or Gilt Finish.

**JEN SET**

Rich Gold, also Art Nouveau Bronze. Height of Clock, 11 inches. Height of Vases, 8 inches.

List Price of Clock, $6.00 — List Price of Set, $12.00

$95.00 — $150.00

List Price of Vases, per pair, $6.00

$150.00

**NIGHT CLOCK "A"**

$150.00

**NIGHT CLOCK "B"**

Gilded Cast Metal Frame with Easel Back to Stand, or can be attached to Gas Burner. 5 inch Opal Glass Dial. One Day Lever Time. Height, 7¼ inches.
List, $5.00

$50.00

**SERENADE**

Rich Gold, also Art Nouveau Bronze.
Height, 8½ inches.
List, $4.60

$45.00

**NAN**

Rich Gold Finish.
Height, 7¼ inches.
List, $3.70

$50.00

**ABE**

Rich Gold, also Art Nouveau Bronze.
Height, 8½ inches.
List, $4.60

**EAGLE**

Rich Gold.
Height, 7 inches.
List, $3.50

$45.00

$50.00

**IVAN**

Rich Gold Finish.
Height, 8¼ inches.
List, $4.30

$50.00

**BUNGALOW**

Rich Gold, also Art Nouveau Bronze. Height, 7¼ inches.
List, $5.00

# 1910 SETH THOMAS NOVELTIES

## METAL NOVELTIES

One Day, 2 inch Porcelain Dials.

$85.00

**GOETHE**

Bronzed Top and Base. Rich Gold Pillars and Center. Height, 8 inches. List, $5.50

$85.00

**MOZART**

Bronze Top and Base, Rich Gold Pillars and Center. Height, 8 inches. List, $5.50

$85.00

**SCHILLER**

Bronze Top and Base, Rich Gold Pillars and Center. Height, 8 inches. List, $5.50

$85.00

**SHAKESPEARE**

Bronze Top and Base, Rich Gold Pillars and Center. Height, 8 inches. List, $5.50

$85.00

**WAGNER**

Bronze Top and Base, Rich Gold Pillars and Center. Height, 8 inches. List, $5.50

$65.00

**DIMPLE**

Rich Gold Finish. Height, 8¾ inches. List, $6.20

$60.00

**QUAINT**

Rich Gold Finish. Height, 9 inches. List, $4.70

$65.00

**FLORIZEL**

Rich Gold Finish. Height, 8¾ inches. List, $5.00

$50.00

**CIS**

Rich Gold Finish, also Art Nouveau Bronze. Height, 8 inches. List, $4.00

$85.00

**VEVA**

Rich Gold Finish. Height, 8½ inches. List, $7.50

$100.00

**PIPER**

Rich Gold Finish and Highly Burnished. Height, 9¾ inches. List, $7.50

$50.00

**LOLA**

Rich Gold Finish. Height, 7½ inches. List, $4.30

$115.00

**JOE**

Rich Gold, also Art Nouveau Bronze. Height, 12½ inches. List, $9.30

$35.00

**JACKET**

Bronze, also Yellow. Height, 3 inches. List, $3.00

$65.00

**JESS**

Rich Gold, also Art Nouveau Bronze. Height, 9 inches. List, $6.20

# 1910 SETH THOMAS NOVELTIES

$65.00

**PETITE**

Metal Case.
Gold Plated. Also Bronze Finish. One Day, Lever Time.
Height. 3½ inches. List, $2.00

$45.00

**SYLVIA**

Adamantine Finish in White, Black, Mahogany, Cocobolo and Green. One Day, Lever Time. Gold Plated Feet. Height, 3 inches. List, $2.00

$55.00

**TRAVELER**

Made in Red, Blue, Green and Maroon Glazed Leather. One Day, Lever Time. Height, 3½ inches.
List, $3.50

**TRAVELER**

$50.00

**FOUNTAIN**

Rich Gold.
Height, 8½ inches.
List, $5.50

$80.00

**COLIN**

Rich Gold and Bronze Art Nouveau
Height, 6 inches.
List, $4.30

$85.00

**PADDOCK**

Rich Gold and Bronze Art Nouveau.
Height, 7½ inches.
List, $5.00

$75.00

**TICK TICK**

Rich Gold and Art Nouveau Bronze.
Height, 6½ inches.
List, $6.50

$75.00

**DORRIT**

Rich Gold.
Height, 5 inches.
List, $7.50

$65.00

**CYRIL**

Rich Gold Finish.
Height, 9½ inches.
List, $6.20

$65.00

**BETH**

Rich Gold Finish, also Art Nouveau Bronze.
Height, 9½ inches.
List, $6.20

$50.00

**COLONIAL**

Rich Gold Finish, also Art Nouveau Bronze.
Height, 8¾ inches.
List, $5.00

$50.00

**FLOSS**

Rich Gold Finish, also Art Nouveau Bronze.
Height, 8½ inches.
List, $5.00

$45.00

**DAINTY**

Rich Gold.
Height, 6½ inches.
List, $3.50

# 1910 SETH THOMAS NOVELTIES

## METAL NOVELTIES

One Day, 2 inch Porcelain Dials.

$45.00

**SCHOOL GIRL**
Rich Gold also Gold Clock with Bronze Girl.
Height, 5 inches.
List, $3.80

$45.00

**NATTY**
Rich Gold.
Height, 6½ inches.
List, $3.40

$45.00

**ELEPHANT**
Rich Gold.
Height, 5 inches.
List, $3.30

$45.00

**ADA**
Rich Gold Finish.
Height, 4 inches.
List, $3.00

$45.00

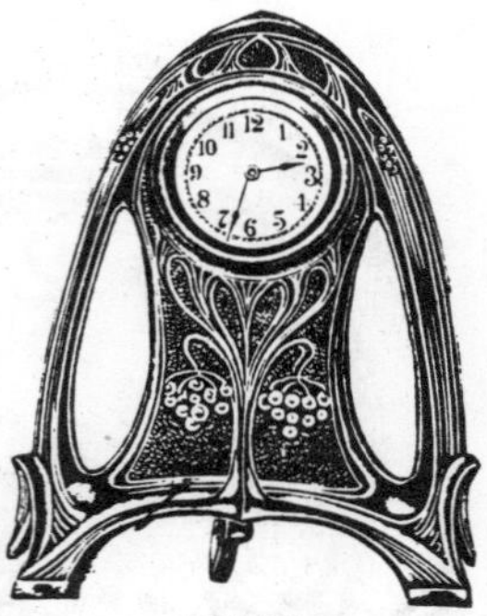

**HOLLY**
Rich Gold.
Height, 7½ inches
List, $3.30

$100.00

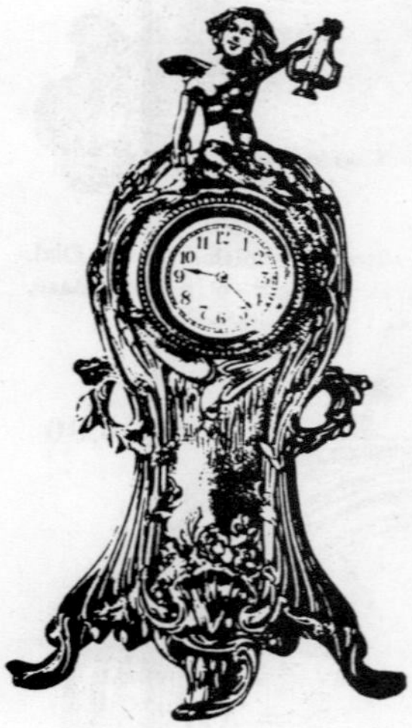

**ELFIN**
Rich Gold Finish and Highly Burnished.
Height, 11 inches.
List, $8.50

$45.00

**BOUQUET**
Rich Gold.
Height, 7 inches.
List, $3.50

$55.00

**SCHOOL DAYS**
Rich Gold, also Gold Clock with Bronze Girl.
Height, 7½ inches.
List, $5.50

$50.00

**CORINNA**
Rich Gold Finish.
Height, 5¾ inches.
List, $3.00

$55.00

**BONA**
Rich Gold Finish.
Height, 5¾ inches.
List, $3.70

$50.00

**GRAPPLE**
Rich Gold Finish.
Height, 7¼ inches.
List, $5.00

$55.00

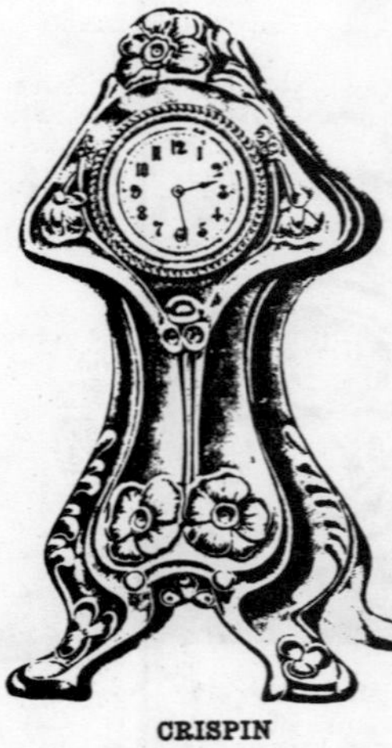

**CRISPIN**
Rich Gold, also Art Nouveau Bronze.
Height, 10¼ inches.
List, $5.00

$60.00

**ALICE**
Rich Gold, also Art Nouveau Bronze.
Height, 10 inches.
List, $5.50

$75.00

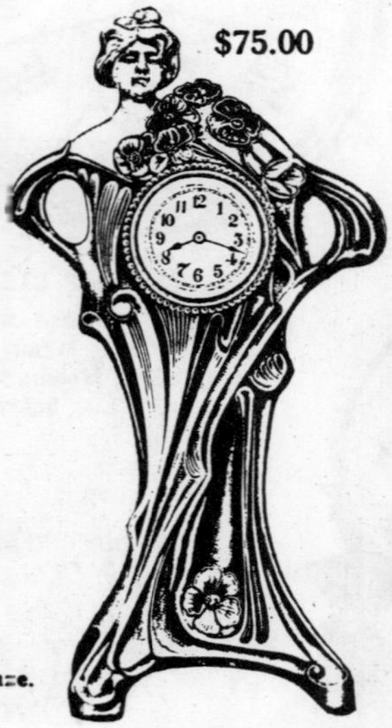

**VANITY**
Rich Gold, also Art Nouveau Bronze.
Height, 12½ inches.
List, $7.50

$45.00

**DIDO**
Rich Gold Finish.
Height, 5¾ inches.
List, $4.00

$40.00

**CHERUBS**
Rich Gold Finish.
Height, 5¾ inches.
List, $2.30

# 1910 SETH THOMAS SHIP'S BELL CLOCKS

## "EIGHT BELL" CLOCKS

The five patterns shown here (and Harbor are Boat Clocks striking "Eight Bells" same as they are rung on board ship. These are called one day movements, but will run for about 3 days. Are strong and simple in construction and do not easily get out of order.

$275.00

**BOAT CLOCK**

Metal Case, Polished Brass, Bronze or Nickel Plated. 6 inch Etched Metal Dial, Silvered. Hinged Bezel. Width of front and back, 8 inches.

List, $20.00

$475.00

**SHIP'S BELL**

Metal Case, Polished Brass or Nickel Plated. 6 inch Dial, Hinged Bezel. Height, 10½ inches. List, $11.00

$225.00

**CABIN**

Metal Case, Bronze Finish. 6 inch Dial. Hinged Bezel. Height, 9 inches. Base, 7½ inches.

List, $15.00

$285.00

**HARBOR**

Metal Case, Bronze Finish. One Day Strike. 6 inch Etched Metal, Silvered Dial, Convex Beveled Glass. For sailors' use, striking "Eight Bells," same as they are rung on board ship. Height, 9 inches. Base, 8½ inches.

List, $25.00

$800.00

**YACHT CLOCK**

Metal Case, Polished Brass or Nickel Plated. 6 inch Metal Dial, Silvered. Hinged Bezel. Height, 10½ inches.

List, $15.70

$400.00

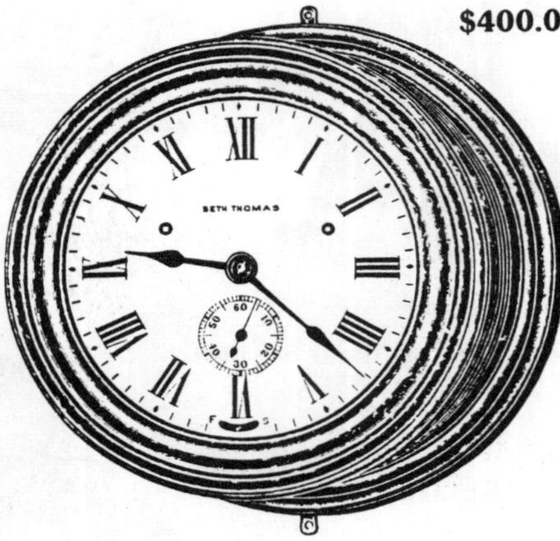


**WARDROOM**

Wood Case, Mahogany or Golden Oak. 5½ inch White Porcelain Dial. Depth of Case, 4¾ inches. Width at back, 8⅝ inches. A fine Cabinet Case.

List, $18.50

**MONTAUK**

*Ship's Bell Strike*

This unusual design has captured the full beauty of swirling Atlantic breakers. Base in cast bronze fitted with finely made, fully jeweled ship's bell strike movement. Height 12¼ inches. Base 22 inches. $175.00

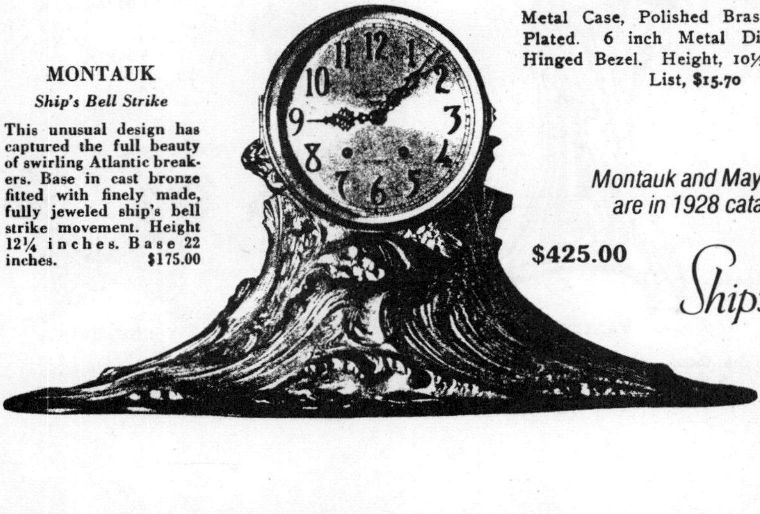

$425.00

*Montauk and Mayflower are in 1928 catalog.*

*Ship's Bell*

$350.00

**Mayflower**

Seven-jewel ships strike movement No. 115. Cast brass head with screw bezel. 4½ inch silvered dial. Mahogany base. Length 13½ inches. Height 9¼ inches.

**Corsair**

Same with seven-jewel No. 112 time movement.

# 1936 SETH THOMAS SHIP'S BELL CLOCKS

**$475.00**

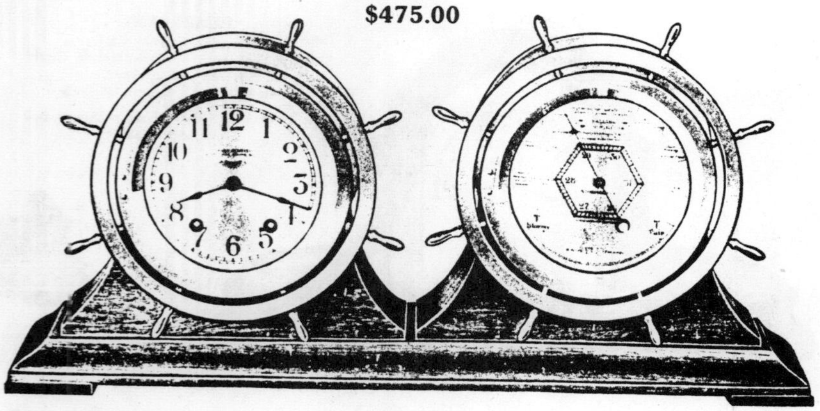

*FINE MARINE CLOCKS*

**Wheel Bezel**

Polished cast brass case with back flange drilled for screws to attach to wall. Screw type moistureproof bezel. Etched silver-plated dial. 7-jeweled fine lever movement. Packed singly.

*8-Day Ship's Bell Strike*

| | Dial Size | Movement No. | Diameter | Retail | Shipping Weight | |
|---|---|---|---|---|---|---|
| No. 44 Wheel | 4 " | 115 | 8¼" | $55.00 | 12 lbs. | **$300.00** |
| No. 66 Wheel | 5½" | 115 | 10⅞" | 62.50 | 16 lbs. | **$350.00** |

*SHIPMATES* ***$125.00***

**Ship's Bell Clock—Barometer Combination**

The pride of all mariners whether seafaring is a vocation or an avocation. An excellent gift and presentation number. Barometer—same as Herald except has hinged bezel and back flange is not drilled for attaching to wall. Clock—has polished cast brass case with hinged wheel type bezel. 4-inch silvered metal dial etched in black. Fitted with 7-jeweled fine lever movement No. 115 that strikes the ship's bells. Both units are mounted on a single base of hand polished solid mahogany. Height 9-inches. Width 18½-inches. Thickness 3⅜-inches. Packed singly. Shipping weight 24 pounds.

**$300.00**

*HELMSMAN* ***$50.00***

*8-Day Ship's Bell Strike*

Ship's wheel circles and cabinet are mahogany with front of diagonally matched mahogany veneer. Spokes of wheel, hinged sash and metal feet are in polished goldplate. Has 4⅛-inch silverplated dial. Fitted with 8-day fine lever 7-jeweled movement No. 115. Strikes the ship's bells. Height 8¼ inches. Thickness 4 inches. Width 13½ inches. Packed singly. Shipping weight 11 pounds.

*This is a modification of earlier model shown on the preceding page. Note the change in the thickness of the bezel and the difference in the bell, and without a wood backboard.*

**$350.00**

*MONITOR* ***$18.50***

*1-Day Ship's Bell Strike*

Polished brass or nickelplated case. Flanged front and back. Hinged bezel. 5¼-inch silverplated dial. 1-day ship's bell strike movement No. 7. Height 10½ inches. Width 7 inches. Packed singly. Shipping weight 7 pounds.

**$235.00**

*MERRIMAC* ***$18.50***

*1-Day Ship's Bell Strike*

Same as Engine Lever but fitted with 1-day ship's bell strike movement No. 7.

# 1915-1921 SETH THOMAS SHIP'S BELL CLOCKS

**$285.00**

PEMBROKE

**PEMBROKE**
Mahogany Cabinet Case.
Silvered Dial, with 4½-inch Minute Circle Silvered and attached.
Height, 10 inches. Length, 18 inches.

List Price with Lever Strike or Ship's Bell Movement, **$75.00.**

**$250.00**

**TAMBOUR No. 4**
Mahogany Cabinet Case.
Four-inch Porcelain or Silvered Convex Dial, Dull Gold Cast Sash with Silvered Mat, and Bevelled Convex Glass. Height, 7½ inches. Base, 15 inches.

List Price with Lever Strike or Ship's Bell Movement, **$50.00.**
With Raised Bronze Numerals, **$5.00** extra.

**$250.00**

**TAMBOUR No. 9**
Mahogany Cabinet Case.
Lever "Ship's Bell" Strike Movement, No. 115. Four-inch Porcelain Dial.
Height, 6½ inches. Length, 14½ inches.

List Price with No. 115 Movement, **$55.00.**

**$315.00**

**CELTIC**
Bronze Finish.
Six-inch Silvered Convex Dials with Raised Bronze Numerals and Cut Serpentine Hands.
Height, 9 inches. Base, 11 inches.

8-Bell Ship's Strike Movement.
List Price, **$85.00.**

**$300.00**

DENTON

**DENTON**
Bronze Finish.
Six-inch Silvered Convex Dials with Raised Bronze Numerals and Cut Serpentine Hands.
Height, 8½ inches. Base, 8½ inches.

With Lever 7-Jewel, 8-Bell Ship's Strike Movement. List Price, **$75.00.**

**$250.00**

RIDEAU

5 inch silvered Dial, Height 9 inches
Width 21½ inches. 8 Day 7 Jewel lever
Ship's Bell Strike. Movement No. 115

**$275.00**

TAMBOUR No. 8

TAMBOUR No. 8

4 inch silvered Dial. Height 7½ inches
Width 18 inches - 8 Day 7 Jewel lever
Ship's Bell Strike - Movement No. 115

# SECTION III

## PUBLIC CLOCKS

# 1928-1930 SETH THOMAS BRACKET CLOCKS

*These clocks are extremely rare and customers for them are limited. However, if they are in good condition you can name your own price. Customers for these clocks include wealthy collectors and owners of shopping malls who prefer a touch of antiquity in their decor.*

**$12,000.00**

*Silk Center Bldg., N. Y. City.*
A most unique bracket clock with moving figures performing every hour, representing the magic development of the raw silk to the finished product.

**$5,500.00**
***With Brackets***

*Ornamental Sheet Bronze Bracket Clock.*

*To receive top prices, clocks must be in original mint condition and in good running order — A replacement dial or missing parts will reduce prices accordingly.*

**$4500.00**

*24" dials mounted in sheet bronze case.*

***Four Dial Bracket Clock***

**$6,650.00**
***With Brackets***

1928 Catalog.

# 1886-1910 SETH THOMAS STREET CLOCKS

*These clocks are extremely rare and customers for them are limited. However, if they are in good condition you can name your own price. Customers for these clocks include wealthy collectors and owners of shopping malls who prefer a touch of antiquity in their decor.*

**$25,000.00**

**FOUR DIAL POST CLOCK**

**Dials, 36 inches in diameter. Iron Dials with Black or Gilt Numerals. Base of Column, 24 inches square. May be fitted with Glass Dials for illuminating if desired. Height, from sidewalk to center of dial, 13 feet 3 inches.**

**$25,000.00**

FOUR-DIAL STREET CLOCK.

Movement to be placed in building, and connection made under sidewalk.

IRON COLUMN.

15 feet to center of Dial. Dial 36 inches.

**$17,000.00**

**TWO-DIAL STREET CLOCK.**

***Movement to be placed in building, and connection made under sidewalk.***

**IRON COLUMN.**

***15 feet to center of Dial. Dial 40 inches.***

**$17,000.00**

**Dials 30 inches in diameter, of Wood, Black, with raised Gilt Numerals. Iron Column, with Head all complete, with Movement and Weight inside Column.**

TWO DIAL STREET CLOCK.

Requires no vault under the sidewalk.

From sidewalk to centre of dials, 12 feet.

**TWO DIAL POST CLOCK**

***40 Inch Dials*** **$17,000.00**

***36 Inch Dials*** **$15,000.00**

**With Iron Column and Iron Head. 40-inch Dials. Height, 13 feet 10 inches. We make Post Clocks in Three Sizes: A Two-Dial Clock having 30-inch Dials, a Two-Dial Clock having 40-inch Dials, and a Four-Dial Clock having 36-inch Dials. Write for Special Post Clock Circulars if interested.**

THESE clocks are unsurpassed in quality of material and workmanship, and in accuracy of time-keeping. They are constructed upon the rules of the highest authorities in horology and no care or expense is spared by us in producing clocks that will, in every way, meet the most exacting and critical requirements.

As proof of this statement, we point to the fact that we have received the highest awards at all expositions in which we have been represented, both in this country and abroad.

In addition to this expert testimony, we also have the testimony of hundreds of individuals and municipalities for whom we have installed clocks, giving unqualified praise to their excellence in all particulars.

We take pleasure in printing, in the latter part of this catalog, a few expressions of praise and satisfaction received from various parties.

In writing for prices and information, please state the diameter and number of dials required, whether they are to be of glass, wood or iron, the weight of the bell or bells (if striking clock is desired), the interior dimensions of the tower and the amount of fall to be had for the weights.

With this information we will be in a position to know what the exact requirements are in each individual case, and will be enabled to make an intelligent estimate for the clock boxed and delivered in New York City, or delivered and installed complete.

Every clock that leaves our factory has been run and thoroughly tested and is guaranteed against original and mechanical defects for a period of five years.

In offering our product to the public, it is not our intent or desire to put on the market clocks that are lower priced than those of other makers. Appreciating that there can be no better advertisement than a clock that performs satisfactorily, our aim will be always to produce clocks which (quality, durability and perfection of construction considered) will prove to be the cheapest on the market.

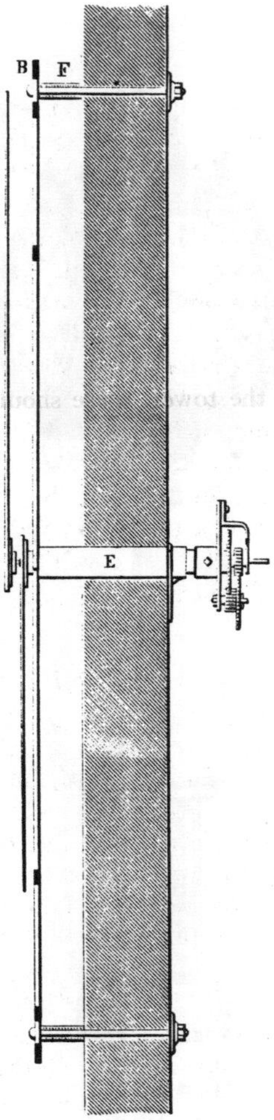


FIG. 1

A—Wall
B—Skeleton Dial
F—Bolt Holding Dial
E—Tube for the Hand Spindles

## NUMERALS

The numerals we furnish for wood or iron dials are made of cast metal (not wood). The face and sides are covered with the best of gold leaf to prevent tarnishing and are fastened on with brass screws.

## DIALS

To look well and show plainly, dials should be one foot in diameter for every ten feet of elevation and be set out flush with or close to the line of the building or tower. Frequently the space available for dials is limited by reason of the architecture of the tower, so this rule cannot be strictly observed, but these proportions should be adhered to as nearly as possible.

## IRON DIALS

Iron dials can be made of galvanized iron with a wood backing an inch or more thick, the full size of the dial. The numerals, etc., are the same as for wood dials. We can furnish iron dials made in this way of any size.

## SKELETON METAL DIALS

FIG. 2 ARABIC

## WOOD DIALS

Wood dials should be made of well-seasoned wood, not over four inches wide, tongued and grooved, from one-half to seven-eighths inch thick, in two layers, with grain crossed to prevent shrinkage, and firmly screwed together, smoothed, painted and smalted. Paint the joints, when putting them together, with paint the same color the face of the dial is to be when finished. This prevents joints showing in case of shrinkage. The joints of the wood on face side of dials should run up and down so that rain will run off. Should dial be placed on the outside of building, it makes a better finish to put a moulding around the outside edge on the front. For conveniently getting at hands at any time from inside the tower, there should be a piece from twelve to fourteen inches square cut from each dial-plate, the lower edge about four inches above and directly over center of dial. This piece can be beveled at top and bottom to keep the rain out; the sides can be straight. Screw a cleat over this piece on the inside to keep it in place. If the room be dark, one of these openings can be glazed to give light. We will furnish numerals and minute marks, or the dial complete, if desired.

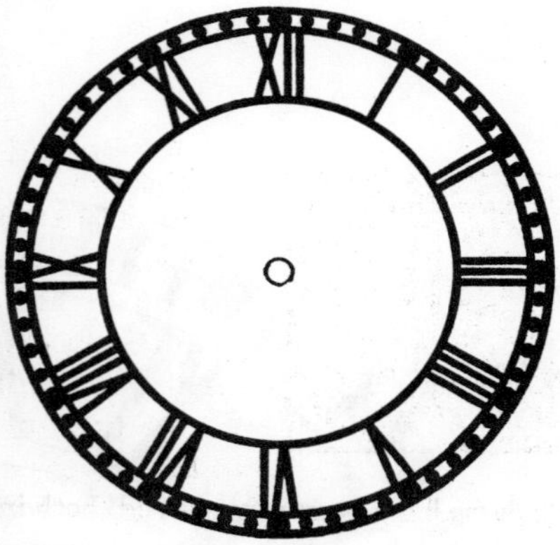

Fig. 3 Roman

## SKELETON METAL DIALS

These dials may be varied in design and made of different metals. Two patterns are illustrated on page 9; one with Arabic figures (minute dots omitted), the other Roman. Our observation is, that it is better for seeing the time, to leave the center or space within the inner ring entirely clear. Skeleton dials should not be placed more than three or four inches from the face of the wall, for they require a backing to show to best advantage. The color of the dials and hands should be considered in connection with the color of the wall or background, with a view to obtaining the contrast desired.

## WHOLE PLATE GLASS DIALS

These may be furnished with safety up to five feet in diameter. The numerals and minute marks are lettered on the surface in black or gold leaf of a durable finish. The backs are stippled with French zinc.

## GLASS

The glass is three-eighths of an inch thick, ground on both sides. The center-piece and segments are of the same thickness and quality throughout, thus insuring an evenly diffused light when they are illuminated.

## ILLUMINATION OF DIALS

It is not satisfactory to illuminate an opaque dial by throwing the light on the surface from the outside of the building. The dials should be of glass and the lights placed behind them. There is no fixed rule for the amount of candle-power required to illuminate any given diameter, for this depends upon the construction of the tower. So as not to cast a shadow, cross-beams and other obstructions should be kept back as far from the inner surface of the dial as possible. The dialroom should be painted white.

## AUTOMATIC LIGHT SWITCH

The switch is controlled by the tower clock and is adjustable to suit the different seasons. It automatically turns on and off the lights to illuminate the dials for whatever hours desired.

## AUTOMATIC WINDING ATTACHMENT

Whenever it is desired, we can furnish motors to wind both the strike and time sides of our clocks automatically. These are so arranged as to make the clocks practically self-winding.

## DIALS IN POSITION WITH FASTENINGS

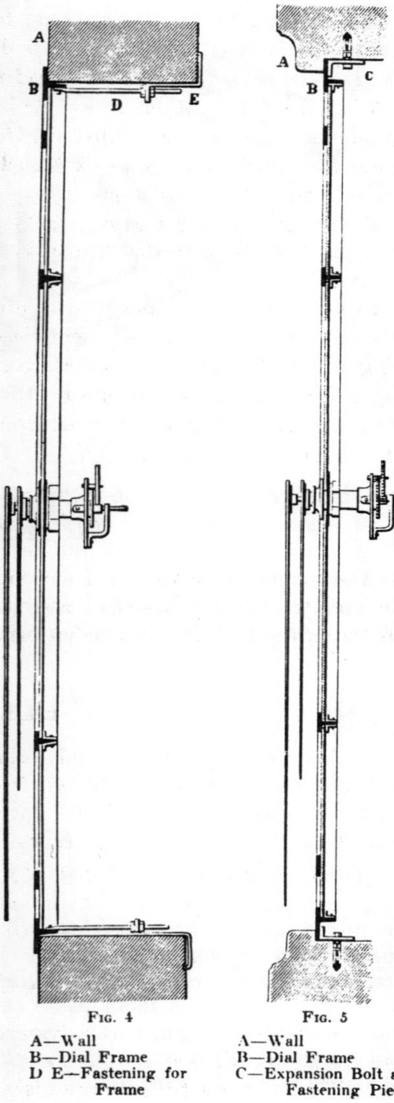


Fig. 4
A—Wall
B—Dial Frame
D E—Fastening for Frame

Fig. 5
A—Wall
B—Dial Frame
C—Expansion Bolt and Fastening Piece

SECTIONAL GLASS AND IRON DIALS

DIALS IN POSITION WITH FASTENINGS

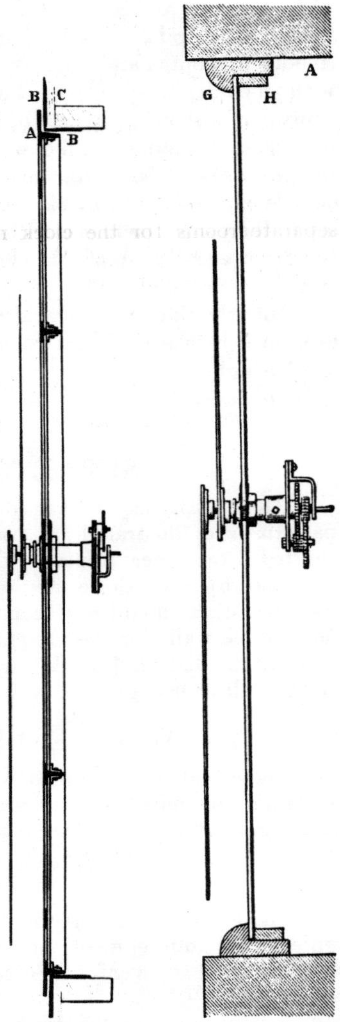


FIG. 6

A—Dial Frame
B B—Metal Covering Turned in
C—Sheathing

FIG. 7

A—Wall
G—Moulding
H—Fastening

## SECTIONAL GLASS DIALS

The openings for design illustrated above should be made in the multiple of six inches. The frame, numerals and minute marks are made of iron or bronze. The dial is divided into sections, which are rebated in the back to received the glass. The glass is held in place with brass clamps. The joints of the frame are fastened together with iron bolts and filled with lead, making the dial, when installed, absolutely watertight.

## ANGELUS

In the year 1326, Pope John XXII. requested that the church bells be rung three times a day as an invitation to prayer, and the custom is kept at the present time. The Angelus striking is as follows: At six o'clock in the morning, again at noon, and at six o'clock in the evening—thus, three strokes, three times in succession, with a pause after each third stroke, and then nine strokes in succession, without pause between. This may be given on a bell which is used for the hours also. Sometimes a separate bell is provided to be used for the Angelus only. Our clock mechanism can be made to cover the requirements for this purpose.

## BELLS

To allow free egress for the sound, the openings in the belfry or bell-chamber ought to be wide and extend, whenever feasible, to top and bottom of the room. If entirely open it is an advantage. When louvres or slats are used, it should be seen that they are not put too closely together, and are set at an angle which will no more than just overlap.

The ceiling of the room should be placed as nearly as possible over the top of the openings, and the floor or bell deck be covered with copper or tin, or constructed of weather-proof material.

A bell should be so suspended that its mouth (bottom part) will be above the level of the base of the openings.

For reasons which may be readily understood, it is preferable to have separate rooms for the clock movement and the bell; the bellroom requiring to be open for the sound, the clockroom closed (except for light) to better protect the movement from exposure to the weather.

How far a bell can be heard is problematical. The chief things are to get a good clock and bell; have a properly constructed belfry, and the sound will carry to its full capacity. It is affected, of course, by obstructions (if any exist), by the lay of the land and direction of the wind.

DIAMETER OF BELLS (AT THEIR MOUTH) AND THEIR WEIGHT

| | | | | | |
|---|---|---|---|---|---|
| 200 lbs. | 22 in. | 750 lbs. | 33 in. | 2,300 lbs. | 48 in. |
| 225 " | 23 " | 800 " | 34 " | 2,500 " | 50 " |
| 250 " | 24 " | 900 " | 35 " | 2,800 " | 52 " |
| 300 " | 25 " | 1,000 " | 36 " | 3,000 " | 54 " |
| 350 " | 26 " | 1,100 " | 37 " | 3,400 " | 55 " |
| 400 " | 27 " | 1,200 " | 38 " | 3,800 " | 57 " |
| 450 " | 28 " | 1,300 " | 39 " | 4,200 " | 59 " |
| 500 " | 29 " | 1,400 " | 40 " | 5,000 " | 62 " |
| 550 " | 30 " | 1,500 " | 41 " | 5,500 " | 64 " |
| 600 " | 31 " | 1,600 " | 42 " | 6,200 " | 66 " |
| 650 " | 32 " | 1,800 " | 44 " | 7,000 " | 72 " |
| 700 " | 33 " | 2,000 " | 46 " | | |

## QUARTER STRIKE CLOCKS

### WESTMINSTER CHIME

A quarter-strike clock consists of a time train to operate the hands, the hour-strike train and the quarter-strike parts. They are practically independent, except that the striking parts are controlled by the time train. To contain the whole, a larger frame is required than for a timepiece only, or for an hour strike. Three weights are required; the fall and amount of weights needed are about the same for both hour and quarter-strike parts.

For the Westminster Chime four bells are required and a fifth may be added for the hour. The complete chime is struck as follows: four blows at the first quarter, eight at the second, twelve on the third. sixteen on the hour, and the hour on the heaviest bell.

Our clocks can be made to strike different arrangements for the quarters.

## FOR TWO BELLS

It is necessary to have two bells of sufficient difference in weight and tone so that the quarter can be distinguished from the hour. One blow is generally struck for the first quarter, two at the half, three at the third quarter and then the hour; or, two at the first quarter, four at the half, etc., which produces a ding-dong effect. A third bell may be added for the hour strike.

## AUTOMATIC SILENCING DEVICE

This silencing device is attached to the tower clock and it will automatically silence and release the striking trains for whatever hour desired.

## HAND SILENCING DEVICE

This hand switch is only used where it is necessary to stop the clock striking for special occasions.

## DIRECTIONS

MOVEMENT.—Tower Clock movements should always be placed upon a firm floor or platform, and in such a position that connections can be made with the dials, either above or below—*above* is the best. The connection to the dial gearing is made from the upright shaft on the movement with rods and the universal joints. The universal joints enable the connecting rods to be run on an angle, avoiding bell or other obstruction. *See illustrations of clocks in position.*

THE CONNECTING RODS.—The rods should be of such length as to allow end shake in the sliding couplings, in case of contraction or expansion of the rods or building. Gas pipe is most frequently used if the rods are short or medium length, we sending couplings to correspond. When required in long lengths they can be of tin tubing soldered on to brass hubs, fitted to universal joints.

DIALS.—See page **130**

WEIGHTS.—See page **142**

PENDULUM.—In case the pendulum bob is below the floor it should be enclosed in a close case sufficiently large to allow plenty of room for vibration. It should be six to eight inches in the clear, on all sides (pendulum swinging), to allow the air to circulate when removed by action of the bob.

PULLEYS.—It is preferable to place the leading or first pulley as far from the movement as possible—the cord works much better. The pulleys should be so secured that their grooves will be in line with the cord so as to prevent friction or chafing of the cord in passing over.

CORD.—The wire rope on the striking train usually requires two coils across the cylinder to run eight days. The time train requires but one coil on the cylinder.

HAMMER.—For a bell which is mounted to swing, the hammer should be placed so as to strike the bell through the standard or support opposite the wheel. It should be set on such an angle as to suit the shape of the bell at the sound bow.

CHAIN GUIDE PULLEYS.—Where possible the connections from the hammer-tail lever on the clock movement to the bell hammer should be in a straight line, but we send guide pulleys that may be used to carry the connections where the clock or bell has to be set at one side, or where the connections need to pass obstructions in the tower.

OIL.—Use tower clock oil on all the work except the pulleys and hammer connections, and for them it is preferable, unless you have good sperm or machine oil. Oil every part about the clock where there is friction, excepting gear and pinion teeth, and, as often as necessary to keep the oil limpid. Put some on the pallets of the escapement. Wipe off the old oil with waste.

# SETH THOMAS TOWER CLOCKS

**TO PUT THE CLOCK IN BEAT.**—Make the beats equal by the two screws back of pendulum on the crutch piece, which connects with the pendulum rod; turning both to the right or to the left, as required, until the pallets are equal in beat on each side of the escape wheel. Then fasten both screws firmly against the crutch piece to hold pallets in place.

**SETTING TO TIME.**—All our clocks larger than Nos. 4 and 14 are made with a new device for setting the hands on the tower, and it also sets the hour and minute hands on small dial on the movement.

There is placed on the time hour wheel shaft a worm wheel with sixty teeth, and fitting it a screw carried by a sleeve on the same shaft. This screw can be turned by a small key sent with the clock; one full turn of screw moves the hands one minute, moving both the hands on the tower and on the movement by the same action. Be careful to take the key off after being used.

**WINDING.**—Wind the clock carefully once a week. To prevent over-winding, have the cord marked to indicate where to stop.

**REGULATING.**—This is done by the screw on top of the movement. Turning to the right shortens the pendulum for faster, and to the left lengthens it to go slower. Make but a slight change at a time.

## CLOCKS IN POSITION

The illustration on the next page represents a tower clock in position inside the tower. In this case the movement is placed below the dials, and the bell is above the dials. We can, however, arrange the dial gearing so as to connect and carry the hands if the dials are above the bell.

It is *always best* to place the movement *below* the bell, if there is room, as there is less vibration in the lower part of the tower.

The cut shows the most usual way of suspending the weights. One side has a "leading pulley" placed to suit the cord cylinder on the movement, and the cord can then pass over the proper pulleys in the corner of the tower. Other illustrations follow on next pages.

FIG. 8

CLOCKS IN POSITION

# 1875 SETH THOMAS CLOCK MOVEMENT DESCRIPTIONS

**$1500.00 w/Motion Works for Dials**

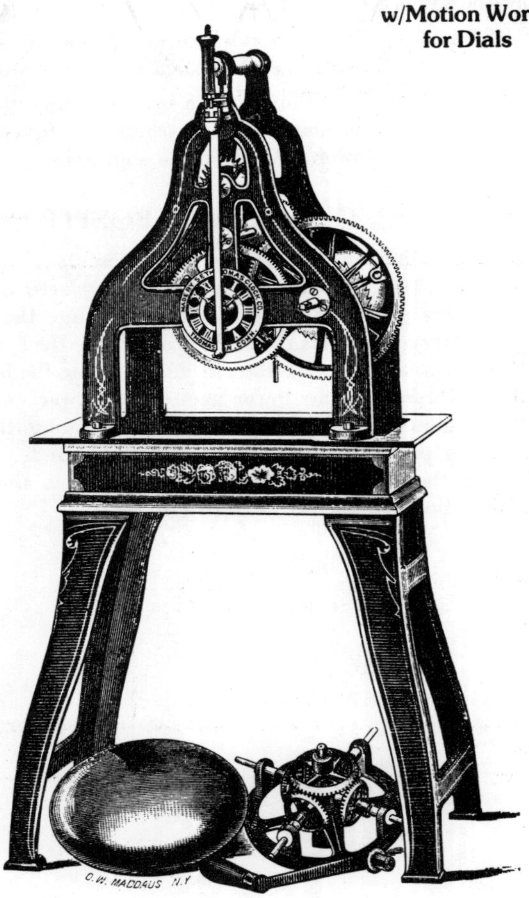


## No. 0—TIME—8 DAY.

20 inches wide, 17 inches deep, 42 inches high.

7 inch Main Wheel, of ample power for a Single Dial five feet or less.

4 foot pendulum. Ball weighs 25 lbs.

## No. 1—TIME—8 DAY.

25 inches wide, 21 inches deep, 47 inches high.

9 inch Main Wheel, of ample power for Single Dial up to six feet, or four Dials of five feet or less.

4 foot pendulum. Ball weighs 50 lbs.

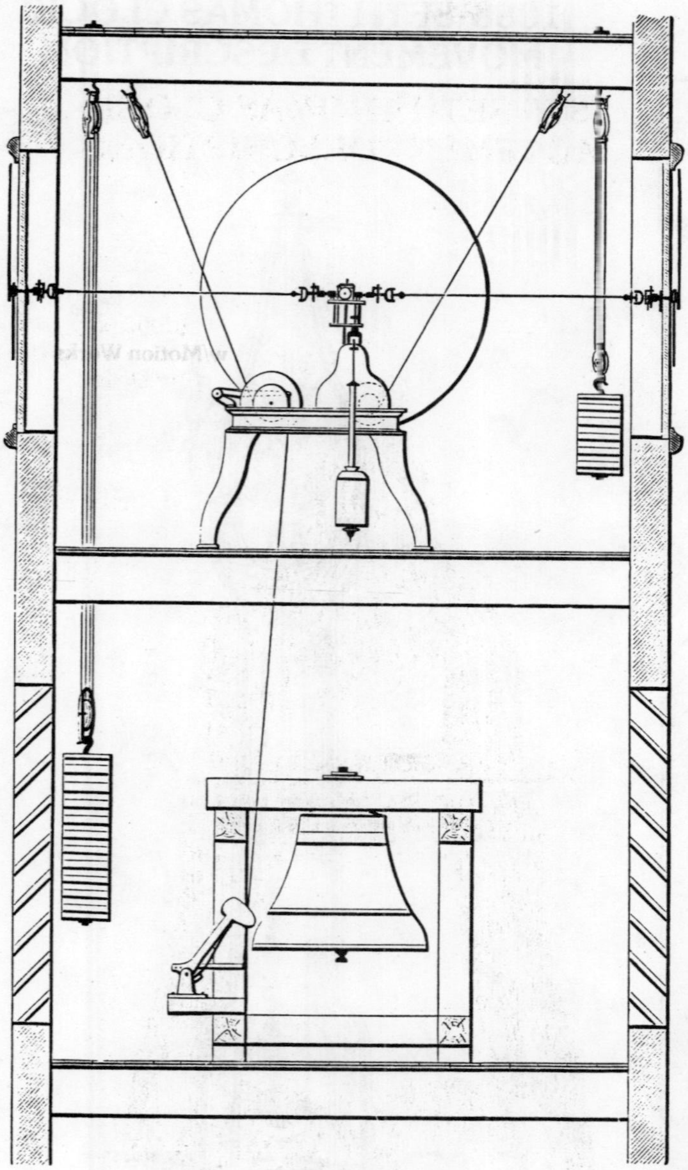

FIG. 9

# 1875 SETH THOMAS CLOCK MOVEMENT DESCRIPTION

No. 2—TIME—8 DAY.

30 inches wide, 27 inches deep, 60 inches high.

12 inch Main Wheel, of ample power for single Dial up to ten feet, or for four Dials from five to eight feet.

5 foot pendulum. Ball weighs 50 lbs.

**$1750.00 w/Dial Works**

# 1888 SETH THOMAS CLOCK MOVEMENT DESCRIPTION

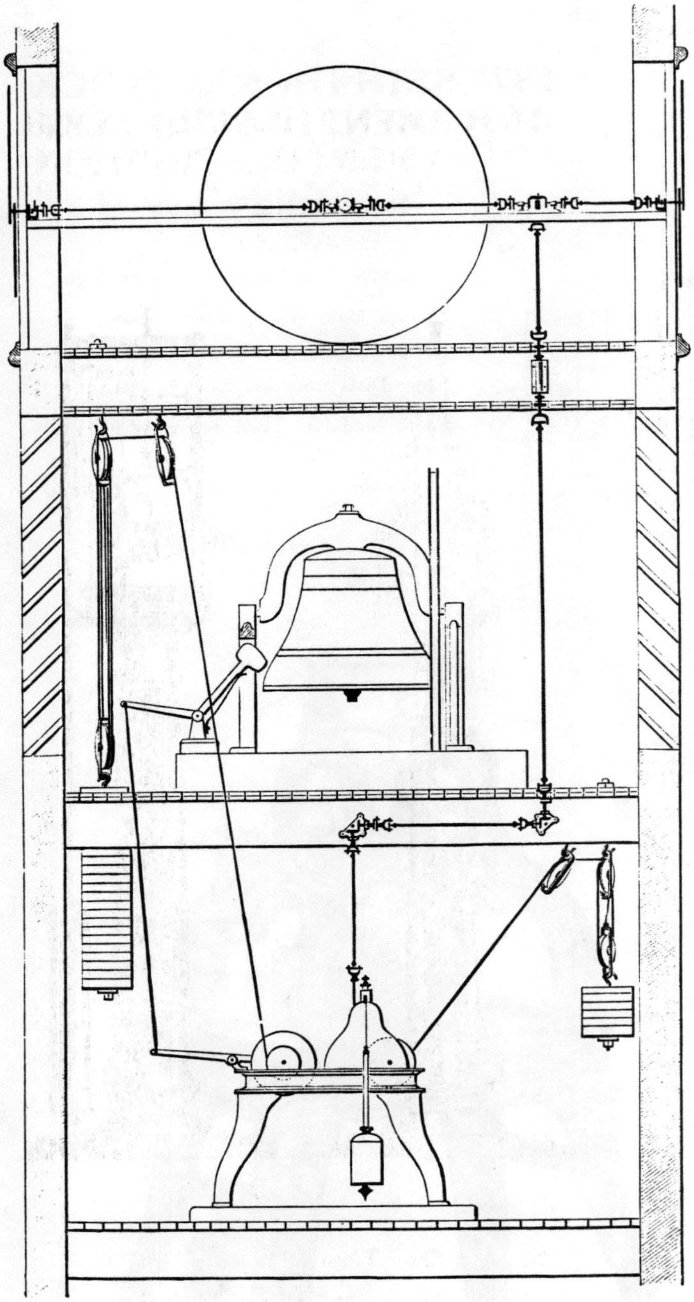

Fig. 10

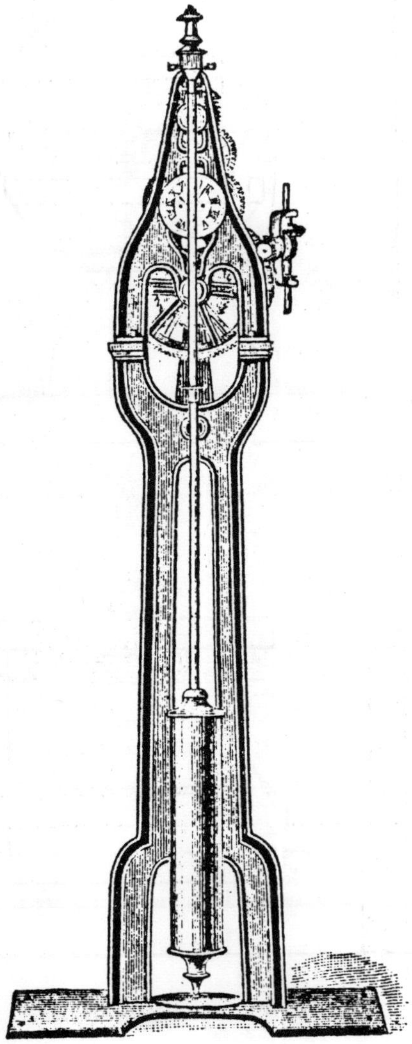

8 Day, No. 4, Time
Width, 9 inches
Depth, 11½ inches
Hight, 51 inches
Main Wheel, 7 inches
Pendulum, 4 feet

**$1250.00 w/Stand**

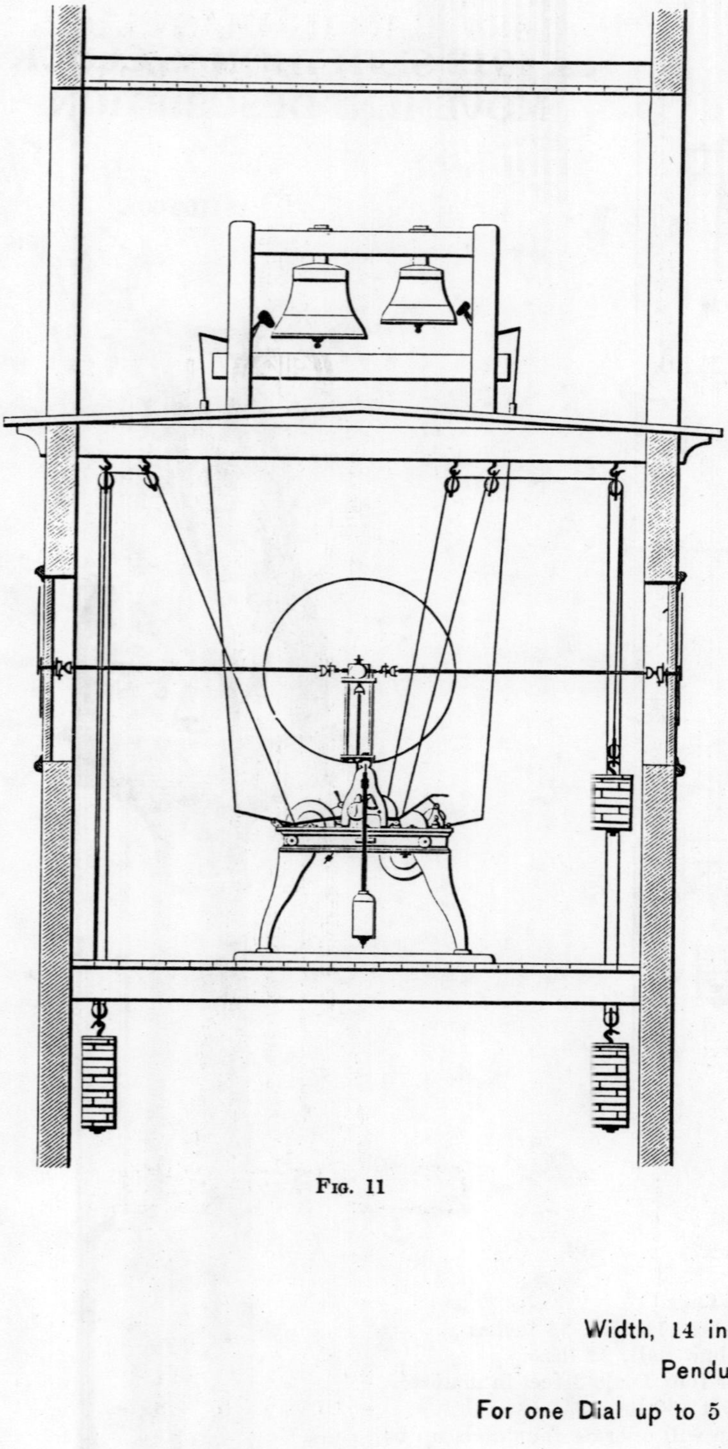

FIG. 11

# 1896 SETH THOMAS CLOCK MOVEMENT DESCRIPTION

**$1250.00**

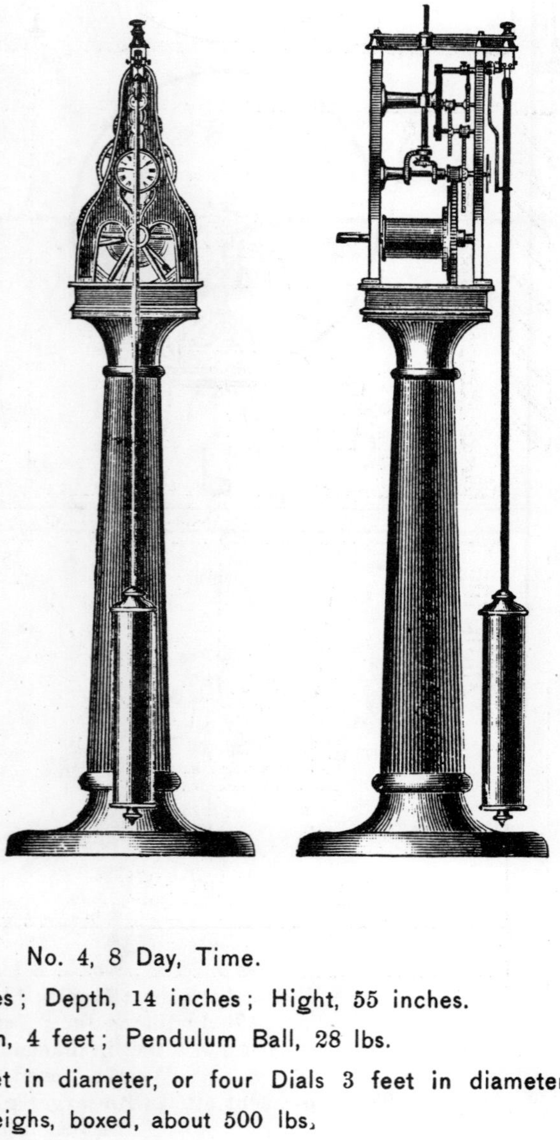

No. 4, 8 Day, Time.

Width, 14 inches; Depth, 14 inches; Hight, 55 inches.

Pendulum, 4 feet; Pendulum Ball, 28 lbs.

For one Dial up to 5 feet in diameter, or four Dials 3 feet in diameter

Weighs, boxed, about 500 lbs.

# ca. 1912 SETH THOMAS CLOCK MOVEMENT DESCRIPTION

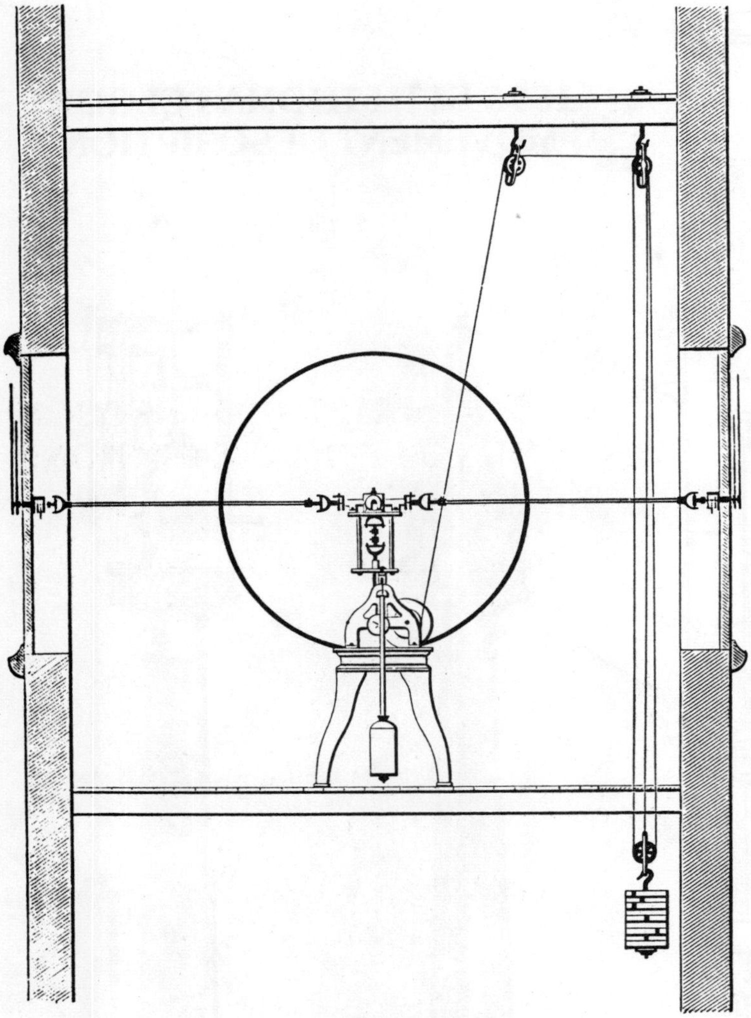

FIG. 12

$1100.00

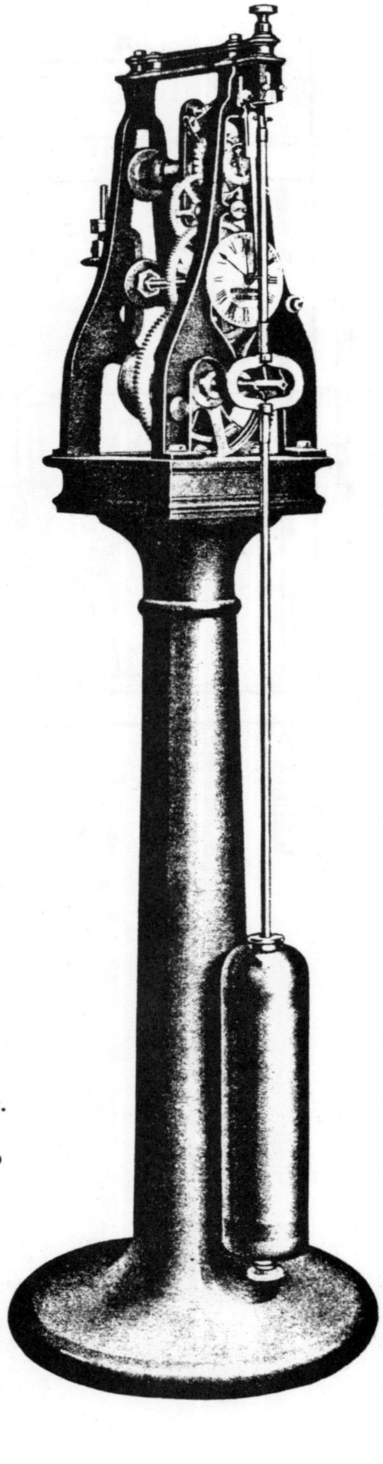

No. 4, 8-Day, Time.
Width, 14 inches; Depth, 14 inches; Height, 55 inches.
Pendulum, 4 feet; Pendulum Ball, 28 lbs.
For one Dial up to 5 feet in diameter, or four Dials 3 feet in diameter.
Weighs, boxed, about 500 lbs.
This movement fitted with larger pinion will operate four dials up to 4 feet in diameter.

SETH THOMAS TOWER CLOCKS.

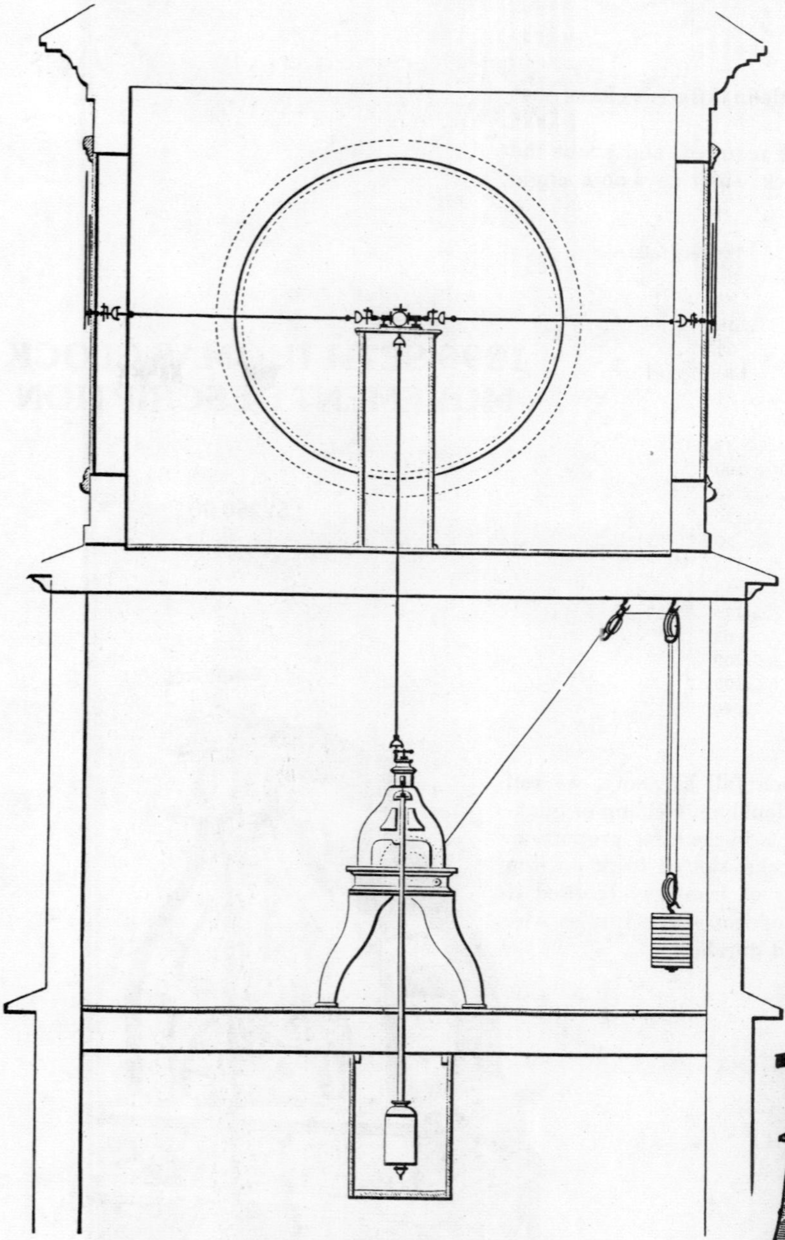

FIG. 13

# 1896 SETH THOMAS CLOCK MOVEMENT DESCRIPTIONS

$2000.00

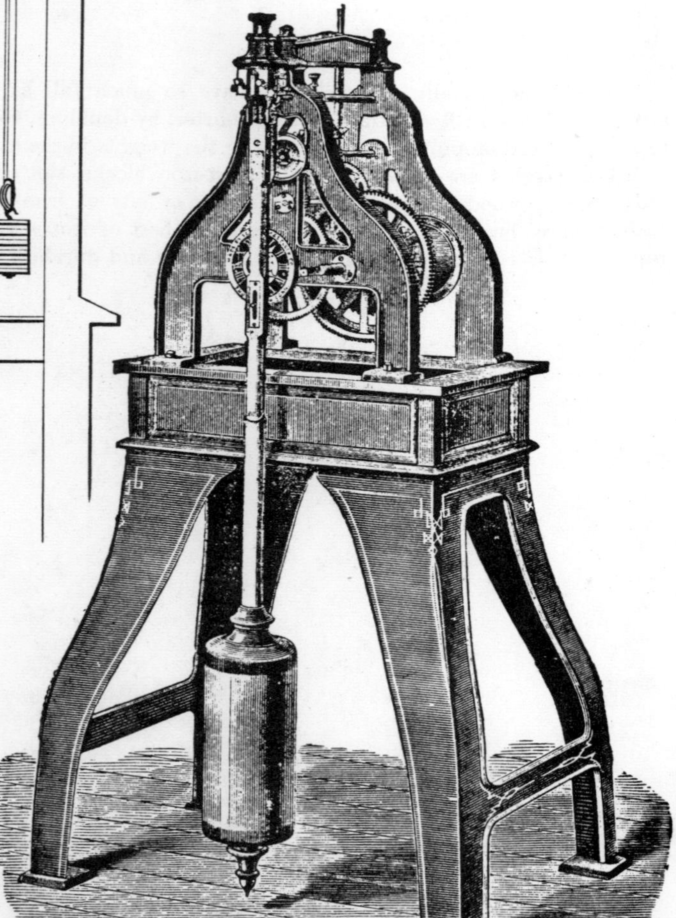

No. 5, 8 Day, Time.
Width, 28 inches; Depth, 24 inches; Hight, 57 inches.
Pendulum, 4 feet; Pendulum Ball, 75 lbs.
For one Dial up to 7 feet, or four Dials 5 feet or less.
Weighs, boxed, about 500 lbs.

No. 6, 8 Day, Time.
Width, 35 inches; Depth, 26 inches; Hight, 60 inches.
Pendulum, 4 feet or 8 feet; Pendulum Ball, 135 lbs.
For one Dial up to 8 feet, or four Dials 7 feet or less.
Weighs, boxed, about 900 lbs.

## WEIGHTS

The *exact* amount of weight needed varies a little in different clocks, and can only be determined by experiment.

We give below an approximation of the weight required, and about the number of feet needed for its fall, to run these clocks eight days on a single cord:

| | | | |
|---|---|---|---|
| No. 4 | | 35 lbs | 24 feet fall. |
| No. 5 | | 75 " | 32 " " |
| No. 6 | | 90 " | 45 " " |
| No. 7 | | 120 " | 58 " " |
| No. 8 | | 175 " | 68 " " |
| No. 9 | | 225 " | 100 " " |
| No. 14 | Time side | 35 " | 24 " " |
| | Strike side | 100 " | 80 " " |
| No. 15 | Time side | 75 " | 32 " " |
| | Strike side | 175 " | 100 " " |
| No. 16 | Time side | 90 " | 45 " " |
| | Strike side | 225 " | 120 " " |
| No. 16 B | Time side | 100 " | 50 " " |
| | Strike side | 275 " | 165 " " |
| No. 17 | Time side | 120 " | 58 " " |
| | Strike side | 300 " | 175 " " |
| No. 18 | Time side | 175 " | 68 " " |
| | Strike side | 375 " | 200 " " |
| No. 19 | Time side | 225 " | 100 " " |
| | Strike side | 500 " | 240 " " |

# 1896 SETH THOMAS CLOCK MOVEMENT DESCRIPTION

**$2250.00**

As it is not usually convenient to have so much fall as above, we can arrange with one-half, one-third or one-quarter, by doubling, trebling or quadrupling the cord on pulleys, and increasing the weights in the same proportion.

The weights are usually made of cast-iron blocks, slotted to fit on iron rods. Strong wooden boxes filled with stones or bits of iron may be used if room permits, but the weights are neatest and best of iron. We furnish wire rope which is constructed with a view to safety and durability.

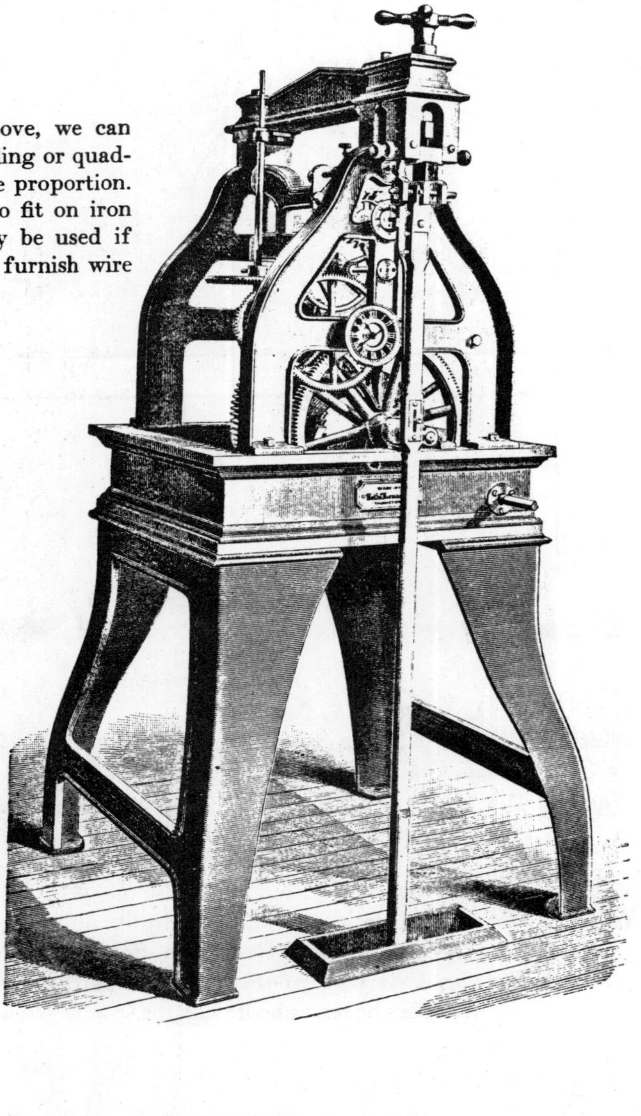

No. 7, 8 Day, Time.
Width, 36 inches; Depth, 37 inches; Hight, 72 inches.
Pendulum, 8 feet: Pendulum Ball, 200 lbs.
Also made with 14 foot Rod and 325 lb. Ball.
For one Dial up to 13 feet, or four Dials of 10 feet or less.
Weighs, boxed, about 1,800 lbs.

## ca. 1912 SETH THOMAS CLOCK MOVEMENT DESCRIPTIONS

No. 7, 8-Day, Time.
Width, 36 inches; Depth, 37 inches; Height, 72 inches.
Pendulum, 8 feet; Pendulum Ball, 200 lbs.
Also made with 14-foot Rod and 325-lb. Ball.
For one Dial up to 13 feet, or four Dials of 10 feet or less, with Gravity Escapement.
For one Dial up to 11 feet, or four Dials of 9 feet or less, with Graham Escapement.
Weighs, boxed, about 1,800 lbs.

$2000.00

## 1890 SETH THOMAS CLOCK MOVEMENT DESCRIPTION

$2500.00

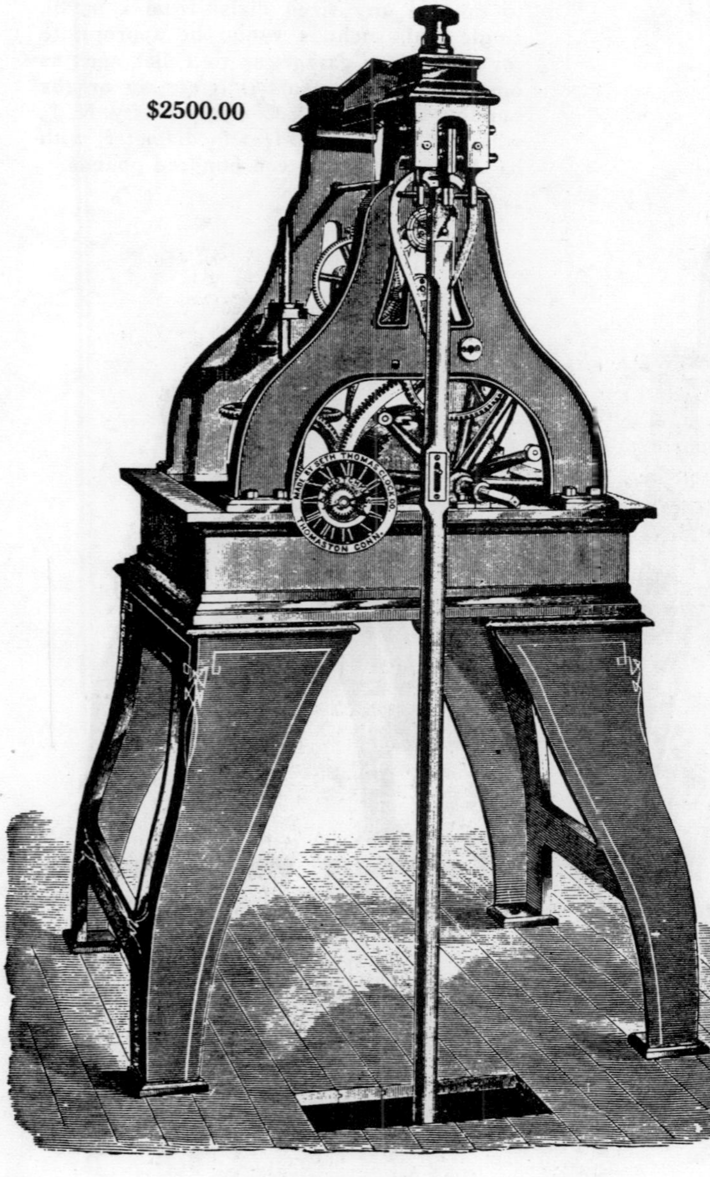

8 Day, No. 7, Time.
Width, 38 inches; Depth, 33 inches; Hight, 65 inches.
Main Wheel, 12 inches; Pendulum, 8 feet.

## 1896 SETH THOMAS CLOCK MOVEMENT DESCRIPTION

**$3500.00**

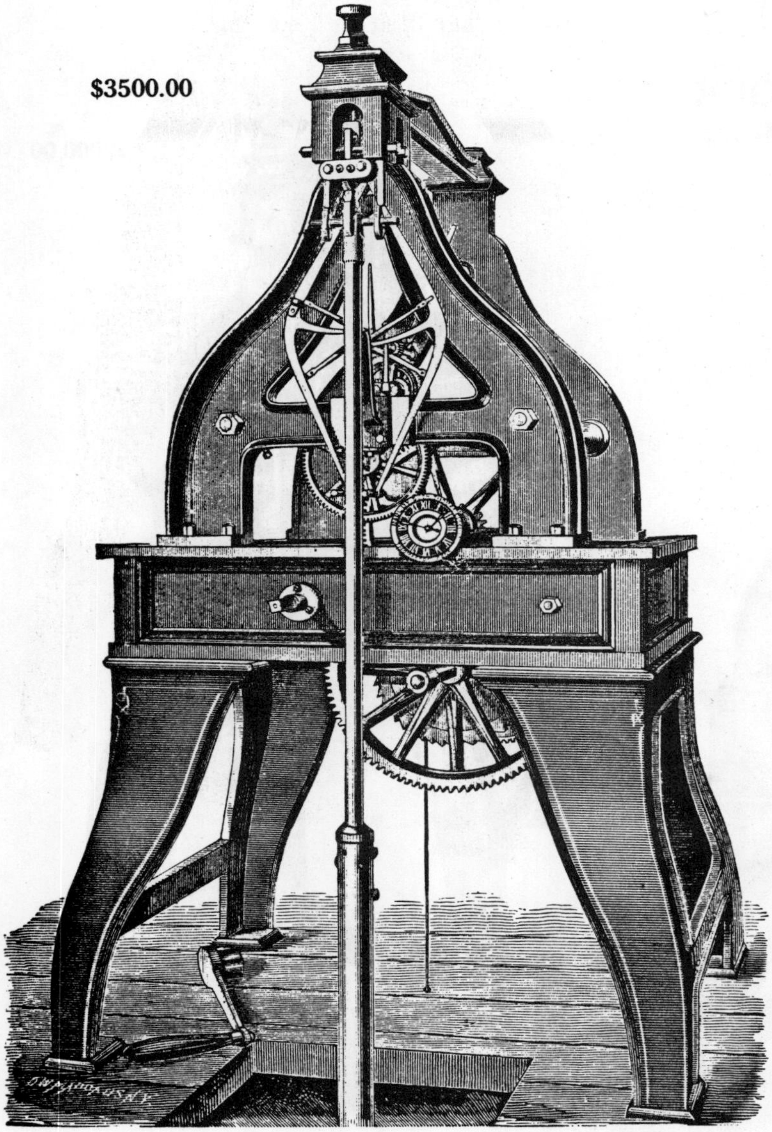

No. 8, 8 Day, Time.
Width, 38 inches; Depth, 38 inches; Hight, 72 inches.
Pendulum, 14 feet; Pendulum Ball, 325 lbs.
For large Dials Weighs, boxed, about 2,300 lbs.

No. 9, 8 Day, Time.
Width, 47 inches; Depth, 43 inches; Hight, 76 inches.
Pendulum, 14 feet. Pendulum Ball, 550 lbs.
For very large Dials. Weighs, boxed, about 2,800 lbs.

## 1915 SETH THOMAS CLOCK MOVEMENT DESCRIPTION

### TOWER CLOCK TIME MOVEMENT

The movement illustrated below is one of many sizes we manufacture, and is a time movement only. These clocks are designed to operate any sized dials—from a small, single dial, such as would be appropriate for a stable or garage, up to a dial, such as one of our movements is operating on the building of Colgate & Co., Jersey City, N. J., which is thirty-eight feet in diameter, with hands weighing fourteen hundred pounds.

SETH THOMAS TOWER CLOCKS.

# ca. 1912 SETH THOMAS CLOCK MOVEMENT DESCRIPTIONS

No. 9, 8-Day, Time.
Width, 45 inches; Depth, 41 inches; Height, 76 inches.
Pendulum, 14 feet; Pendulum Ball, 550 lbs.
For one Dial up to 22 feet, or four Dials of 18 feet or less.
Weighs, boxed, about 2,800 lbs.

**$3500.00 w/Winding Mechanism**

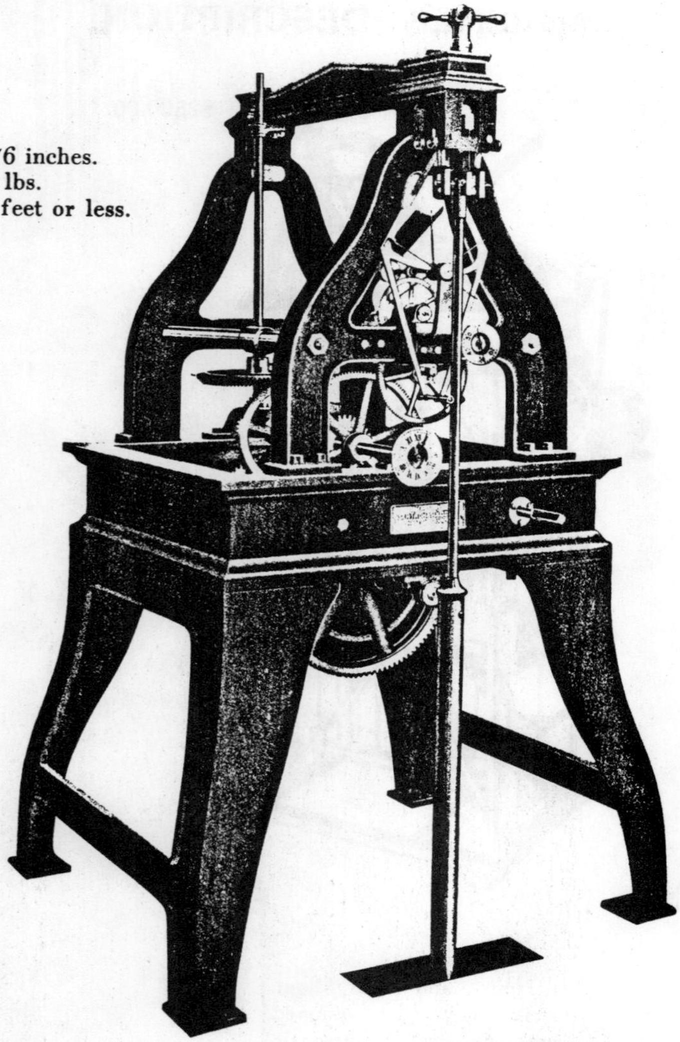

**$2750.00**

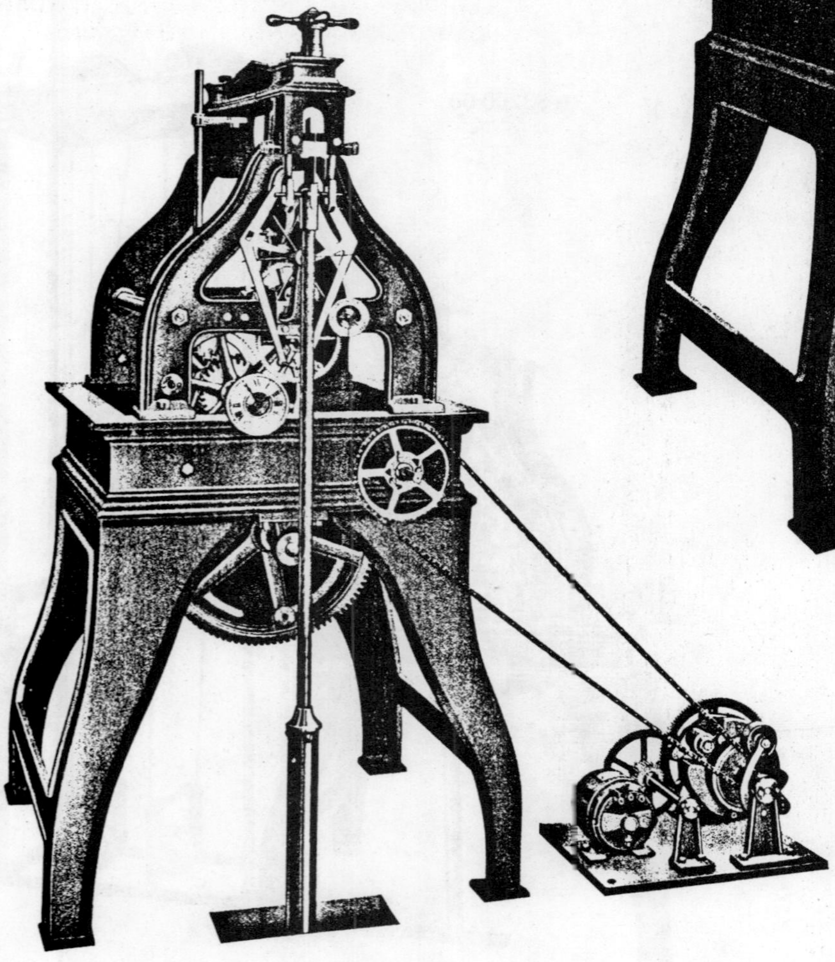

No. 8, 8-Day, Time.
Width, 38 inches; Depth, 39 inches; Height, 72 inches.
Pendulum, 14 feet; Pendulum Ball, 325 lbs.
For one Dial up to 18 feet, or four Dials of 15 feet or less.
Weighs, boxed, about 2,300 lbs.
May be Equipped with Power Wind as in Illustration.

## 1922 SETH THOMAS CLOCK MOVEMENT DESCRIPTION

$2500.00

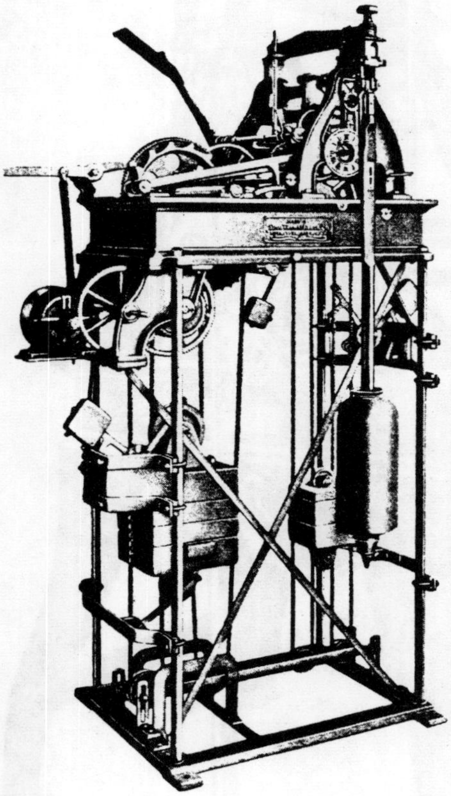

Hour and half hour strike movement equipped with motors for automatic winding. This movement is designed for one dial up to 8 feet or four dials of 5½ feet or less and to operate a bell not over 1000 lbs.

## ca. 1912 SETH THOMAS CLOCK MOVEMENT DESCRIPTION

$3200.00

No. 10, 8-Day Time, Chain Wind.
Width, 52 inches; Depth, 48 inches; Height, 80 inches.
Pendulum, 8 feet or 14 feet; Pendulum Ball, 330 lbs.
For one Dial up to 32 feet, or 4 Dials up to 25 feet.
Weight, boxed, about 8,500 lbs.
Weight gear frame made to suit conditions.

# 1875 SETH THOMAS CLOCK MOVEMENT DESCRIPTION

**$3200.00 w/Dial Works**

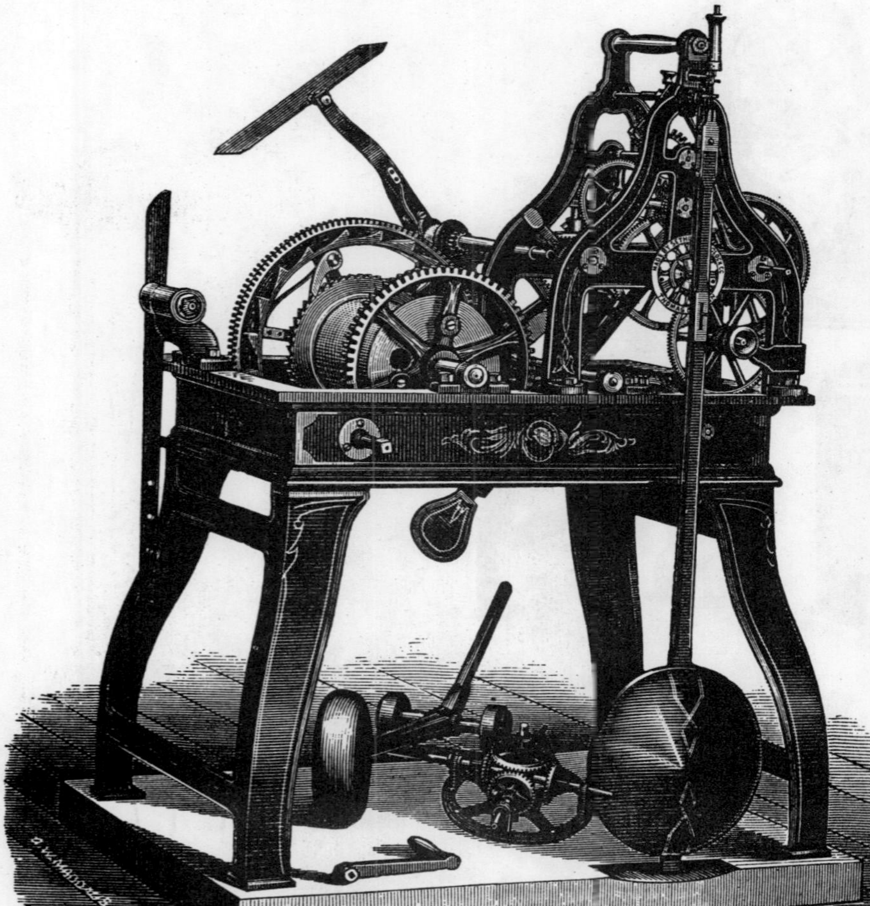

No. 11—STRIKE—8 DAY.

40 inches wide, 42 inches deep, 51 inches high.

9 inch Main Wheel, 16 inch Strike Wheel. For same Dials as No. 1 Time, and striking part of ample power for Bells of 1000 lbs. or less.

5 foot pendulum. Ball weighs 50 lbs.

# 1875 SETH THOMAS CLOCK MOVEMENT DESCRIPTIONS

### No. 12. A—STRIKE—8 DAY. $3200.00

48 inches wide, 42 inches deep, 54 inches high.

12 inch Main Wheel, 20 inch Strike Wheel, with Rollers. For any sized Dial, and to strike Bell up to 1500 lbs.

8 foot pendulum. Ball weighs 75 lbs.

### No. 12. B—STRIKE—8 DAY. $3200.00

Same as above, but striking part made heavier and ample for Bell up to 3,000 lbs.

### No. 13—STRIKE—8 DAY. $3200.00

62 inches wide, 50 inches deep, 64 inches high.

16 inch Main Wheel; 24 inch Strike Wheel, with Rollers. For any sized Dials, and for very heavy Bells. All the Wheels made of Bronze, except the main Strike.

8 foot pendulum. Ball weighs 75 lbs.

### No. 14—STRIKE—8 DAY. $3200.00

72 inches wide, 50 inches deep, 70 inches high.

18 inch Main Wheel, 30 inch Strike Wheel, with Rollers. For any sized dial, and for very heavy bells. *All the wheels made of bronze.*

13 foot pendulum. Ball weighs 300 lbs.

## 1890 SETH THOMAS CLOCK MOVEMENT DESCRIPTIONS

No. 14, 8 Day, Hour Strike

Width, 30 inches; Depth, 28 inches; Hight, 54 inches.

Pendulum, 4 feet; Pendulum Ball, 28 lbs.

For one Dial 5 feet or less in diameter, or four Dials 3 feet in diameter.

Strikes on a fine large gong weighing about 40 lbs., attached to the movement.

Weighs, boxed, about 700 lbs.

Suitab e for large hall or room, also for stables, porter's lodges, etc.

## 1896 SETH THOMAS CLOCK MOVEMENT DESCRIPTIONS

**$4000.00**

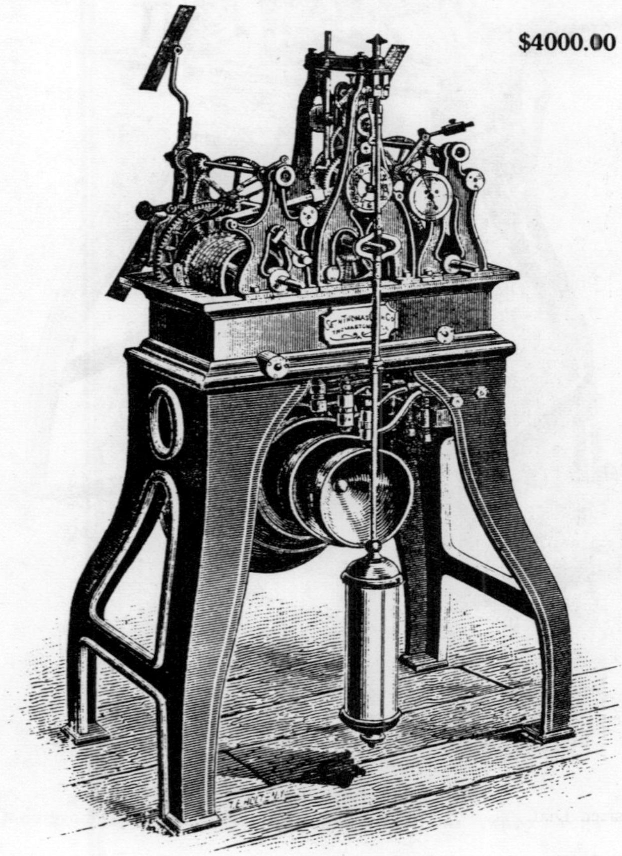

No. 14, 8 Day, Quarter Chime.

Width, 36 inches; Depth, 28 inches; Hight 54 inches.

Pendulum, 4 feet; Pendulum Ball, 28 lbs.

For one Dial 5 feet or less in diameter, or four Dials 3 feet in diameter:

Plays Westminster chime on four accurately attuned gongs at each quarter, and strikes hours on a larger gong.

Weighs, boxed, about 800 lbs.

Suitable for towers and buildings on private estates, country residences, etc.

**$2000.00**

# 1896 SETH THOMAS CLOCK MOVEMENT DESCRIPTIONS

$3000.00

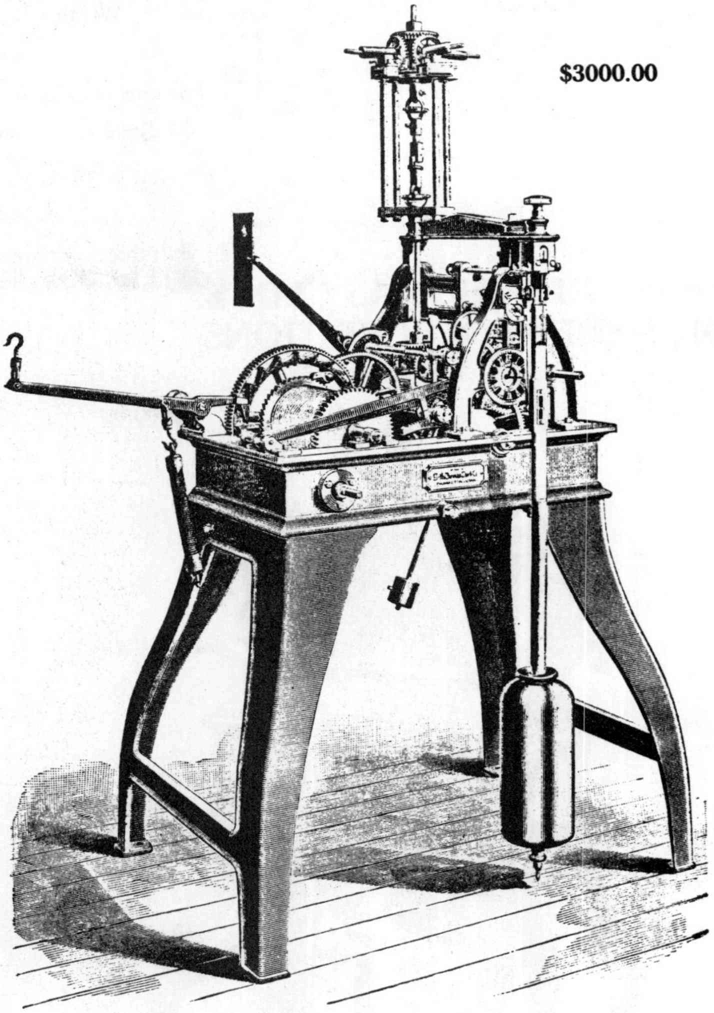

No. 15, 8 Day, Strike.
Width, 40 inches; Depth, 33 inches; Hight, 58 inches.
Pendulum, 4 feet; Pendulum Ball, 75 lbs.
For one Dial up to 7 feet, or four Dials 5 feet or less.
For Bell of 1,000 lbs. or less.
Weighs, boxed, about 1,000 lbs.
Cut shows central dial works when placed on top of a movement.

$3000.00

No. 14, 8 Day, Hour Strike.
Width, 36 inches; Depth, 25 inches; Hight, 54 inches.
Pendulum, 4 feet; Pendulum Ball, 28 lbs.
For one Dial 5 feet or less in diameter, or four Dials 3 feet in diameter.
For a large bell-metal gong weighing about 90 lbs., attached to an adjustable iron frame, as shown above, or a bell up to 200 or 300 lbs. weight, suspended in the usual way.
Clock weighs, boxed, about 700 lbs.
Gong and Frame, boxed, weigh about 220 lbs.
Suitable for towers and buildings on private estates, country residences, etc.

# 1880 SETH THOMAS CLOCK MOVEMENT DESCRIPTIONS

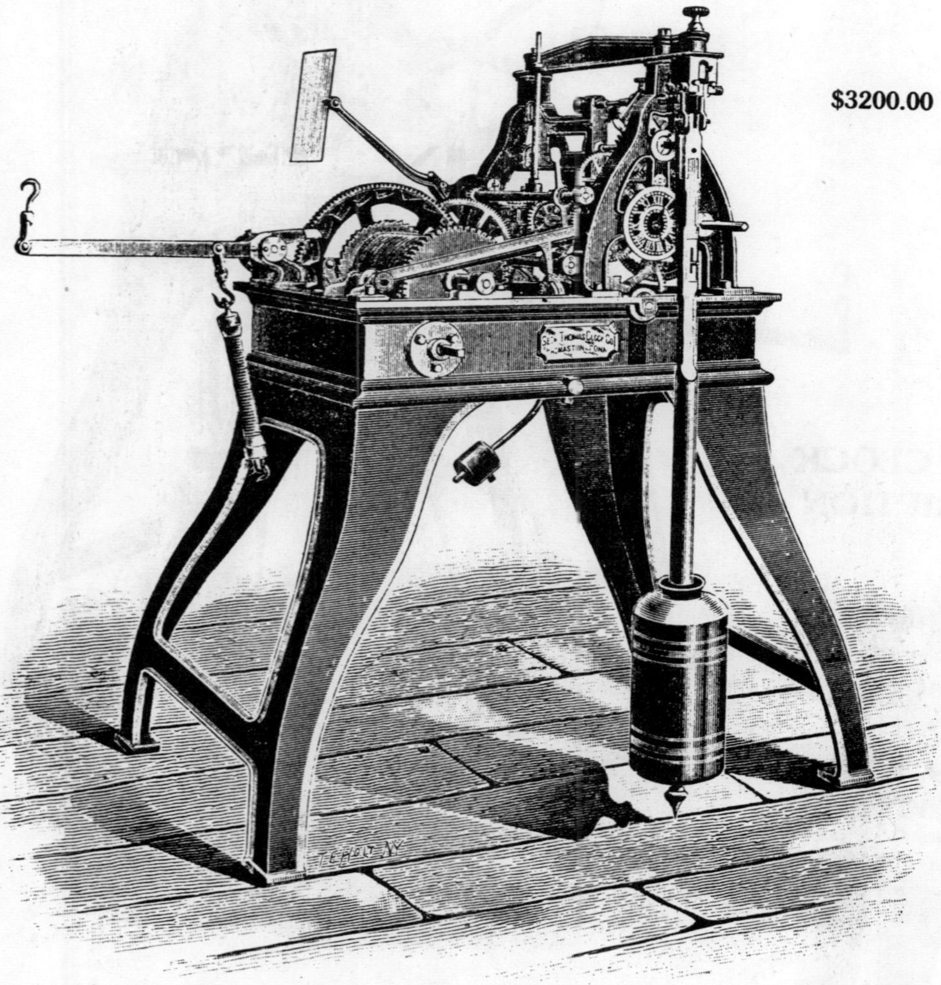


$3200.00

No. 15, 8 Day, Strike.

Width, 40 inches; Depth, 33 inches; Hight, 58 inches.

Pendulum, 4 feet; Pendulum Bal , 75 lbs.

For one Dial up to 7 feet, or four Dials 5 feet or less.

For Bell of 1,000 lbs. or less.

Weighs, boxed, about 1,000 lbs.

## 1910 SETH THOMAS CLOCK MOVEMENT

**$2500.00**

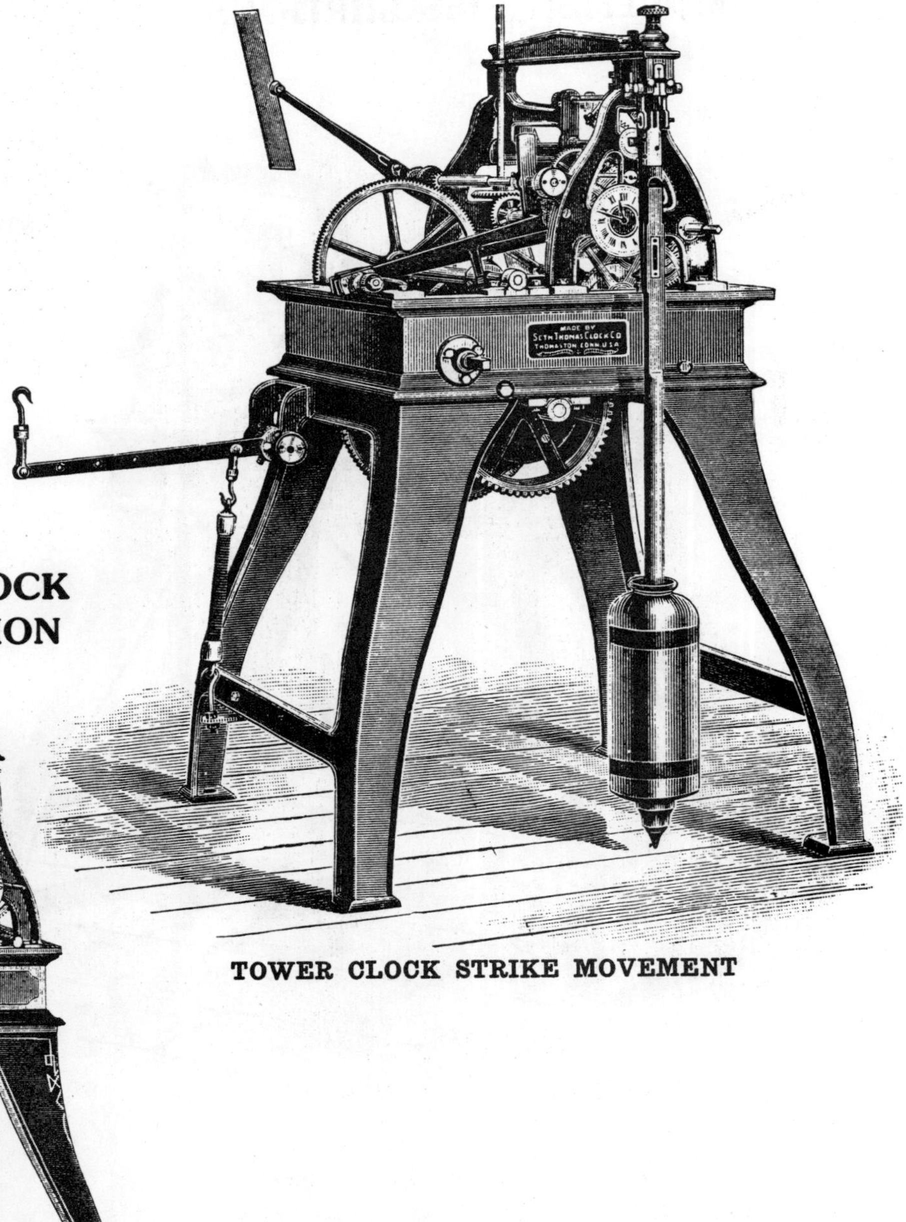

**TOWER CLOCK STRIKE MOVEMENT**

## 1890 SETH THOMAS CLOCK MOVEMENT DESCRIPTION

**$2700.00**

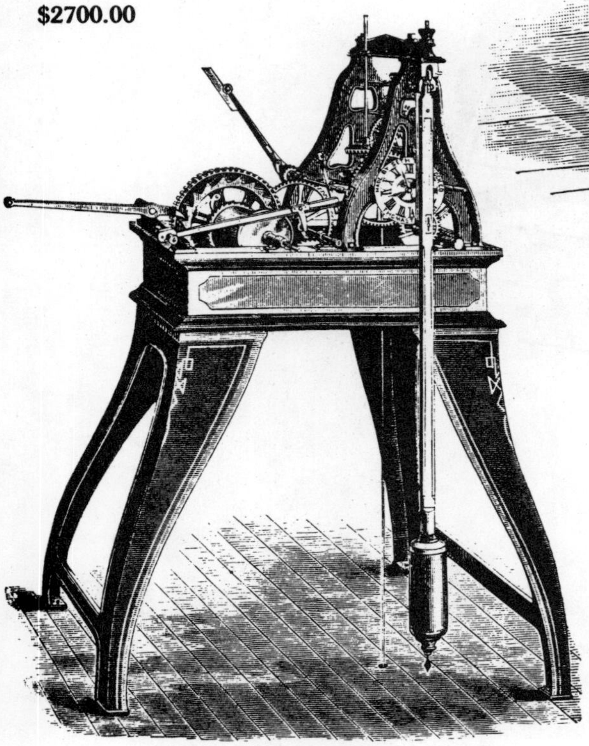

8 Day, No. 15, Strike.
Width, 38 inches; Depth, 33 inches; Hight, 57 inches.
Main Wheel, 7 inches; Pendulum, 4 feet.
10 inch Strike Wheel.

# 1890 SETH THOMAS CLOCK MOVEMENT DESCRIPTION

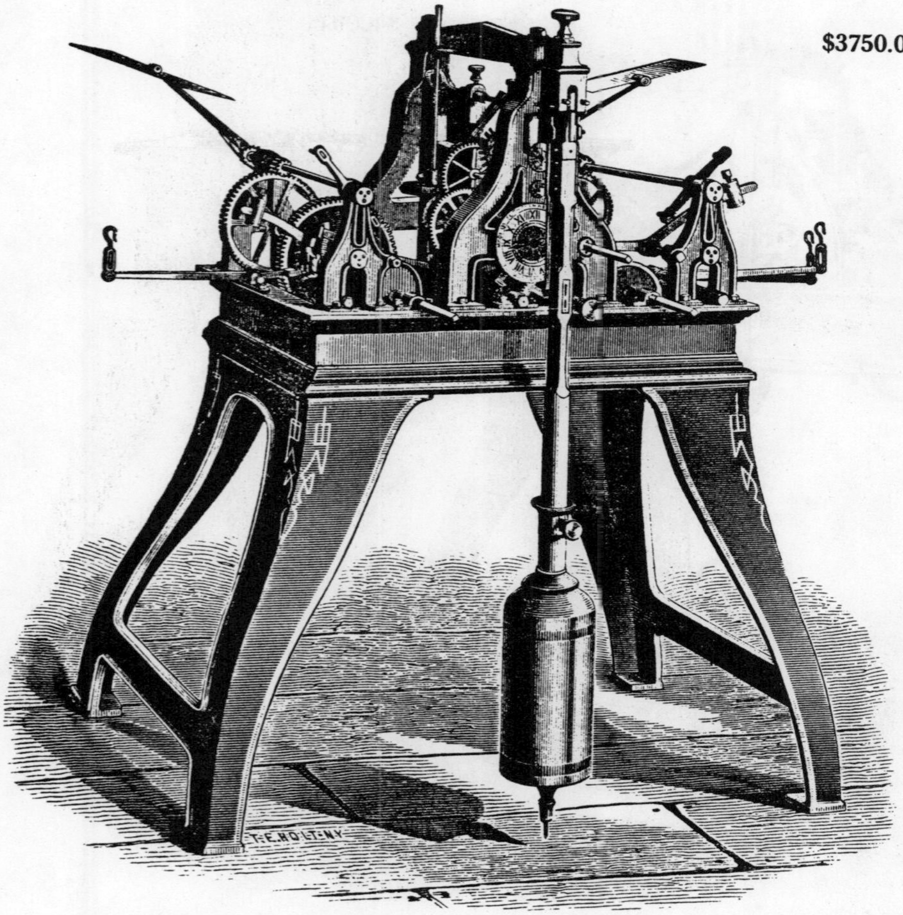


$3750.00

No. 15, 8 Day, Quarter Strike.

Width, 60 inches; Depth, 39 inches; Hight, 58 inches.

Pendulum, 4 feet; Pendulum Ball, 75 lbs.

For one Dial up to 7 feet, or four Dials 5 feet or less.

For three or four Bells, the largest of which can be 1,000 lbs.

Weighs, boxed, about 1,300 lbs.

# 1896 SETH THOMAS CLOCK MOVEMENT DESCRIPTION

**$4000.00**

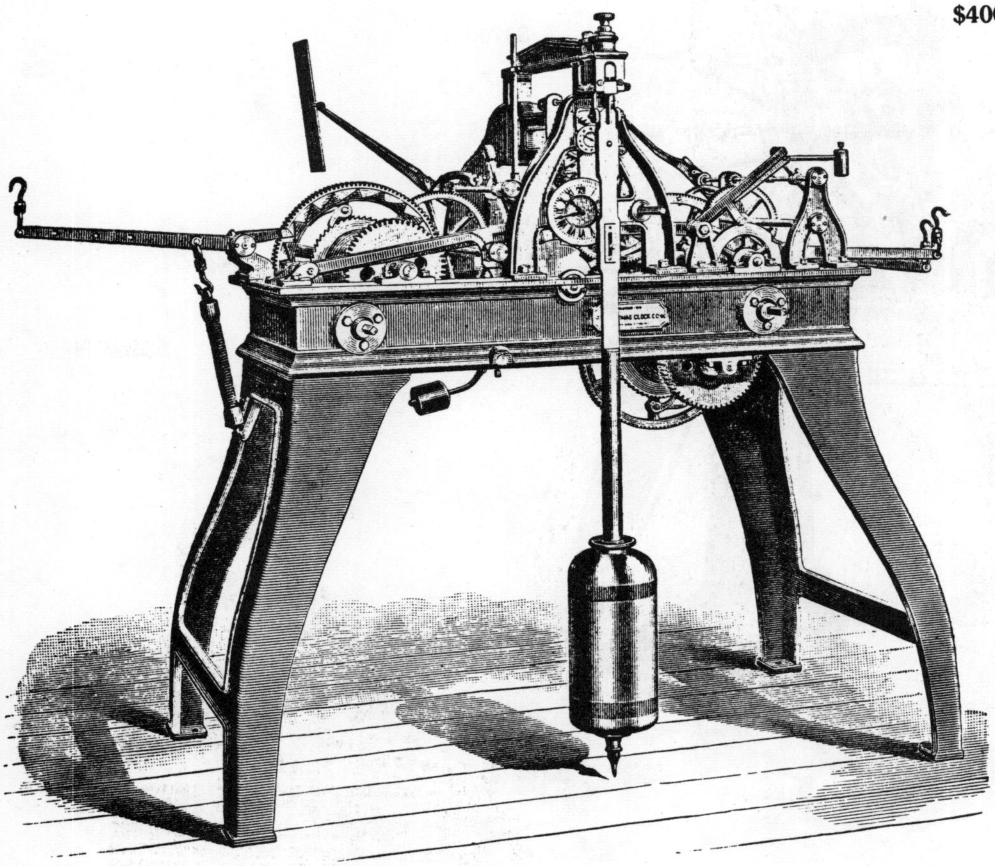

No. 15, 8 Day, Quarter Strike.

Width, 60 inches; Depth, 39 inches; Hight, 60 inches.

Pendulum, 4 feet; Pendulum Ball, 75 lbs.

For one Dial up to 7 feet, or four Dials 5 feet or less.

For peal of Bells, the largest of which can be 1,000 lbs.

Weighs, boxed, about 1,500 lbs.

# 1886 SETH THOMAS CLOCK MOVEMENT DESCRIPTIONS

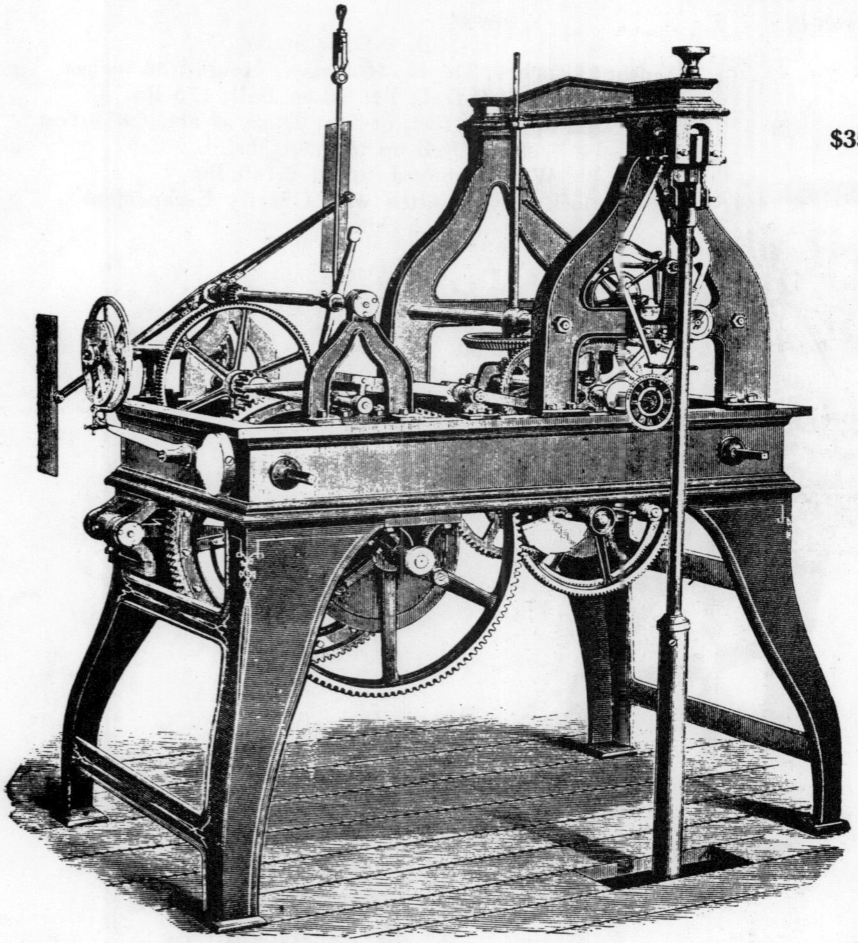

$3500.00

No. 16, 8 Day, Strike
Pendulum, 4 feet, as shown on "No. 6, Time," page 123; Pendulum Ball, 75 lbs.
For one Dial up to 7 feet, or four Dials of 6 feet or less.
No. 16 B, 8 Day, Strike.
Pendulum, 4 feet; Pendulum Ball, 150 lbs.
For one Dial up to 7 feet, or four Dials of 6 feet or less.
For Bell up to 2,500 lbs.
No. 17, 8 Day, Strike.
Pendulum, 8 feet; Pendulum Ball, 124 lbs.
For four 8 foot Dials and Bell up to 3,000 lbs.
No. 18, 8 Day, Strike.
Pendulum, 14 feet; Pendulum Ball, 300 lbs.
For large Dials and heavy Bell.
No. 19, 8 Day, Strike.
Pendulum, 14 feet; Pendulum Ball, 500 lbs.
For very large Dials and very heavy Bell.

$3200.00

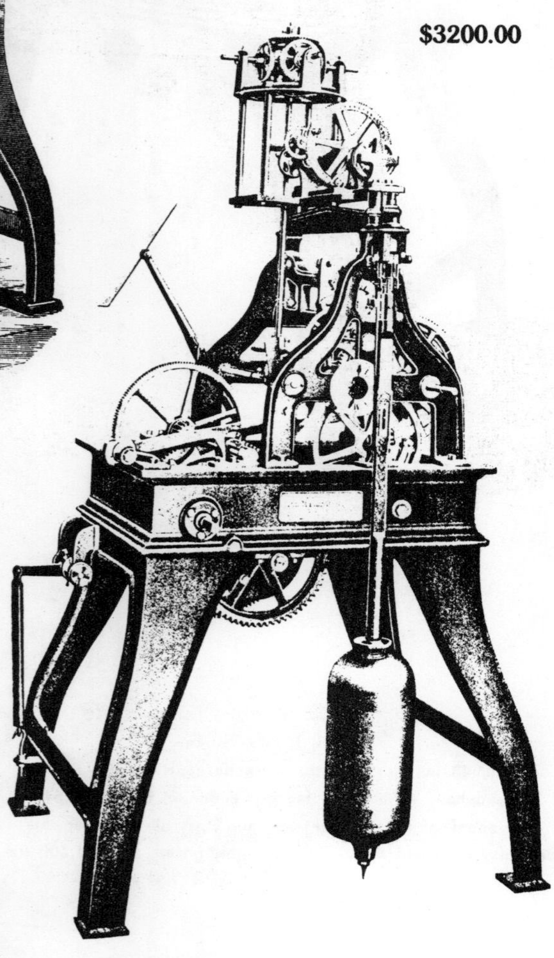

## SETH THOMAS CLOCK MOVEMENT DESCRIPTION

No. 16, 8-Day, Strike.
Width, 54 inches; Depth, 36 inches; Height, 58 inches.
Pendulum, 4 feet or 8 feet; Pendulum Ball, 135 lbs. or 175 lbs.
For one Dial up to 10 feet, or four Dials of 8½ feet or less.
Bell may be about 2,800 lbs.
Weighs, boxed, about 1,500 lbs.

# 1886 SETH THOMAS CLOCK MOVEMENT DESCRIPTIONS

No. 16B, 8-Day, Strike.
Width, 72 inches; Depth, 36 inches; Height, 58 inches.
Pendulum, 8 feet; Pendulum Ball, 175 lbs.
For one Dial up to 10 feet, or four Dials of about 8½ feet.
For Bell up to 3,500 lbs.
Weighs, boxed, about 2,100 lbs.
This clock may be fitted with Gravity Escapement.

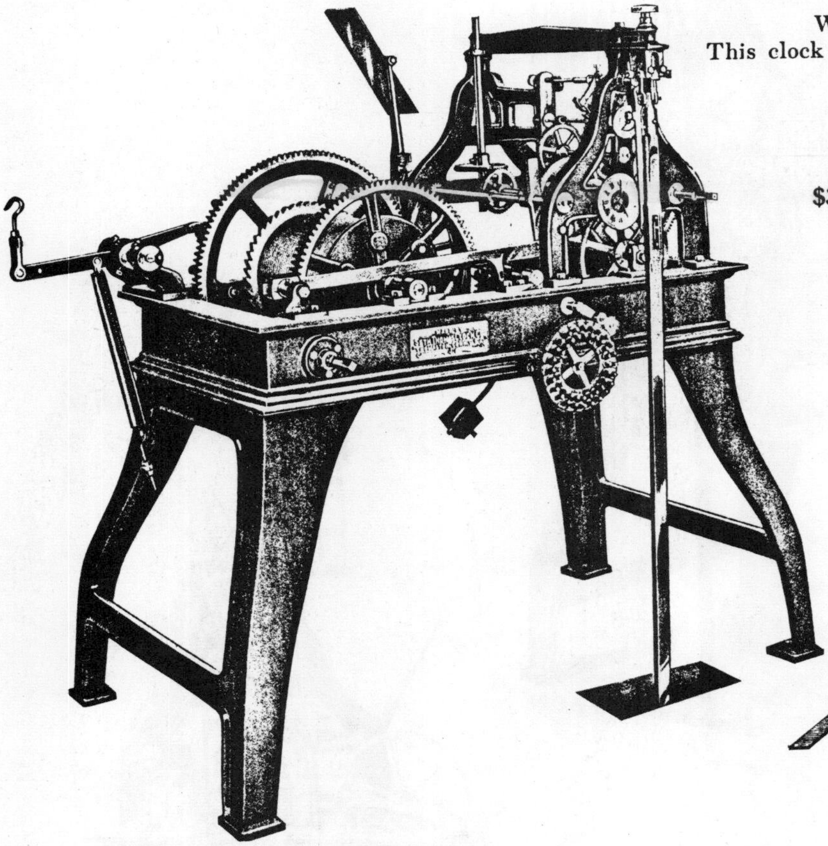

**$3500.00**

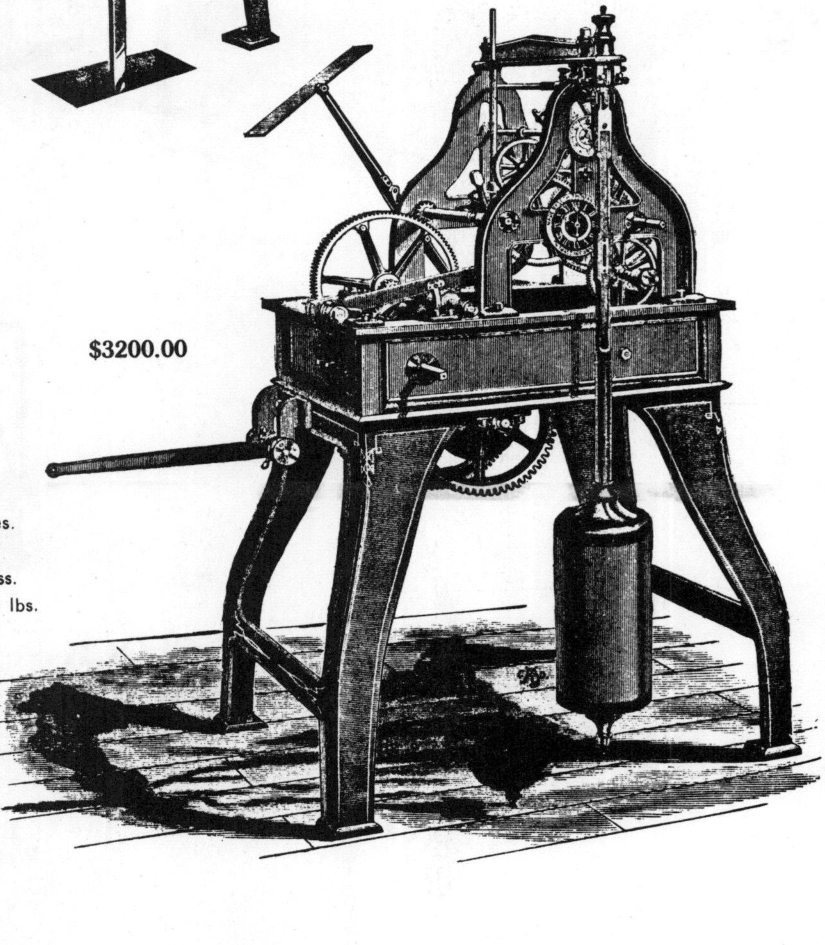

**$3200.00**

No. 16, 8 Day, Strike.
Width, 43 inches; Depth, 36 inches; Hight, 58 inches.
Pendulum, 4 feet or 8 feet; Pendulum, Ball 135 lbs.
For one Dial up to 8 feet, or four Dials of 7 feet or less.
Bell may be about 2,000 lbs. Weighs, boxed, about 1,500 lbs.

# 1886 SETH THOMAS CLOCK MOVEMENT DESCRIPTIONS

$3500.00

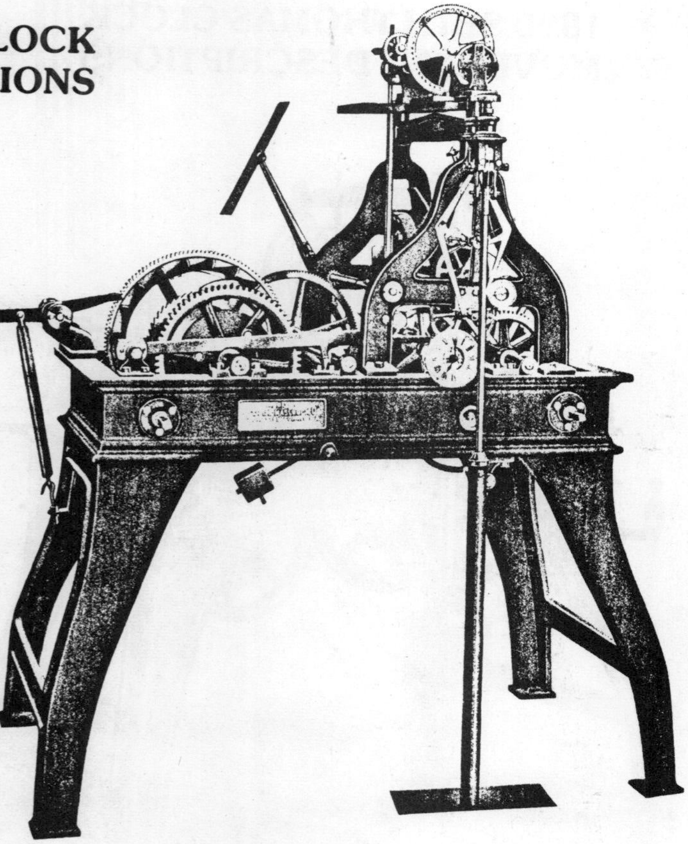

No. 16A Gravity Escapement, 8-Day, Strike.
Width, 70 inches; Depth, 36 inches; Height, 58 inches.
Pendulum, 8 feet or 14 feet; Pendulum Ball, 135 lbs. or 175 lbs.
For one Dial up to 11 feet, or four Dials of 9 feet or less.
Bell may be about 2,800 lbs.
Weighs, boxed, about 1,600 lbs.

$3500.00

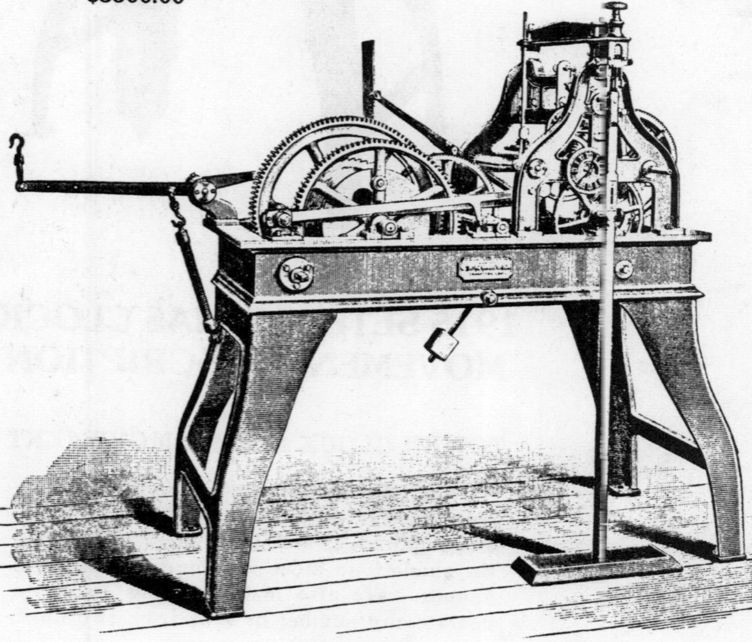

No. 16 B, 8 Day, Strike.
Width, 58 inches; Depth, 36 inches; Hight, 58 inches.
Pendulum 8 feet; Pendulum Ball, 175 lbs.
For one Dial up to 10 feet, or four Dials about 8 feet.
For Bell up to 3,500 lbs. Weighs, boxed, about 2,100 lbs.

# 1890 SETH THOMAS CLOCK MOVEMENT DESCRIPTIONS

**$4000.00**

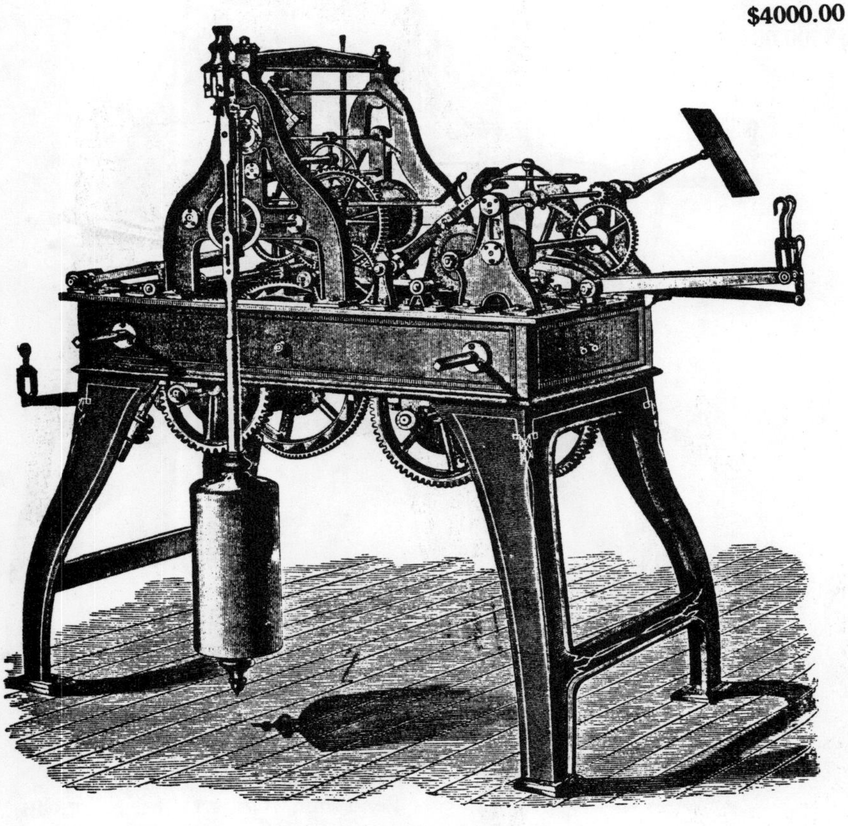

No. 16, 8 Day, Quarter Strike.
Width, 77 inches; Depth, 39 inches; Hight 58 inches.
Pendulum, 4 feet; Pendulum Ball, 135 lbs.
For one Dial up to 8 feet, or four Dials of 7 feet or less.
For three or four Bells, the largest of which can be 1,800 lbs.
Weighs, boxed, about 1,800 lbs.

**$2000.00 w/Motion Works**

# 1915 SETH THOMAS CLOCK MOVEMENT DESCRIPTION

### TOWER CLOCK STRIKE MOVEMENT

These clocks are built to operate dials from two feet in diameter to practically any size required, as well as to strike bells from the smallest to those weighing thousands of pounds. We also make chiming clocks to operate any number of bells from two up.

# ca. 1912 SETH THOMAS CLOCK MOVEMENT DESCRIPTIONS

No. 16, 8-Day, Quarter Strike.
Width, 88 inches; Depth, 36 inches; Height, 58 inches.
Pendulum, 4 feet or 8 feet; Pendulum Ball, 135 lbs. or 175 lbs.
For one Dial up to 10 feet, or four Dials of 8½ feet or less.
For peal of Bells, the largest of which may be about 2,800 lbs.
Weighs, boxed, about 2,300 lbs.
This clock may be fitted with Gravity Escapement.

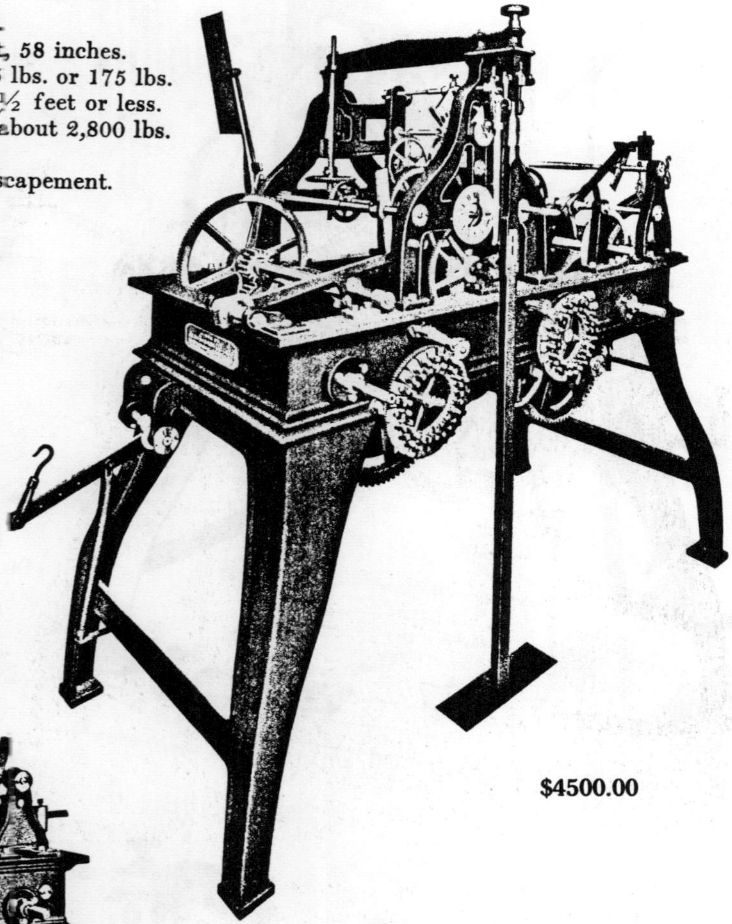

$4000.00

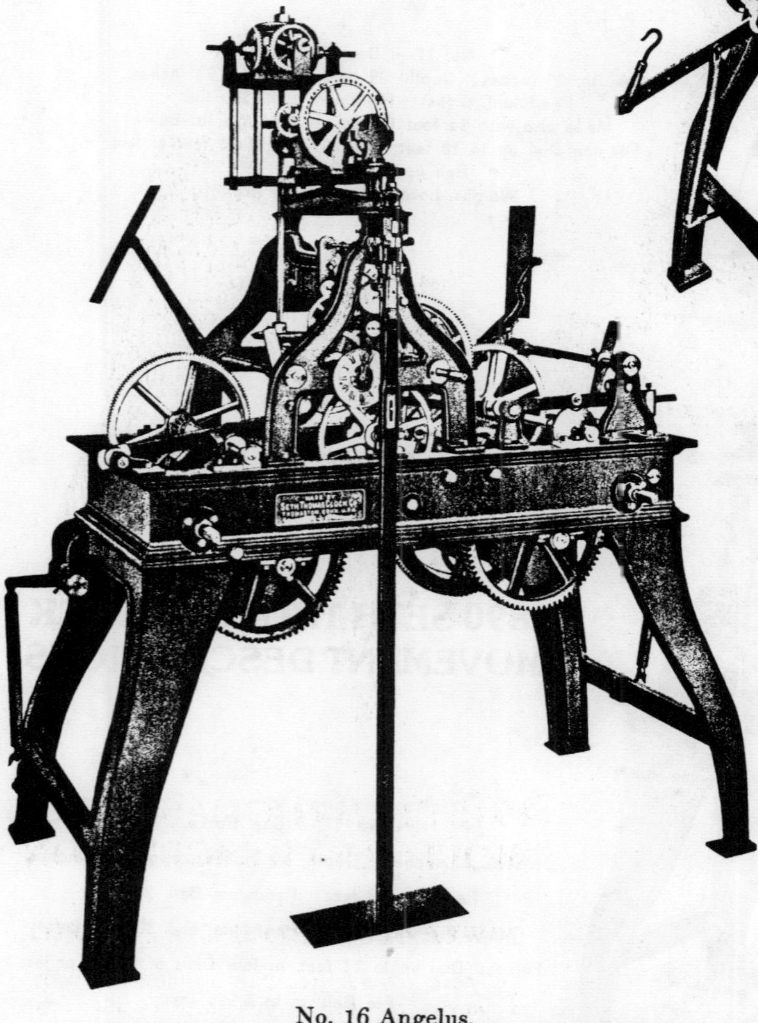

$4500.00

No. 16 Angelus.
Width, 72 inches; Depth, 36 inches; Height, 58 inches.
Pendulum, 8 feet; Pendulum Ball, 175 lbs.
For one Dial up to 10 feet, or four Dials about 8½ feet.
For Bell up to 3,000 lbs.
Weight, boxed, about 2,200 lbs.

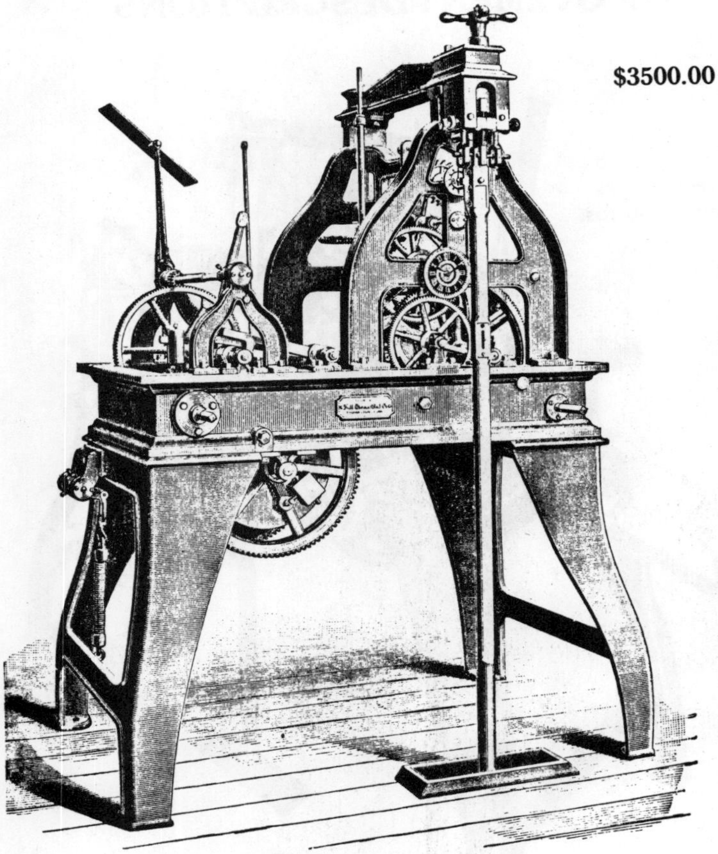

$3500.00

# 1896 SETH THOMAS CLOCK MOVEMENT DESCRIPTIONS

No. 17, 8 Day, Strike.
Width, 65 inches; Depth, 39 inches; Hight, 72 inches.
Pendulum, 8 feet; Pendulum Ball, 200 lbs.
Made also with 14 foot Pendulum, and 325 lb. Ball.
For one Dial up to 13 feet, or four Dials of 10 feet or less.
Bell up to 4,500 lbs.
Weighs, boxed, about 2,800 lbs.

$4000.00

# 1890 SETH THOMAS CLOCK MOVEMENT DESCRIPTIONS

No. 17, 8 Day, Strike.
Width, 53 inches; Depth, 39 inches; Hight, 65 inches.
Pendulum, 8 feet; Pendulum Ball, 200 lbs.
Made also with 14 foot Pendulum, and 300 lb. Ball.
For one Dial up to 11 feet, or four Dials of 9 feet or less.
For Bell up to 3,500 lbs.
Weighs, boxed, about 2,800 lbs.

# SETH THOMAS TOWER CLOCKS.

## 1896 SETH THOMAS CLOCK MOVEMENT DESCRIPTIONS

**$3500.00**

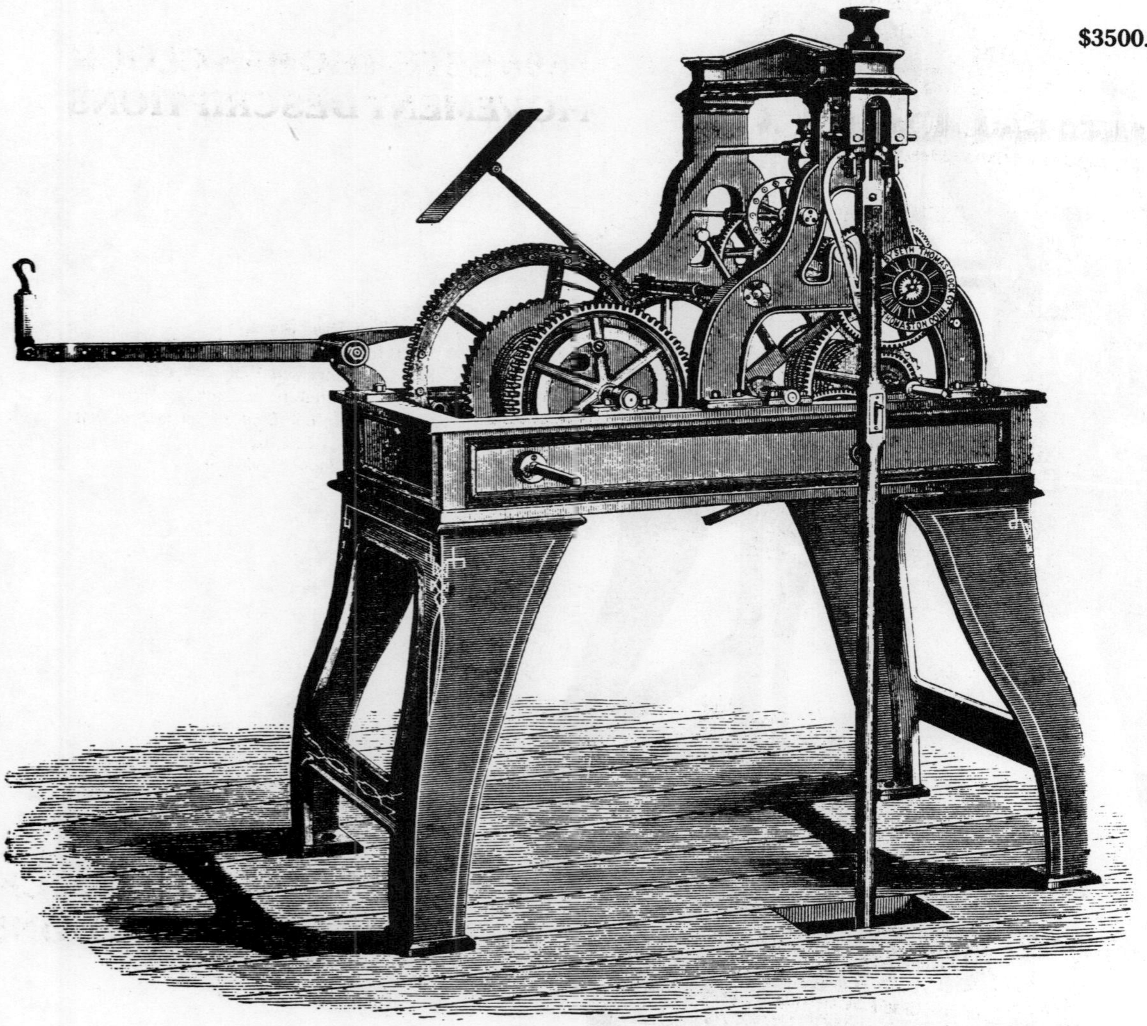

No. 17, 8 Day, Strike.

Width, 53 inches; Depth, 39 inches; Hight, 65 inches.

Pendulum, 8 feet; Pendulum Ball, 200 lbs.

# ca. 1912 SETH THOMAS CLOCK MOVEMENT DESCRIPTIONS

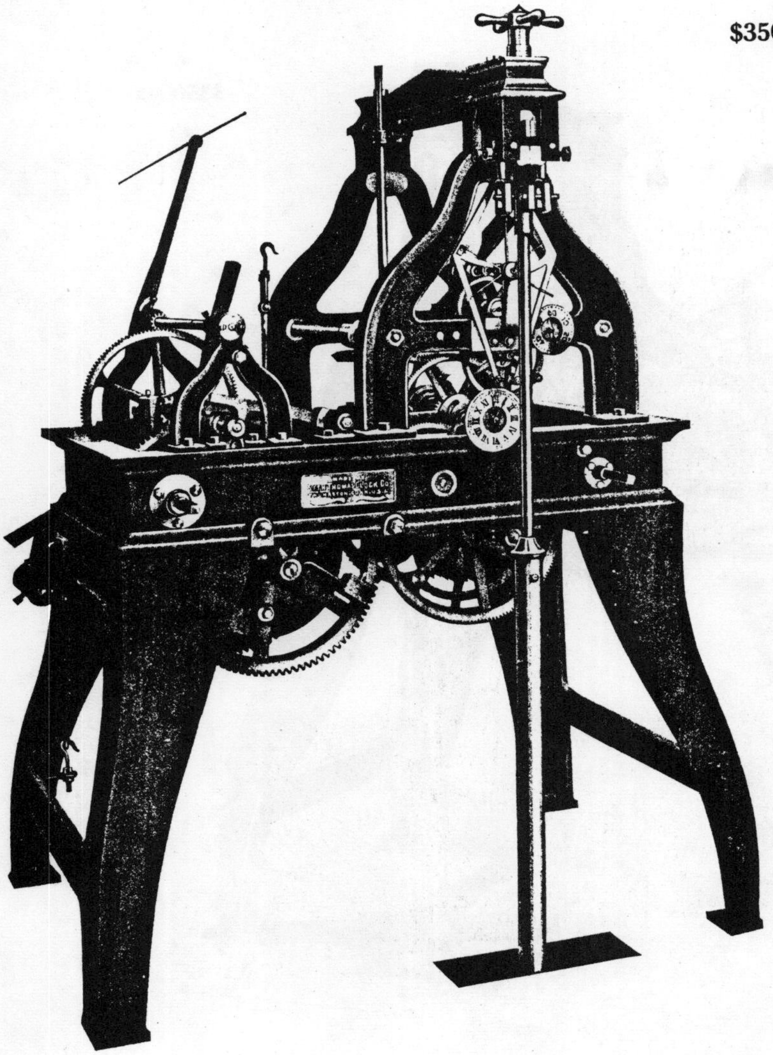

**$3500.00**

No. 17 Gravity, 8-Day, Hour Strike.
Width, 68 inches; Depth, 39 inches; Height, 72 inches.
Pendulum, 8 feet, or 14 feet; Pendulum Ball, 325 lbs.
For one Dial up to 14 feet, or four Dials up to 11 feet.
Bell about 4,500 lbs.
Weight, boxed, about 2,800 lbs.

# 1890 SETH THOMAS CLOCK MOVEMENT DESCRIPTIONS

$7000.00

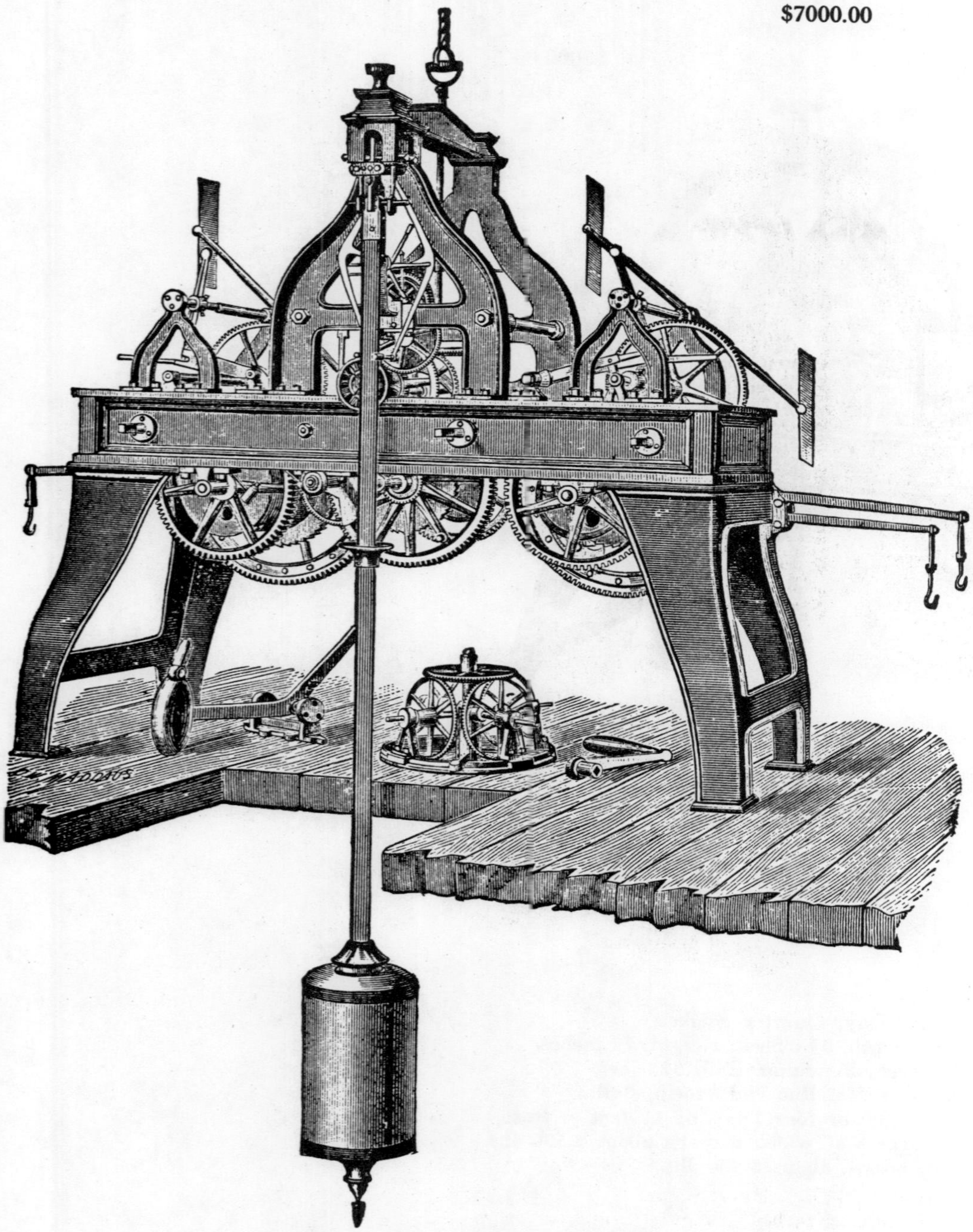

No. 17, 8 Day, Quarter Strike.
Width, 86 inches; Depth, 37 inches; Hight, 72 inches.
Pendulum, 8 feet; Pendulum Ball, 325 lbs.
Also made with 14 foot Roc and 325 lb. Ball.
For one Dial up to 14 feet, or four Dials of 11 feet or less.
For peal of Bells, the largest of which may be about 4,500 lbs.
Weighs, boxed, about 3,800 lbs.

# ca. 1912 SETH THOMAS CLOCK MOVEMENT DESCRIPTIONS

**$5000.00**

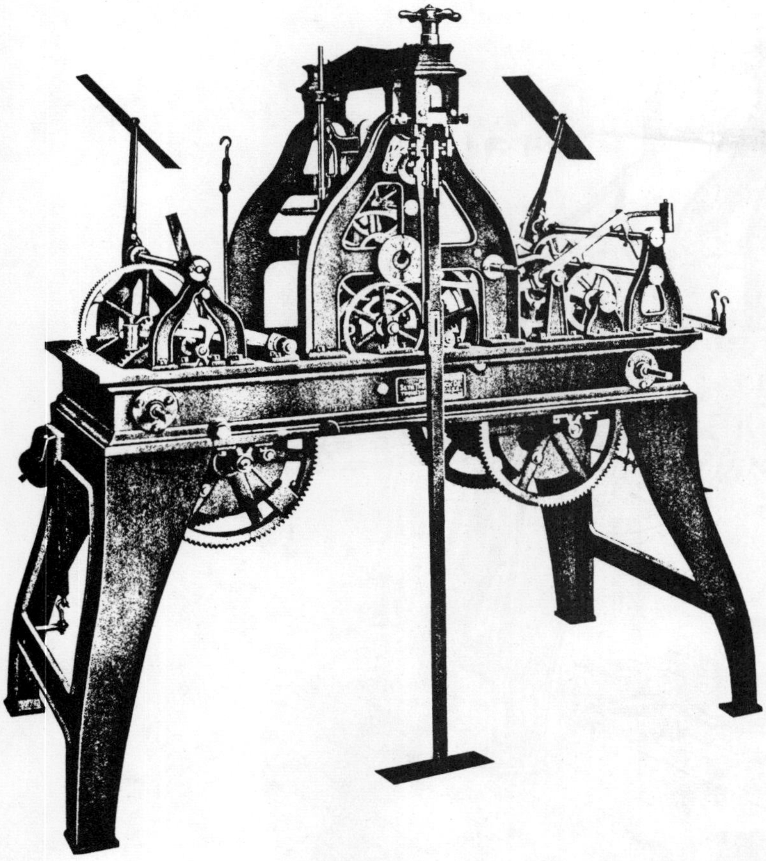

No. 17, 8-Day, Quarter Strike.
Width, 98 inches; Depth, 39 inches; Height, 72 inches.
Pendulum, 8 feet; Pendulum Ball, 325 lbs.
Also made with 14-foot Rod and 325-lb. Ball.
For one Dial up to 14 feet, or four Dials of 11 feet or less.
For peal of Bells, the largest of which may be about 4,500 lbs.
Weighs, boxed, about 3,800 lbs.

# 1890 SETH THOMAS CLOCK MOVEMENT DESCRIPTIONS

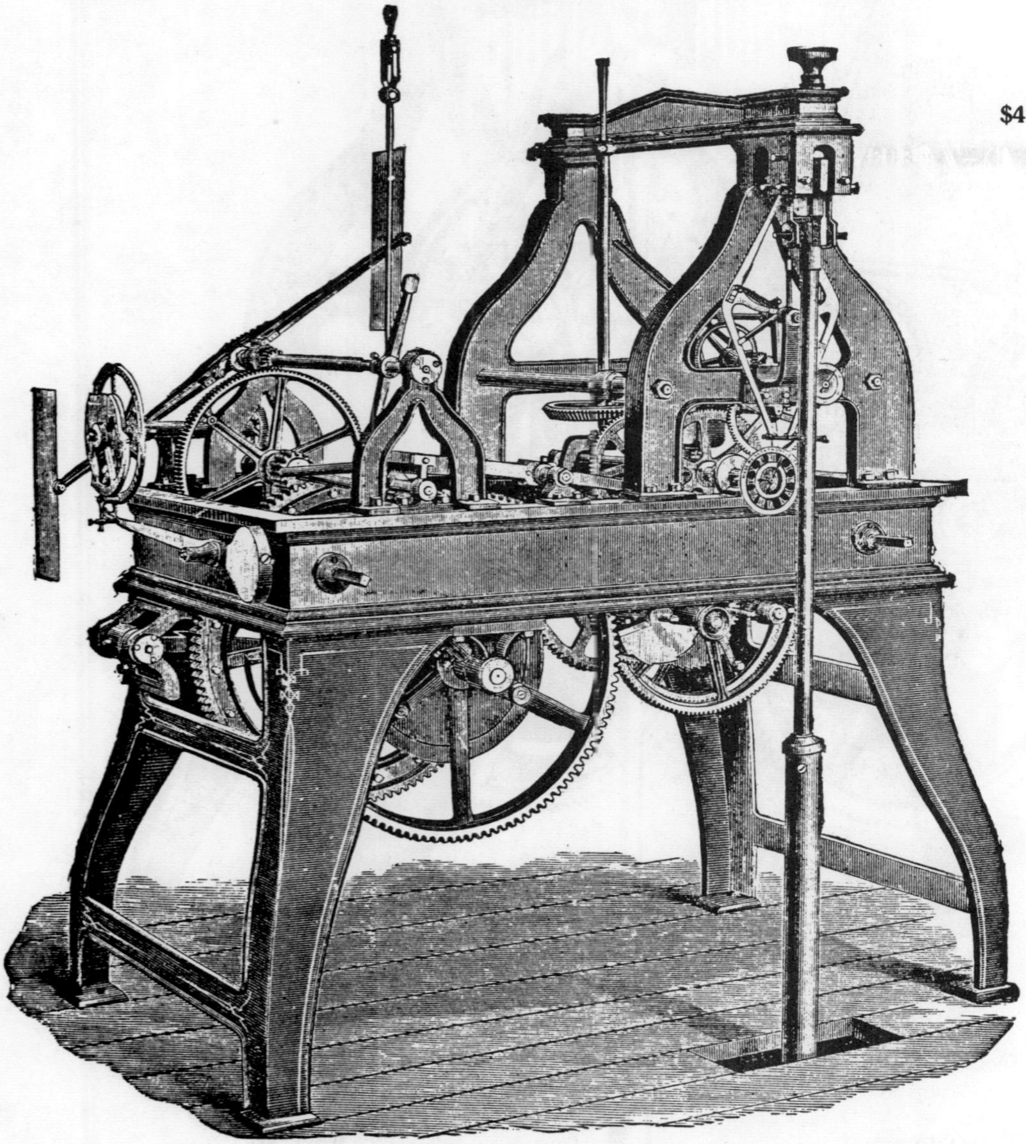

$4500.00

No. 18, 8 Day, Strike.
Width, 72 inches; Depth, 42 inches; Hight, 72 inches.
Pendulum, 14 feet; Pendulum Ball, 325 lbs.
For large Dials and Heavy Bell.
Weighs, boxed, about 3,300 lbs.

No. 19, 8 Day, Strike.
Width, 85 inches; Depth 50 inches; Hight, 76 inches.
Pendulum, 14 feet; Pendulum Ball, 550 lbs.
For very large Dials and very heavy Bell.
Weighs, boxed, about 4,900 lbs.

SETH THOMAS TOWER CLOCKS.

# Ca. 1912 SETH THOMAS MOVEMENT DESCRIPTIONS

**$4500.00**

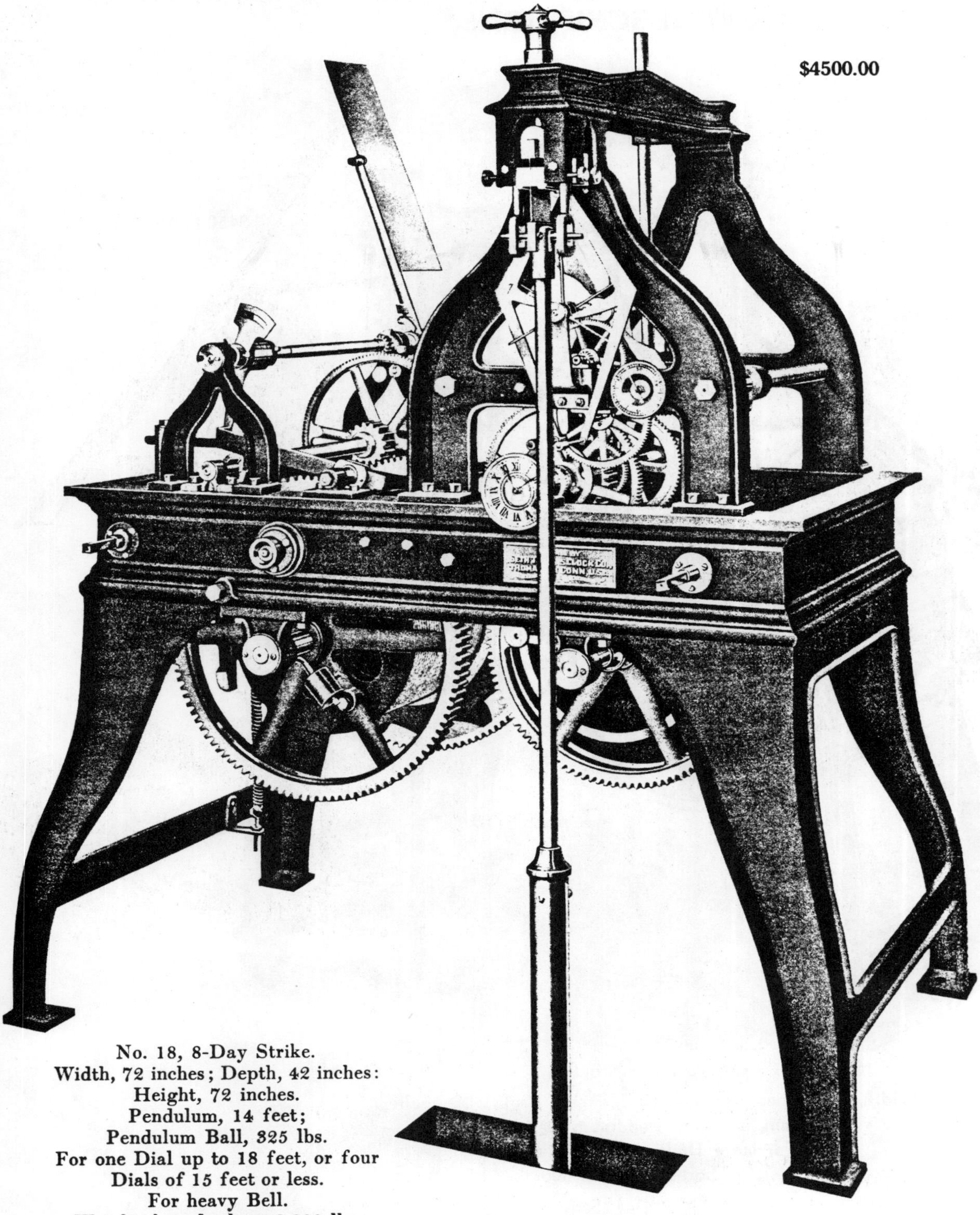

No. 18, 8-Day Strike.
Width, 72 inches; Depth, 42 inches:
Height, 72 inches.
Pendulum, 14 feet;
Pendulum Ball, 325 lbs.
For one Dial up to 18 feet, or four
Dials of 15 feet or less.
For heavy Bell.
Weighs, boxed, about 3,300 lbs.

No. 19, 8-Day, Strike.
Width, 85 inches; Depth, 50 inches; Height, 76 inches.
Pendulum, 14 feet; Pendulum Ball, 550 lbs.
For one Dial up to 22 feet, or four Dials of 18 feet or less.
For very heavy Bell.
Weighs, boxed, about 4,900 lbs.
Quarter-strike clocks of the above numbers are also made.

# SETH THOMAS TOWER CLOCKS.

## 1896 SETH THOMAS CLOCK MOVEMENT DESCRIPTIONS

$9000.00

No. 18, 8 Day, Quarter Strike.
Width, 104 inches; Depth, 42 inches; Hight, 75 inches.
Pendulum, 14 feet; Pendulum Ball, 325 lbs.
For peal of Bells, the largest of which can be about 6,000 lbs.
Weighs, boxed, about 4,500 lbs.

No. 19, 8 Day, Quarter Strike.
Width, 129 inches; Depth, 47 inches; Hight, 77 inches.
Pendulum, 14 feet; Pendulum Ball, 550 lbs.
For very large Dials and Bells.
Weighs, boxed, about 8,500 lbs.

No. 20, 8 Day, Quarter Strike.
Width, 132 inches, Depth, 60 inches; Hight, 96 inches.
Pendulum, 14 feet; Pendulum Ball, 550 lbs.
For Dials and Bells of any size.
Weighs, boxed, about 12,000 lbs.

SETH THOMAS TOWER CLOCKS.

$1000.00 to $3500.00

# ca. 1930 SETH THOMAS CLOCK MOVEMENT DESCRIPTIONS

## *Automatically Wound Time Movements*

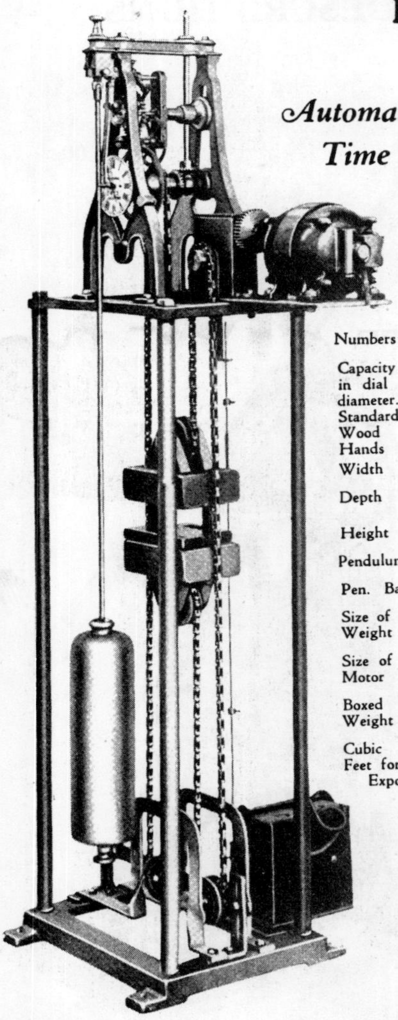

## *Specifications*

| Numbers | 4 | 5 | 6 | 7 | 8 | 9 | 10 |
|---|---|---|---|---|---|---|---|
| Capacity in dial diameter. Standard Wood Hands | 1 -5½′<br>2 -4½′<br>3 -4′<br>4 -3½′ | 1 -8′<br>2 -7′<br>3 -6′<br>4 -5½′ | 1 -11′<br>2 -10′<br>3 -9½′<br>4 -9′ | 1 -15′<br>2 -14′<br>3 -13′<br>4 -12′ | 1 -20′<br>2 -18′<br>3 -17′<br>4 -16′ | 1 -26′<br>2 -23′<br>3 -21′<br>4 -20′ | 1 -35′<br>2 -30′<br>3 -27′<br>4 -25′ |
| Width | 1′-10″ | 2′-0″ | 2′-8″ | 3′-3″ | 3′-3″ | 3′-6″ | 4′-8″ |
| Depth | 1′-8″ | 1′-10″ | 2′-6″ | 3′-4″ | 3′-4″ | 3′-6″ | 4′-6″ |
| Height | 5′-4″ | 5′-10″ | 7′-1″ | 7′-9″ | 1½ sec.<br>9′-6″ | 11′-0″ | 12′-0″ |
| Pendulum | 3′-9″ | 3′-10″ | 5′-9″ | 5′-9″ | 1½ sec.<br>8′3″ | 8′-3″ | 8′-3″ |
| Pen. Ball | 28 lbs | 75 lbs | 135 lbs | 175 lbs | 330 lbs | 330 lbs | 330 lbs |
| Size of Weight | 18 lbs to 30 lbs | 30 lbs to 40 lbs | 57 lbs to 75 lbs | 190 lbs to 240 lbs | 190 lbs to 275 lbs | 238 lbs to 285 lbs | 268 lbs to 315 lbs |
| Size of Motor | 1/10 H.P. | 1/10 H.P. | 1/8 H.P. | 1/8 H.P. | 1/8 H.P. | 1/6 H.P. | 1/4 H.P. |
| Boxed Weight | 375 | 600 | 850 | 2400 | 2600 | 2900 | 3500 |
| Cubic Feet for Export | 25 | 35 | 40 | 50 | 50 | 60 | 80 |

$850.00 to $2500

## *Time Movements with Electric Release for Operation from Master Clock*

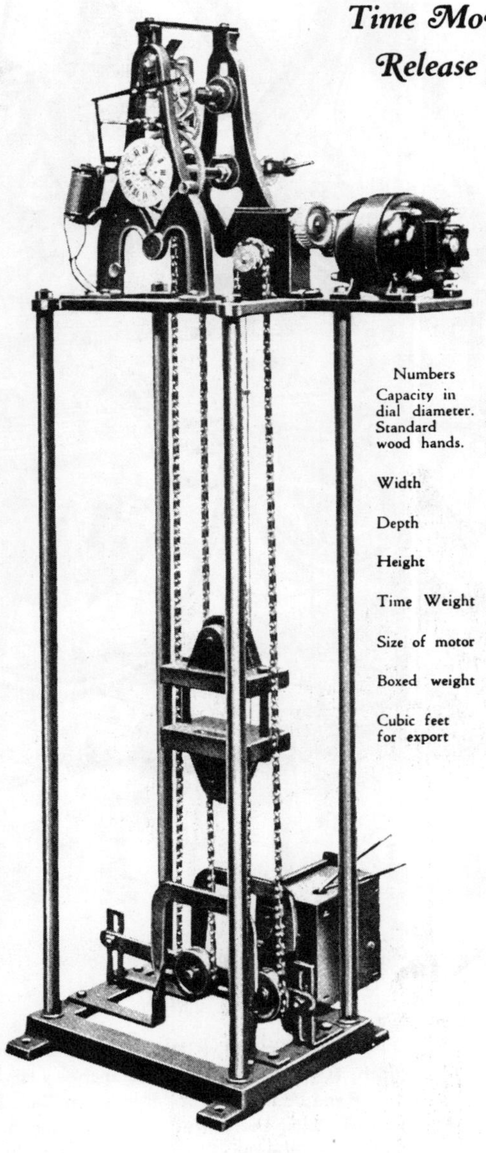

### *Specifications*

| Numbers | 4 | 5 | 6 | 7 | 8 | 9 |
|---|---|---|---|---|---|---|
| Capacity in dial diameter. Standard wood hands. | 1 -6′<br>2 -5½′<br>3 -5′<br>4 -4′ | 1 -8′<br>2 -7′<br>3 -6′<br>4 -5½′ | 1 -12′<br>2 -11′<br>3 -10′<br>4 -9½′ | 1 -15′<br>2 -14′<br>3-13′<br>4 -12′ | 1 -20′<br>2 -18′<br>3-17′<br>4 -16′ | 1 -30′<br>2 -26′<br>3 -24′<br>4 -22′ |
| Width | 1′-10″ | 2′-0″ | 2′-8″ | 3′-3″ | 3′-3″ | 3′-4″ |
| Depth | 1′-8″ | 1′-8″ | 2′-4″ | 3′-0″ | 3′-0″ | 3′-2″ |
| Height | 4′-8″ | 5′-9″ | 5′-8″ | 7′-9″ | 9′-0″ | 6′-8″ |
| Time Weight | 18 to 40 lbs. | 30 to 50 lbs. | 57 to 93 lbs. | 190 to 265 lbs. | 190 to 295 lbs. | 238 to 335 lbs. |
| Size of motor | 1/10 H.P. | 1/10 H.P. | 1/8 H.P. | 1/6 H.P. | 1/6 H.P. | 1/4 H.P. |
| Boxed weight | 340 | 525 | 750 | 2300 | 2400 | 2600 |
| Cubic feet for export | 24 | 32 | 37 | 45 | 45 | 55 |

## *Automatic Winders for Old Clocks*

If you have an old style hand wound 8-day Tower Clock we strongly recommend that you convert it into the modern type by using the new Seth Thomas Automatic Winders. These eliminate the old style cables and heavy weights, always a hazard.

# ca. 1930 SETH THOMAS CLOCK MOVEMENT DESCRIPTIONS

See following page for specifications

**$2500.00**

***Automatically Wound Time and Strike Movements***

Furnished Optionally to Strike Hours Only or Hours and Half Hours. Similiar Type Movements to Strike the Angelus. Also to Strike the Ship's Bells.

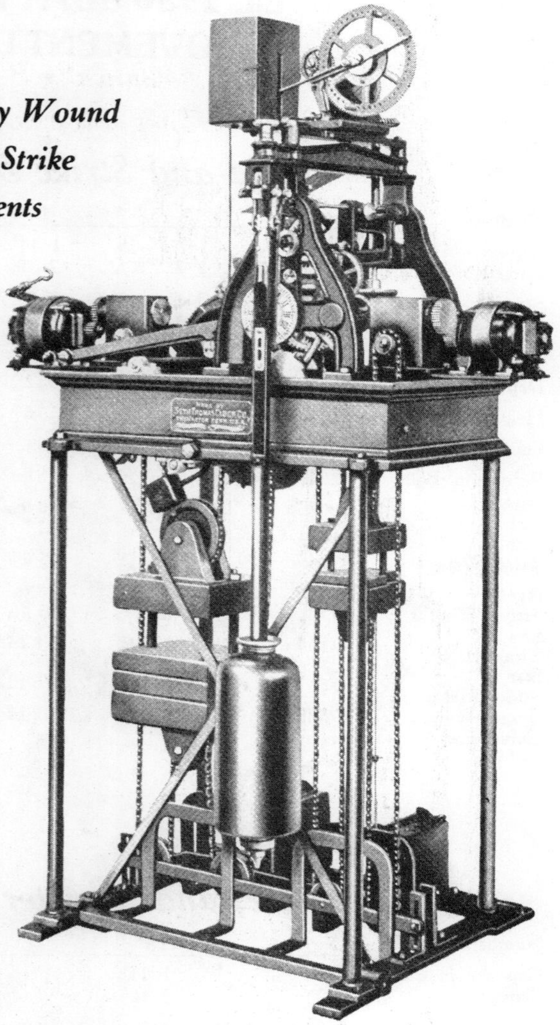

**$4000.00**

These movements can be supplied with Westminster or Whitington Chimes or with single or double strike quarters or with ding-dcng arrangement. Any special arrangement can be supplied.

***Automatically Wound Time, Strike and Quarter Strike Movements***

See following page for specifications

# ca. 1930 SETH THOMAS CLOCK MOVEMENT DESCRIPTIONS

## *Time and Strike Movements*

| Numbers | 14 | 15 | 16 | 17 | 18 | 19 |
|---|---|---|---|---|---|---|
| Capacity in dial diameters. | 1 -5½′<br>2 -4½′ | 1 -8′<br>2 -7′ | 1 -11′<br>2 -10′ | 1 -15′<br>2 -14′ | 1 -20′<br>2 -18′ | 1 -26′<br>2 -23′ |
| Standard Wood Hands | 3 -4′<br>4 -3½′ | 3 -6′<br>4 -5½′ | 3 -9½′<br>4 -9′ | 3 -13′<br>4 -12′ | 3 -17′<br>4 -16′ | 3 -21′<br>4 -20′ |
| Width | 4′-0″ | 4′-6″ | 5′-2″ | 6′-6″ | 7′-0″ | 8′-0″ |
| Depth | 2′-2″ | 2′-6″ | 3′-2″ | 3′-4″ | 3′-8″ | 4′-4″ |
| Height | 5′-7″ | 5′-7″ | 7′-6″ | 7′-9″ | 10′-6″ | 11′-6″ |
| Pendulum | 45″ | 46″ | 69″ | 69″ | 99″ | 99″ |
| Pendulum Ball | 28 lbs | 75 lbs | 135 lbs | 175 lbs | 330 lbs | 330 lbs |
| Largest Bell | 500 lbs | 1200 lbs | 3000 lbs | 4000 lbs | 5000 lbs | 7000 lbs |
| Size of Time Weight | 18 to 30 lbs | 30 to 40 lbs | 57 to 75 lbs | 190 to 240 lbs | 190 to 275 lbs | 238 to 285 lbs |
| Size of Strike Weight | 125 to 175 lbs | 170 to 220 lbs | 238 to 285 lbs | 238 to 317 lbs | 273 to 320 lbs | 355 to 500 lbs |
| Size of Time Motor | 1/10 H.P. | 1/10 H.P. | 1/8 H.P. | 1/6 H.P. | 1/6 H.P. | 1/6 H.P. |
| Size of Strike Motor | 1/8 H.P. | 1/6 H.P. | 1/6 H.P. | 1/6 H.P. | 1/4 H.P. | 1/2 H.P. |
| Boxed Weight | 875 lbs | 1400 lbs | 2550 lbs | 2800 lbs | 3500 lbs | 4500 lbs |
| Cubic Feet for Export | 25 | 40 | 50 | 75 | 85 | 110 |

## *Time, Strike, and Quarter Strike Movements*

| Numbers | 15 | 16 | 17 | 18 | 19 |
|---|---|---|---|---|---|
| Capacity in dial diameters. | 1 -8′<br>2 -7′ | 1 -11′<br>2 -10′ | 1 -15′<br>2 -14′ | 1 -20′<br>2 -18′ | 1 -26′<br>2 -23′ |
| Standard Wood Hands | 3 -6′<br>4 -5½′ | 3 -9½′<br>4 -9′ | 3 -13′<br>4 -12′ | 3 -17′<br>4 -16′ | 3 -21′<br>4 -20′ |
| Width | 5′-8″ | 8′-0″ | 9′-3″ | 10′-0″ | 12′-0″ |
| Depth | 3′-0″ | 3′-8″ | 3′-10″ | 3′-8″ | 4′-4″ |
| Height | 5′-7″ | 7′-6″ | 7′-9″ | 10′-6″ | 11′-6″ |
| Pendulum | 46″ | 69″ | 69″ | 99″ | 99″ |
| Pendulum Ball | 75 lbs | 135 lbs | 175 lbs | 330 lbs | 330 lbs |
| Largest Bell | 1200 lbs | 3000 lbs | 4000 lbs | 5000 lbs | 7000 lbs |
| Size of Time Weight | 30 to 40 lbs | 57 to 75 lbs | 190 to 240 lbs | 190 to 275 lbs | 238 to 285 lbs |
| Size of Strike Weight | 170 to 220 lbs | 238 to 285 lbs | 238 to 317 lbs | 273 to 320 lbs | 355 to 500 lbs |
| Size of Quarter Strike Weight | 170 to 220 lbs | 238 to 285 lbs | 238 to 317 lbs | 273 to 320 lbs | 355 to 585 lbs |
| Size of Time Motor | 1/10 H.P. | 1/8 H.P. | 1/6 H.P. | 1/6 H.P. | 1/6 H.P. |
| Size of Strike Motor | 1/6 H.P. | 1/6 H.P. | 1/6 H.P. | 1/4 H.P. | 1/2 H.P. |
| Size of Quarter Strike | 1/6 H.P. | 1/6 H.P. | 1/6 H.P. | 1/4 H.P. | 1/2 H.P. |
| Boxed Weight | 2200 | 3100 | 3500 | 4900 | 7500 |
| Cubic Feet for Export | 75 | 100 | 150 | 195 | 280 |

# ca. 1930 SETH THOMAS CLOCK MOVEMENT DESCRIPTIONS

**$1250.00**

## *Miniature Tower Clock*

## *Miniature Tower Clock in Unfinished Wood Cabinet With 12″ Pilot Dial No. 2*

No. 2 Movement

Overall dimensions
Mounted in Cabinet

| | |
|---|---|
| Width | 20″ |
| Depth | 12″ |
| Height | 38″ |

Capacity: 1 dial 5′; 2-4′; 3-3′; 4-2′-6″

Pendulum: 25″

Pendulum Ball: 2 lbs.

Size of weight: From 14 to 22 lbs.

Size of motor: 1/20 H.P.

Boxed weight: 240 lbs. with trimmings.

This movement is of the same high quality workmanship and material and carries the same guarantee as all other tower clock movements of our manufacture.

It is especially adapted for street electric sign use, for bracket clocks and for use over entrances and doorways. Due to an eighty beat pendulum its regulation is only guaranteed within one minute per month if properly maintained. It can, however, be hung on the wall of a store or even mounted inside of a bracket clock if necessary.

It is also furnished without cabinet, but mounted on an iron frame where required, and is particularly adapted to the replacement of obsolete street clock movements of other makes.

It is moderately priced. Standard equipment includes two pairs of hands, dialworks, couplings, and shafting. Due to short pendulum it will perform satisfactorily in locations where vibrations would interfere with operations of movements with longer pendulums.

***To receive top prices, movements must be complete and original, and they must be in good running or repairable condition. Damaged, worn-out or missing parts will reduce prices accordingly.***

# The National Association of Watch & Clock Collectors

The National Association of Watch & Clock Collectors is a non-profit scientific and educational corporation founded in 1943 to bring together people who are interested in timekeeping in any form or phase. This worldwide organization, with in excess of 32,000 members, meets the needs of its membership through its bi-monthly 128 page *Bulletin* and its bi-monthly *Mart* advertising magazines, its Answer Box service staffed by over one hundred volunteer experts, its 130 chapters around the world, and its Museum, Reference and Lending Libraries located in Columbia, PA. Dues are $25 in the U.S. and $32 for members residing outside of the U.S. The Museum is open year-round, Monday through Friday from 9:00-4:00 and Saturdays from 9:00-5:00. For further information write to the above address or call (717) 684-8261.

**THE NATIONAL ASSOCIATION OF WATCH & CLOCK COLLECTORS, INC.**

514 Poplar Street
Columbia, PA 17512

# SECTION IV

## MOVEMENTS

# PREFACE

The movements in this section have been compiled from the movement catalogs issued by Seth Thomas, as well as from the regular issues of the company's catalogs, some of which have occasional references to movements.

Using this section, you can:

- Establish the movement numbers that Seth Thomas used for different clock models.
- Identify the proper movement for an authentic match with your case, when you have a case from which the movement has been removed.
- Check the movements and cases on those special finds, when you have reason to suspect they might be "married", or mismatched.

Readers are cautioned to use the information provided in this section judiciously. There are many gaps in the data since the company did not list its entire line of movements in published catalogs, and it is probable that there were also additional catalogs issued that have not yet been located. Also, it was not until 1907 that the company specified which movement went with which case.

You should also keep in mind that marriages between cases and movements occasionally took place in the factory and, as a result, movements sometimes were originally installed in cases other than those for which they were intended.

Now, for the first time, a Movements Table has been prepared for your use in research or as an aid in your efforts to properly match your Seth Thomas movement with the correct case. When you use the table, please remember that discontinuation of a movement in a catalog did not necessarily mean that corresponding clock models were discontinued. Undoubtedly, inventory was worked off, and there is a possibility the discontinued movements were used in newer models.

# Seth Thomas
# MOVEMENTS TABLE

| TYPE/MODEL NAME/ NUMBER | MOVEMENTS USED |
|---|---|
| **ALARM** | |
| Aden, 8-Day | 110 |
| Anvil | 92-C |
| Artist | 94, 94-A |
| Crystal | 94, 94-A |
| Dix | 6 |
| Eight-Day Automatic | 110 |
| Eik | 92-A |
| Joker | 94, 94-A |
| Long | 3, 3-A |
| Mikado | 94, 94-A |
| Nutmeg | 92 |
| Reminder | 89-F |
| Standish, 8-Day | 110 |
| Owl | 3, 3-A |
| **BANJO** | |
| No. 1 | 41, 117, 123 |
| No. 2, Strike | 125 |
| No. 3, Pendulum | 123 |
| No. 5 | 123 |
| Blackstone, Black Case | 103-A |
| Boston, 8-Day Time | 7-Jeweled |
| Brookfield 3 W | 120 |
| Cornwall, Time | 123 |
| Crandall | 120, 120-K |
| Danvers | 103-A |
| Deerfield | 103-A |
| Delaware, Lever | 5100 |
| Dover | 103-A |
| Grenville, Green Case | 103-A |
| Hampton | 103-A |
| Newport | 103-A |
| Pittsfield | 103-A |
| Pomfret, 8-Day | 4-Jeweled |
| Ramgate | 103, 103-A |
| Redress, Red Case | 103-A |
| Southbury | 103-A |
| **BOUDOIR & DESK** | |
| Anita | 5100 |
| Baltic | 5100 |
| Banner No. 1 W | 5100 |
| Banner No. 2 W | 5100 |
| Barnard Green | 102 |
| Barnard Ivory | 102 |
| Bayne | 102 |
| Belmont No. 1 | 5100 |
| Belmont No. 2 | 5100 |

| TYPE/MODEL NAME/ NUMBER | MOVEMENTS USED |
|---|---|
| **BOUDOIR & DESK** | |
| Boudoir No. 1 | 103, 103-A |
| Boudoir No. 2 | 103, 112 |
| Boudoir No. 3 | 103 |
| Boudoir No. 4 | 103, 112 |
| Boudoir No. 5 | 103-A |
| Brayton | 102 |
| Brewster | 102 |
| Burton | 102 |
| Cairo | 103-A |
| Caldwel | 103-A |
| Calgary | 103 |
| Calgary, Bronze | 103 |
| Camden | 103, 103-A |
| Candor | 103 |
| Cannock | 103, 103-A |
| Capel | 103, 103-A |
| Cardiff | 103-A |
| Carthage | 103-A |
| Castine No. 1 | 103-A |
| Castine No. 2 | 103-A |
| Castine No. 3 | 103-A |
| Castine No. 4 | 103-A |
| Celtic | 105 |
| Columbia | 103-A |
| Comax | 103, 103-A |
| Como | 103, 103-A |
| Corinth | 103 |
| Corona | 5100 |
| Crest No. 1 W | 5100 |
| Cromar | 103-A |
| Dallas | 103-A |
| Dale | 103-A |
| Dartmore | 103-A |
| Deane | 103-A |
| Delart | 5100 |
| Delpart | 5100 |
| Delpart No. 2 | 5100 |
| Delpart No. 3 | 5100 |
| Delpart No. 4 | 5100 |
| Delray | 5100 |
| Delrod | 5100 |
| Delta No. 1 | 5100 |
| Delta No. 3 | 5100 |
| Devon | 103 |
| Discus | 5100 |
| Doncaster | 103 |
| Dura | 5100 |
| Earl | 103-A |

| TYPE/MODEL NAME/ NUMBER | MOVEMENTS USED |
|---|---|
| **BOUDOIR & DESK** | |
| Duplex | 5100 |
| Eden | 112 |
| Eliot | 112 |
| Elfin | 112 |
| Elton | 112 |
| Enfield | 112 |
| Ensign | 112 |
| Exeter | 112 |
| Janet | 5100 |
| Lido | 103-A |
| Marquis | 103-A |
| Matin | 5100 |
| Mayfair | 103-A |
| Maywood | 103-A |
| Milan | 103-A |
| Myrna | 5100 |
| Odin | 103-A |
| Petite No. 1 | 102 |
| Petite No. 2 | 102 |
| Petite No. 3 | 102 |
| Primrose | 5100 |
| Regent | 103-A |
| Shamrock | 5100 |
| Sunglow | 5100 |
| Tabor No. 1 W | 5100 |
| Thor | 103-A |
| Tivoli | 103-A |
| Twilight | 5100 |
| Vernon | 103, 103-A |
| Verona | 103-A |
| Vogue | 5100 |
| Windsor | 5100 |
| **CALENDAR** | |
| Baird Drop | 97-A |
| Calendar/Large | |
| German Rolls | 37-C |
| Portuguese Rolls | 37-B |
| Spanish Rolls | 37-A |
| White Rolls | 37 |
| Calendar/Small | |
| Black Rolls | 36-A |
| German Rolls | 36-D |
| Portuguese Rolls | 36-C |
| Spanish Rolls | 36-B |
| White Rolls | 36 |
| Dixie | 34-G, 44-W |
| Drop Octagon | 44-O |

## CALENDAR

| TYPE/MODEL NAME/NUMBER | MOVEMENTS USED |
|---|---|
| Drop Octagon, Strike | 89-F |
| Fashion, New | 85-D |
| Fashion, Old | 85-F |
| Fashion, Southern | 85-G |
| Globe | 44-X |
| Ithica Drop | 41-G |
| Jackson | 44-N |
| Jackson Drop | 34-I |
| Office No. 4 | 41-P |
| Office No. 5 | 41-P, 62-D |
| Office No. 6 | 41-L, 44-P |
| Office No. 7 | 41-L, 44-P |
| Office No. 9 | 62-E |
| Office No. 10 | 55 |
| Office No. 11 | 62-E |
| Office No. 12 | 55-A |
| Office No. 13 | 55-A |
| Parlor No. 1 | 82-B |
| Parlor No. 3 | 44-Q |
| Parlor No. 4 | 44-Q |
| Parlor No. 5 | 44-Q |
| Parlor No. 6 | 44-Q |
| Parlor No. 7 | 44-Q |
| Parlor No. 8 | 44-S |
| Parlor No. 9 | 44-S |
| Parlor No. 10 | 82-B |
| Saturn | 97-H |

## CARRIAGE

| TYPE/MODEL NAME/NUMBER | MOVEMENTS USED |
|---|---|
| Carriage No. 1 | 9 |

## CHIME

| TYPE/MODEL NAME/NUMBER | MOVEMENTS USED |
|---|---|
| Astor | 113 |
| Chime 50 | 124 |
| Chime 52 | 124, 124-B |
| Chime 53 | 124, 124-B |
| Chime 56 | 124-D |
| Chime 57 | 124 |
| Chime 60 | 124 |
| Chime 70 | 113 |
| Chime 71 | 113, 113-A |
| Chime 72 | 113, 113-A |
| Chime 73 | 113, 113-A |
| Chime 74 | 113 |
| Chime 75 | 113 |
| Chime 77 | 113 |
| Chime 78 | 113 |
| Chime 79 | 113 |
| Chime 80 | 113 |
| Chime 81 | 113-E |
| Chime 90 | 124 |
| Chime 91 | 124 |
| Chime 92 | 124 |
| Chime 93 | 124 |
| Chime 94 | 124 |
| Chime 95 | 124 |
| Chime 96 | 124 |
| Chime 97 | 124 |
| Chime 98 | 124 |
| Chime 99 | 124 |

## CHIME

| TYPE/MODEL NAME/NUMBER | MOVEMENTS USED |
|---|---|
| Chime 100 | 124 |
| Chime 101 | 113, 113-B |
| Chime 102 | 113, 113-B |
| Chime 103 | 113, 113-B |
| Chime 116 | 124 |
| Sulgrave | 113 |

## CRYSTAL REGULATOR

| TYPE/MODEL NAME/NUMBER | MOVEMENTS USED |
|---|---|
| Empire | 48-M, 48-N, 51-E, 51-F, 51-I, 51-J |
| Empire No. 40 | 51-G |
| Empire No. 50 | 51-N |
| Empire No. 200 | 120 |
| Empire No. 201 | 120 |
| Empire No. 300 | 120 |
| Empire No. 301 | 120 |
| Orchid | 51-E, 51-H |

## ELECTRIC BANJO

| TYPE/MODEL NAME/NUMBER | MOVEMENTS USED |
|---|---|
| Banjo No. 24 | ETM |
| Brookfield/SSSS | 1602 |
| Canfield | SET |
| Duffield | SSET |
| Fairfield | SSE |
| Mansfield | SSE |
| Penfield | SSE |

## ELECTRIC BOUDOIR & DESK

| TYPE/MODEL NAME/NUMBER | MOVEMENTS USED |
|---|---|
| Azure No. 1 | SSET |
| Axure No. 2 | SSET |
| Banner 1E | SSET |
| Banner 2E | SSET |
| Crest 1E | SSET |
| Dickson | SSET |
| Durango | SET |
| Echo | SSET |
| Floret 1E | SSET |
| Gibson | SET |
| Hour Glass | SET |
| Judson | SET |
| Kitchenette | SET |
| Linnet | SSET |
| Phoenix | SSET |
| Pierson | SET |
| Primrose 2E | SSET |
| Rutland | SSET |
| Sonora | SET |
| Sunset | SSET |
| Tabor 1E | SSET |
| Talson | SSET |
| Upson | SET |

## ELECTRIC CHIME

| TYPE/MODEL NAME/NUMBER | MOVEMENTS USED |
|---|---|
| Marbury/SSWC | 1700 |
| Westbury/SSWC | 1750, 1770 |
| Wickford/SSWC | 1701, 1720, 1721 |

## ELECTRIC HALL

| TYPE/MODEL NAME/NUMBER | MOVEMENTS USED |
|---|---|
| Hall No. 1 | SSWC |
| Hall No. 1E | SSWC |
| Hall No. 3 | SSWC |
| Hall No. 3E | SSWC |
| Hall No. 4E | SSWC |

## ELECTRIC HANGING

| TYPE/MODEL NAME/NUMBER | MOVEMENTS USED |
|---|---|
| Auditor | SET |
| Counsil No. 2 | SSET |
| Executive | SET |
| Judson | 1500 |
| Kitchenette | SET |
| No. 50 | ETM |
| No. 51 | ETM |
| No. 52 | ETM |
| Plaza | SSET |
| President/SSST | 1510 |
| Principal No. 2E | SSET |
| Recorder No. 1 | SSET |
| Recorder No. 2 | SSET |
| Self-winding No. 1 | 86-AF |
| Senate/SSST | 1520 |

## ELECTRIC MANTEL

| TYPE/MODEL NAME/NUMBER | MOVEMENTS USED |
|---|---|
| Abington | SSSE |
| Bennington | SSSE |
| Century No. 2/SSSS | 1604 |
| Century No. 3/SSSS | 1604 |
| Chevron No. 1E | SSET |
| Colton No. 2E | SSE |
| Console No. 1 | 1604 |
| Console No. 2 | 1604 |
| Console No. 4E | SSET |
| Console No. 5E | SSET |
| Dalesbury | SSWC |
| Durango | SSET |
| Electric No. 1 | ETM |
| Electric No. 10 | ETM |
| Electric No. 20 | ETM |
| Electric No. 21 | ETM |
| Electric No. 22 | ETM |
| Electric No. 23 | ETM |
| Falsbury | SSWC |
| Gibson | SSET |
| Judson | SSET |
| Lipton | SSET |
| Ludbury | SSWC |
| Marbury/SSWC | 1700 |
| Medbury/SSWC | 1702 |
| Middleton/SSSS | 1600, 1601 |
| Nanking | SSSE |
| Newton | SSSE |
| Pierson | SSET |
| Rex | SSET |
| Rodson | SSET |
| Sonora | SSET |

ETM = Electric Time Movement
SSE = Self Starting Electric
SET = Synchronous Electric Time
SSET = Self Starting Electric Time
SSSE = Synchronous Self-Starting Electric
SSSS = Synchronous Self-Starting Strike
SSST = Synchronous Self-Starting Time
SSWC = Synchronous Self-Starting Westminster Chime

# ELECTRIC MANTEL

| TYPE/MODEL NAME/ NUMBER | MOVEMENTS USED |
|---|---|
| Roxbury/SSWC | 1700 |
| Stetson | SSET |
| Tilton | SSSE |
| Upson | SSET |
| Westbury/SSWC | 1750 |
| Winton | SSSE |

# HALL

| TYPE/MODEL NAME/ NUMBER | MOVEMENTS USED |
|---|---|
| Hall No. 3W | 113, 113-B |
| Hall No. 4 | 113-E |
| Hall No. 4W | 113 |
| Hall No. 21 | 73 |
| Hall No. 22 | 74-B |
| Hall No. 23 | 72-A |
| Hall No. 24 | 72-A |
| Hall No. 25 | 74-C |
| Hall No. 26 | 74-B |
| Hall No. 27 | 72-B |
| Hall No. 28 | 72-B |
| Hall No. 33 | 72-B |
| Hall No. 34 | 72-B |
| Hall No. 35 | 72-B |
| Hall No. 36 | 72-B |
| Hall No. 37 | 72-B |
| Hall No. 38 | 84 |
| Hall No. 39 | 84 |
| Hall No. 2272C | 72-C |
| Hall No. 2274D | 74-D |
| Hall No. 2784 | 84 |
| Hall No. 2874D | 74-D |
| Hall No. 3374D | 74-D |
| Hall No. 3434D | 74-D |
| Hall No. 3574D | 74-D |
| Hall No. 3684 | 84 |
| Hall No. 3684D | 74-D |
| Hall No. 3774D | 74-D |
| Hall No. 3784 | 84 |
| Hall No. 3974 | 74-D |
| Precision Regulator | 76, 76-A, 100, 100-A |

# HANGING

| TYPE/MODEL NAME/ NUMBER | MOVEMENTS USED |
|---|---|
| Albey | 51 |
| Arcade No. 1/18 inch | 86, 86-T |
| Arcade No. 2/24 inch | 86, 86-T |
| Arctic, Strike | 89-A |
| Arctic, Time | 41, 123-A |
| Assembly | 41 |
| Baird, Strike | 44-U |
| Baird, Time | 97, 97-B, 97-C, 97-D, 97-E, 97-F, 97-G |
| Bradford 8 and 12 inch | 89-B, 89-C |
| Chateau | 105 |
| Chatham | 41, 41-M, 41-AC, 44-V |
| Chatham, Strike | 89-A |
| Chatham, Time | 123-A |
| Dante | 103-A |
| Dauphin | 103-A |
| Diana | 103-A |
| Double Dial/Ceiling | 86 |
| Double Dial/Wall | 86 |
| Drop, 8 inch, Time | 41, 41-S, 123-B |
| Drop Octagon, Time | 41-K |
| Drop Octagon, 8 inch Strike | 89-B |
| Drop Octagon, 8 inch | 41-B, 41-S |
| Drop Octagon, 10 inch Strike | 89-C, 89-D, 89-F |
| Drop Octagon, 10 inch Time | 41, 41-B, 41-AC, 123-C |
| Drop Octagon, 12 inch Strike | 89-D, 89-F |
| Drop Octagon, 12 inch Time | 41, 41-A, 41-AF, 123-D |
| DuBary | 103-A |
| Duchess | 103-A |
| English Office, Time | 41-D, 41-E, 41-F |
| Exton | 112 |
| Flora | 66-A |
| Gallery 18 inch | 86-F |
| Gallery 24 inch | 86, 86-G |
| Globe, Strike | 44-F, 89-E |
| Globe, Time | 41-C, 123-E |
| Hudson | 86, 86-T |
| Juno | 41-N, 41-O |
| Lobby 14 and 18 inch | 86-A, 86-G |
| Lunar | 82-C |
| Lyre | 103-A |
| Marble Dial No. 0 | 86, 86-AE |
| Marble Dial No. 1 | 86, 86-AB |
| Marble Dial No. 2 | 86-AC |
| Marcy | 88-A |
| N.Y. Store | 40, 40-A, 48-A |
| Office No. 1, Strike | 89-A |
| Office No. 1, Time | 41, 41-AC, 123-A |
| Office No. 6, Strike | 89-E |
| Office No. 6, Time | 41, 41-AN |
| Office No. 7 | 41-AN |
| Office No. 9 | 86 |
| Office No. 10 | 86 |
| Office No. 11 | 86 |
| Office No. 14 | 86 |
| Office No. 16 | 86, 86-W |
| Panama | 34-J, 44-E |
| Philadelphia | 41-H |
| Precise | 5100 |
| Principal 2W | 41 |
| Queen Anne | 41-H, 85 |
| Regulator No. 1 | 62-A |
| Regulator No. 2 | 61, 77, 77-A, 77-B |
| Regulator No. 3, Strike | 63, 63-A |
| Regulator No. 3 | 62 |
| Regulator No. 4 | 64, 77-A, 77-B |
| Regulator No. 5 | 64 |
| Regulator No. 6 | 61 |
| Regulator No. 6/Double Time | 61-B |
| Regulator No. 7 | 62, 65 |
| Regulator No. 8 | 62-C |
| Regulator No. 9 | 61 |
| Regulator No. 10 | 70 |
| Regulator No. 12 | 71 |
| Regulator No. 14 | 70 |
| Regulator No. 15 | 71 |
| Regulator No. 16 | 67 |
| Regulator No. 17 | 62-B |
| Regulator No. 17/Double Time | 62-F |
| Regulator No. 18 | 61-A |
| Regulator No. 19 | 67-A |
| Regulator No. 21 | 77-A |
| Regulator No. 30 | 61 |
| Regulator No. 31 | 68 |
| Regulator No. 32 | 69 |
| Regulator No. 70 | 68 |
| Regulator No. 71 | 68 |
| Regulator No. 71/Sweep | 67 |
| Regulator No. 80 | 68 |
| Schultz Watchman | 2-I |
| Senate 3W | 41 |
| Sonant | 120 |
| Swez | 85-B |
| World | 85-E, 85-J, 86-W |
| World with seconds | 86 |

# KITCHEN

| TYPE/MODEL NAME/ NUMBER | MOVEMENTS USED |
|---|---|
| Alba | 41-S |
| Chef | 10 |
| Clair | 103-A |
| Kitchen | 123 |
| Round Nursery | 41-AM |
| Savarin | 41-AM |
| Square Kitchen | 41-AM |
| Square Nursery | 123 |
| White Lever | 10 |
| Willow | 41-AM |

# LEVER

| TYPE/MODEL NAME/ NUMBER | MOVEMENTS USED |
|---|---|
| Anvill | 92-B |
| Arctist | 95-A |
| Colt | 96-F |
| Engine | 10 |
| Joker | 93, 95, 95-A |
| Metal Lever No. 3 | 112, 112-B |
| Metal Lever No. 4 | 112, 112-A |
| Metal Lever No. 6 | 117 |
| Metal Lever No. 80 | 117 |
| Night Clock | 96-J |
| No. 60 | 117 |
| Nutmeg | 91 |
| Pansy | 96 |
| Paper Weight | 96-A, 96-G |
| Pony Porcelain | 96-D |
| Shetland | 96-B |
| Soldier Boy | 96-I |
| Vulcan | 96-C |
| Wardroom | 7-A |
| White Lever | 10 |
| Wood Lever, 6, 8, 12 inch dial | 10 |

# MANTEL

| TYPE/MODEL NAME/ NUMBER | MOVEMENTS USED |
|---|---|
| Abbey | 51 |

ETM = Electric Time Movement
SSE = Self Starting Electric
SET = Synchronous Electric Time
SSET = Self Starting Electric Time
SSSE = Synchronous Self-Starting Electric
SSSS = Synchronous Self-Starting Strike
SSST = Synchronous Self-Starting Time
SSWC = Synchronous Self-Starting Westminster Chime

| TYPE/MODEL NAME/ NUMBER | MOVEMENTS USED |
|---|---|
| **MANTEL** | |
| Alto | 89 |
| Angelus | 34-E, 44-M |
| Arts | 89-G |
| Avon | 120 |
| Baso | 89 |
| Bedford | 89 |
| Belmont No. 1 | 89 |
| Belmont No. 2 | 89 |
| Belmont No. 5 | 89 |
| Belmont No. 8 | 89 |
| Cambridge | 89 |
| Carson | 45 |
| Cello | 89 |
| Celtic | 51 |
| Century No. 2 | 120-M |
| Century No. 3 | 120-M |
| Century No. 4W | 120 |
| Chairman | 89-A |
| Chairman, Time | 41-AC |
| Chateau | 51, 105 |
| Chevron No. 1W | 5100 |
| Clifton No. 2 | 89 |
| Clifton No. 3 | 89 |
| Colonial | 82-D |
| Concord | 89 |
| Console No. 1 | 120-L |
| Console No. 2 | 120-L |
| Console No. 4W | 120 |
| Console No. 5W | 120 |
| Cottage | 89-A |
| Crafts | 89-G |
| Cymbal 1 | 89 |
| Cymbal 2 | 89 |
| Cymbal 3 | 89 |
| Cymbal 4 | 89 |
| Cymbal 5 | 89 |
| Cymbal 6 | 89 |
| Cymbal 8 | 89 |
| Cymbal 10 | 89 |
| Dana 1 | 89 |
| Dana 3 | 89 |
| Dana 8 | 89 |
| Delos | 5100 |
| Denton | 51 |
| Devotional/Musical | 34-B |
| Dunbar | 120, 120-F |
| Dundee | 120 |
| Eichler | 89-B |
| Essex | 120 |
| Franklin | 82-D |
| Frontenac | 120 |
| Gilt Column | 89 |
| Gloucester | 112 |
| Gramercy | 103-A |
| Hecla | 88 |
| Henley | 89 |
| Hotel | 85-C |
| Kingsbury | 89, 89-A |
| Lafayette | 125 |
| Lawrence | 125 |
| Leader 1 | 89 |
| Leader 2 | 89 |
| Leader 3 | 89 |
| Lexington | 125 |

| TYPE/MODEL NAME/ NUMBER | MOVEMENTS USED |
|---|---|
| **MANTEL** | |
| Lincoln | 66 |
| Locarno | 125 |
| Lowell | 125 |
| Lynn | 125 |
| Majestic 1 | 105 |
| Majestic 2 | 105 |
| Mezzo | 89 |
| Modena | 48-R |
| Navarre | 120 |
| New Shasta | 89 |
| New Sucile | 89 |
| Norfolk | 89 |
| Nymph | 51 |
| O.G./Large | 82-A |
| Parma | 48-R |
| Patrician 3 | 112 |
| Patrician 4 | 112 |
| Peer | 105 |
| Pembroke | 51 |
| Plymouth No. 5573 | 5101 |
| Plymouth No. 5584 | 891-L |
| Plymouth No. 5585 | 891 |
| Plymouth No. 5591 | 891-M |
| Plymouth No. 5592 | 891-L |
| Plymouth No. 5593 | 891-M |
| Plymouth No. 5595 | 891 |
| Plymouth No. 5596 | 1120-J |
| Principal, Strike | 89-A |
| Principal, Time | 41-AC |
| Priscilla | 89 |
| Prospect No. 00 | 89 |
| Prospect No. 2 | 89 |
| Prospect No. 15 | 89 |
| Prospect No. 16 | 89 |
| Prospect No. 17 | 89 |
| Prospect No. 18 | 89 |
| Prospect No. 19 | 89 |
| Putnam | 82-D |
| Regal | 89 |
| Rex | 89 |
| Rideau | 105 |
| Ritz | 120 |
| Rival | 89 |
| Riverside | 103-A |
| Rivoli | 51-K |
| Roy | 89 |
| Salem | 89 |
| Saxo | 89 |
| Sentinel 1 | 89 |
| Sentinel 2 | 89 |
| Sentinel 3 | 89 |
| Sentinel 4 | 89 |
| Sentinel 5 | 89 |
| Sentinel 6 | 89 |
| Sentinel 7 | 89 |
| Sentinel 8 | 89 |
| Sentinel 9 | 89 |
| Sentinel 10 | 89 |
| Sentinel 12 | 89 |
| Sentinel 14 | 89 |
| Sentinel 15 | 89 |
| Sentinel 17 | 89 |
| Sentinel 18 | 89 |
| Sentinel 19 | 891 |

| TYPE/MODEL NAME/ NUMBER | MOVEMENTS USED |
|---|---|
| **MANTEL** | |
| Severn | 120 |
| Sharon | 89, 89-A |
| Summit | 34-F |
| Sutton | 103-A |
| Tambour | 51-L |
| Tambour 1 | 120, 120-F |
| Tambour 2 | 120-F |
| Tambour 3 | 120 |
| Tambour 4 | 51 |
| Tambour 5 | 89, 120 |
| Tambour 6 | 89, 120 |
| Tambour 7 | 89, 120 |
| Tambour 8 | 105 |
| Tambour 9 | 105 |
| Tambour 12 | 120 |
| Tambour 18 | 120 |
| Tambour 19 | 120 |
| Tambour 20 | 120 |
| Tambour 21 | 120 |
| Tempo | 89 |
| Thames | 120-F |
| Tory | 48-Q |
| Touraine | 48-R |
| Trent | 105 |
| Tyne | 51 |
| Wales | 48-R |
| Whitby | 120 |
| Windsor No. 3 | 89 |
| Windsor No. 4 | 89 |
| Wye | 120 |
| York 1 | 89 |
| York 2 | 89 |
| Yorktown | 85-H |
| Zephyr | 120 |
| **SHIP'S BELL** | |
| Abbey | 115 |
| Boat Clock | 7, 7-D |
| Celtic | 115 |
| Chateau | 115 |
| Denton | 115 |
| Helmsman | 115 |
| Mayflower | 115, 115-A |
| Merrimac | 7 |
| Monitor | 7 |
| No. 44 | 115, 115-A |
| No. 44 Wheel | 115 |
| No. 66 | 115, 115-A |
| No. 66 Wheel | 115 |
| No. 660, Hinge Bezel | 115 |
| No. 880/8½ inch | 115 |
| Peer | 115 |
| Rideau | 115 |
| Ship's Bell, One Day | 7 |
| Shipmates | 115 |
| Tambour 8 | 115 |
| Tambour 9 | 115 |
| Tyne | 115 |
| Trent | 115 |
| **SHIP'S** | |
| Argos | 10 |
| Celtic | 105 |
| Chateau | 105 |

| TYPE/MODEL NAME/ NUMBER | MOVEMENTS USED |
|---|---|

## SHIP'S

| | |
|---|---|
| Columbia Ship's Wheel | 103-A |
| Corsair | 112 |
| Denton | 105 |
| Engine Lever | 10 |
| Fine Marine | 117 |
| Merrimac | 7-D |
| No. 3, Time | 112 |
| No. 4, Time | 112 |
| No. 4, Wheel | 112 |
| No. 6, Time | 117 |
| No. 6, Wheel | 117 |
| No. 80 | 117 |
| Rideau | 51, 105 |
| Tambour 8 | 105 |
| Tyne | 51 |

## UNSPECIFIED

| | |
|---|---|
| Andrews | 52 |
| Bank Movement | 101 |
| Delmore | 5100 |
| Delkirk | 5100 |
| Drummer | 34-G |
| English, James Stockwell | 41-J |
| Greenwood, J.W. | 40-B |
| Herald | 41-I |
| Kroeber | 89 |
| Mitchel, Vance & Co. | 47 |
| Paper Winder | 99, 99-A |
| Turn Table Movement shows windows | 98 |
| Precise | 5101 |

# Ca. 1877 SETH THOMAS CLOCK MOVEMENTS DESCRIPTIONS

*To receive top prices, movements must be complete and original, and they must be in good running or repairable condition. Damaged, worn-out or missing parts will reduce prices accordingly.*

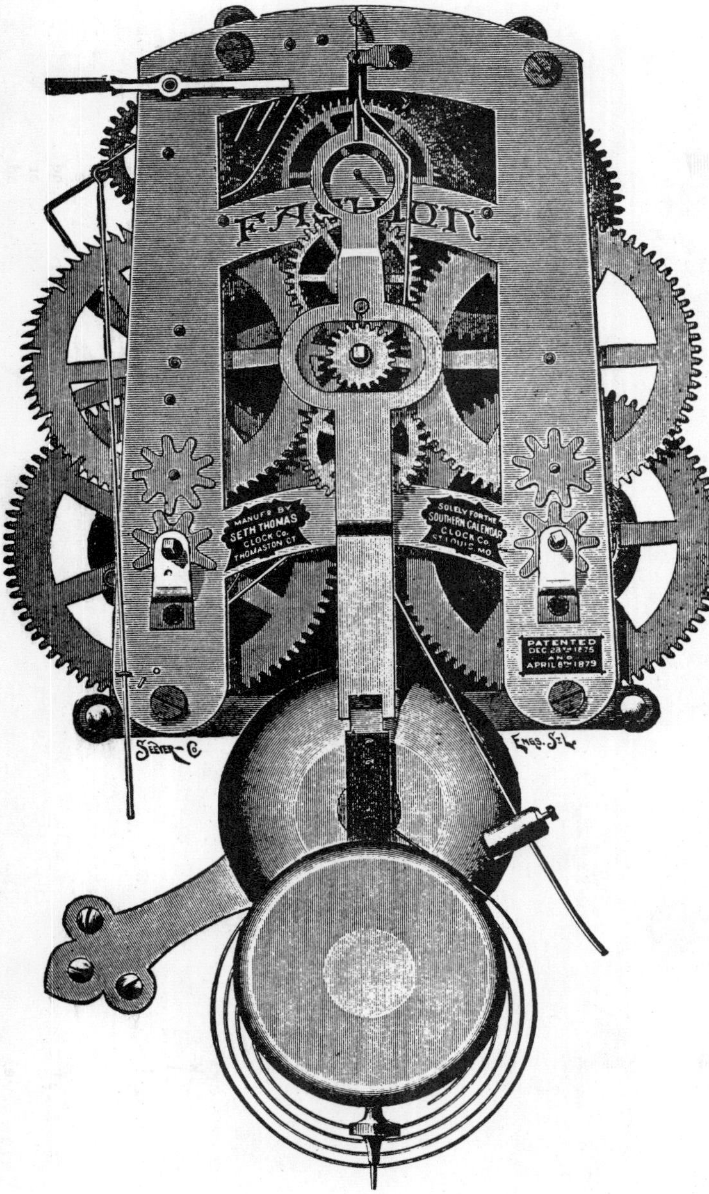


FASHION
Manufactured by Seth Thomas Clock, Co., Thomason, CT.
Solely for the Southern Calendar Clock Co., St. Louis, MO.

**$400.00**

## Directions

The "Fashion" will give the hours, minutes and seconds; month, day of the month, and day of the week; also, 29 days to February in leap year, with perfect accuracy, if the following directions are carefully observed:

First -- It is positively essential that it should be perfectly level, in order to give it a regular beat; that is, the motion of the pendulum must be equal, and the escape as distinct upon one side as upon the other; it must also be plumb, to prevent the pendulum ball from dragging on the rim of lower dial.

Second -- It must be wound either Saturday night or Sunday morning of each week. If allowed to run down it will cause the calendar to change in the day instead of at night, also throws it out of date. If this occurs turn the hands of the time dial forward (in all cases by the minute hand) until the calendar hand changes dates, then hold the minute hand at the figure XII, and turn the hour hand up to the same figure, then set the day and month rolls by placing the point of a knife against them through the openings in the dial and pressing down until the right day of the week and month is indicated, then turn the calendar hand to the day of the month. Now, if it is past noon when starting, turn the hands of the time dial forward by the minute hand 12 hours in addition to the hour of starting. If in the forenoon, turn simply to the hour of the day; this will cause the calendar to change at 12 o'clock at night, instead of 12 o'clock in the day.

Third -- If the calendar hand should miss a date, it may be turned backward or forward, but never backward after the 28th day of the month, and in no case further than to the 1st day.

Fourth -- As is well known by all, leap year comes once in four years, and gives to February 29 days. If the calendar should give 29 days to February before 1884 it has not been set right in starting, and can be easily corrected. If it gives 29 days to February, 1881, turn the month roll backward to March three times (which is three years); if in 1882, twice (which is two years); if in 1883, once (which is one year). This may be done by placing the point of a knife on the month roll and pressing down; then turn the calendar hand to the proper date; this will give to February in 1884 (which is a leap year) 29 days.

Fifth -- To regulate, carefully remove the pendulum ball from the rod; if it runs slow, screw up the tap at the bottom; if fast, lower it.

Sixth -- Never allow the children to wind or interfere with it; let one person attend to winding and keep the key.

Seventh -- To clean the case use a piece of soft flannel with a little linseed oil.

# Ca. 1877 SETH THOMAS CLOCK MOVEMENTS DESCRIPTIONS

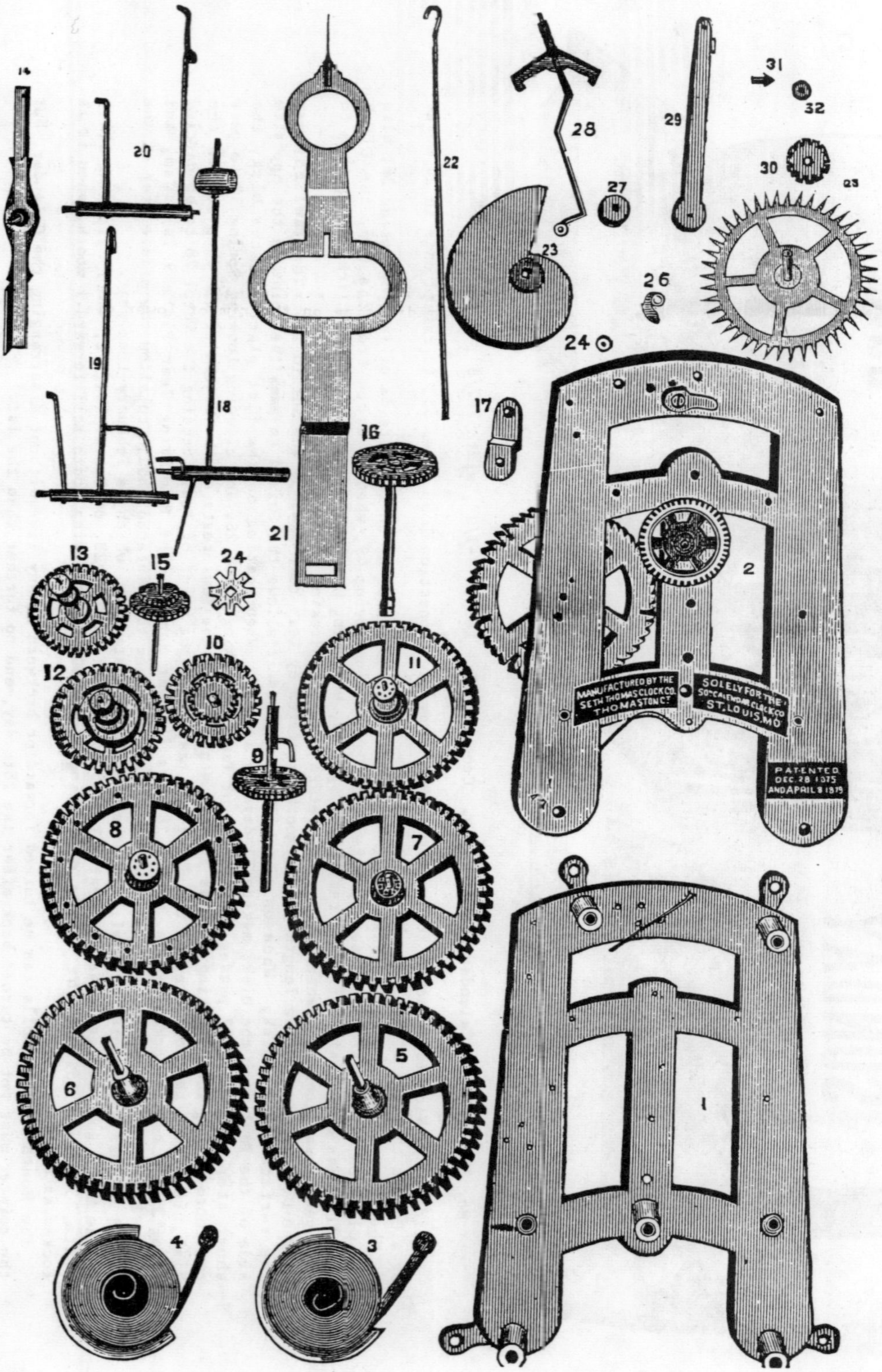


SECTIONAL PARTS OF TIME MOVEMENT.

## Specification of Parts of Time Movement.

1-Back Plate, with Pillow; Brackets for fastening in Case; Connection Guide Wire; Connection Bracket, riveted on; Bushings; Turn Back Spring.
2-Front Plate on which is Escapement Button and Screw; Count Wheel and Stud and Spring; Dial Wheel; Pinion Stud and Washers; Bushings; Pendulum Stud.
3-Main Time Spring.
4-Main Strike Spring.
5-Main Time Arbor, with Wheel Bracket and Hub.
6-Main Strike Arbor, with Wheel Bracket and Hub.
7-Second Time Wheel on First Time Pinion.
8-Second Strike Wheel, containing 12 Pins on First Strike Pinion.
9-Center Shaft, with Wheel.
10-Socket with Gear and Wheel.
11-Third Time Wheel on Second Time Pinion.
12-Third Strike Wheel and Cam on Second Strike Pinion.
13-Fourth Strike Wheel on Third Strike Pinion.
14-Fly Pinion and Fly.
15-Intermediate Shaft with two Wheels.
16-Cam Shaft with Wheel.
17-Ratchet Bridges (two in each movement).
18-Lockwork, Hammer Shaft.
19-Lockwork, Lift Shaft.
20-Lockwork, Count Shaft.
21-Pendulum Reel.
22-Strike Wire.
23-Calendar Cam with square hole Bushing.
24-Ratchet, 4 in each movement.
25-26--Crown Wheel and Click.
27-Pillow Screws (5 in each movement).
28-Verge with Shaft and Crutchwire.
29-Calendar Connection Lever.
30-Center Shaft Gear.
31-Calendar Connection Rivet.
32-Calendar Connection Washer.

# Ca. 1877 SETH THOMAS CLOCK MOVEMENTS DESCRIPTIONS

**$350.00**

**No. 36. -- Small Calendar, white rolls. Center 1/4". Length 3-1/2", width 7-7/8", depth 8-3/17".**

Slide 29, which is attached to Connecting Rod 61 by the Connection Loop Screw 58, is raised once in 24 hours by a Cam at the back of the Time Movement.

When the Slide rises, the Passing Pin on Click Plate 35 passes up on the outside of Passing Spring 38; also the Click on Click Plate 34 passes up outside of Day Locking Spring 49, when it drops from the Cam, the Click and Pin pass down inside of their respective Springs, unlocking Wheels 26 and 28, when the Clicks cause them to be moved forward one day, when the Springs return to place, locking the wheels as before.

The Disk 24 governs the length of the month, assisted by Lea~ Year Pawl 39 and Shield 17. The Pawl follows the surface of the Disk. When on the highest points it allows the Shield to pass the point on the opposite side of the Pawl from the Disk, making 31 days in a month. When resting on the flat places of the Disk, the short step of Shield 17 passes under point of Pawl, unlocking Wheel 28, and allowing Locking Spring 32 to pass one tooth in the wheel, making 30 days in that month. When the Pawl rests on the bottom of the deep cut in the Disk, it allows the long step of the Shield to pass under the point of Pawl, causing the Wheel 28 to be carried along 3 days before locking again, making 28 days in that month. The Revolving Gear Plate 9 is a Cam, and comes before the deep cut once in four years, and shortens the distance enough to allow the middle step of the Shield to pass under point of Pawl, and make 29 days in that month, which is February-Leap year.

The Month Roll Changes when the Center Lift 51 drops off the Cam 23, on Center Shaft 2, when Click on Click Plate 33 passes down on the inside of Locking Spring 50, and carrying Month Roll forward 1 month, when it is locked as before.

The Month and Day Rolls can be turned forward or backward at any time without disarranging the Calendar, but the pointer must not be turned back after the 28th day, and no further than the 1st.

*To receive top prices, movements must be complete and original, and they must be in good running or repairable condition. Damaged, worn-out or missing parts will reduce prices accordingly.*

# Ca. 1877 SETH THOMAS CLOCK MOVEMENTS DESCRIPTIONS

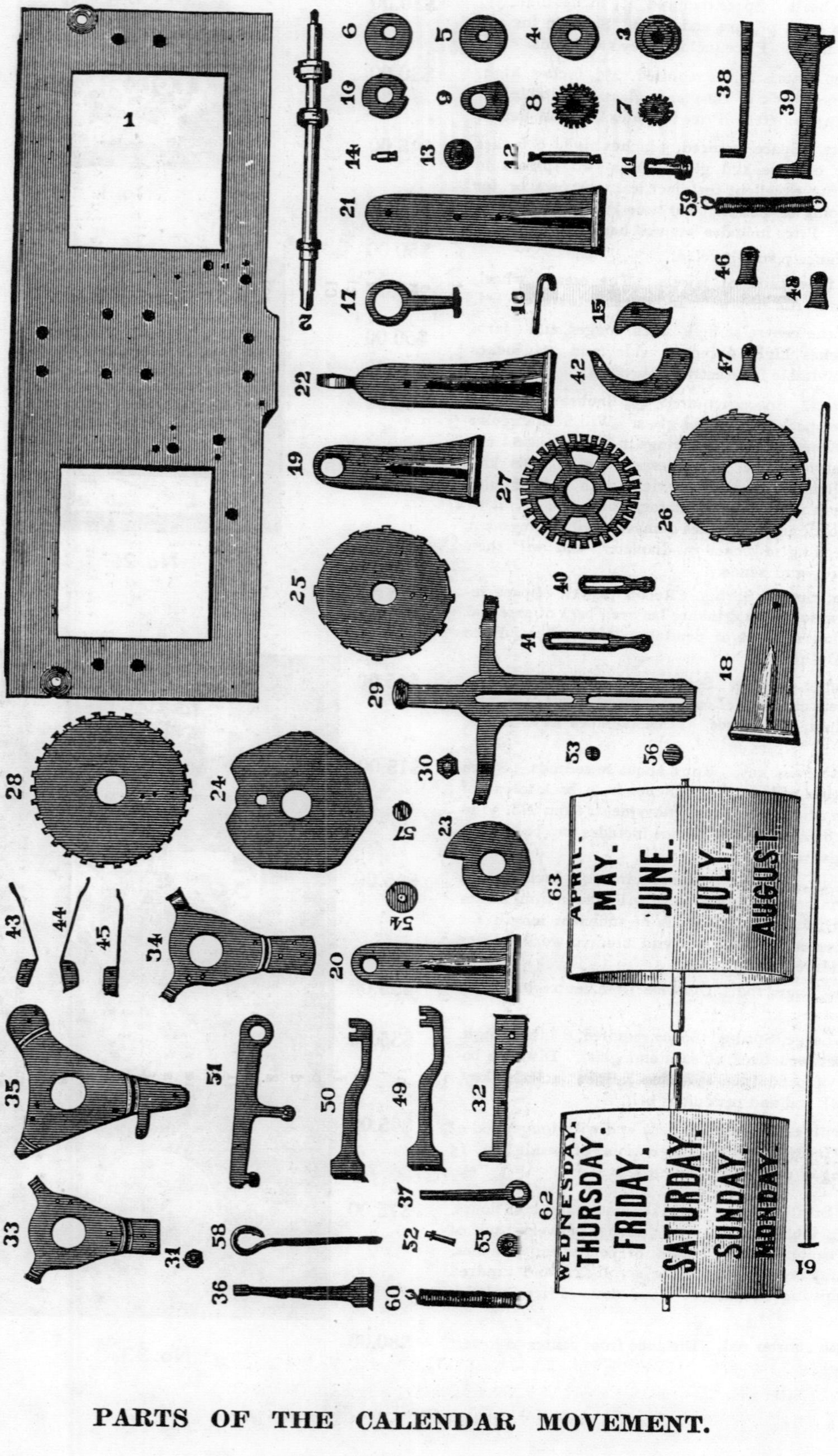


PARTS OF THE CALENDAR MOVEMENT.

**Specification of Parts of Calendar Movement,**

1-Plate.
2-Center Shaft.
3-Disk Hub.
4-Month Wheel Hub.
5-Day Wheel Hub.
6-Center Wheel Hub.
7-Stationary Gear.
8-Revolving Gear.
9-Revolving Gear Plate.
10-Safety Ratchet.
11-Center Lift Hub.
12-Center Lift Hub Stud.
13-Slide Bushing.
14-Connection Loop Pin.
15-Safety Click.
16-Staples for Spiral Springs.
17-Shield.
18-Outside Roll Standards (2 in each Movement).
19-Stationary Gear Standards.
20-Front Center Standards.
21-Back Center Standards.
22-Center Lift and Day Roll Standards.
23-Cam for Center Shaft.
24-Disk.
25-Month Wheel (12 Teeth).
26-Day Wheel (14 Teeth).
27-Friction Center Wheel.
28-Center Wheel (31 Teeth).
29-Slide.
30-Standard Rivets (6 in each Movement).
31-Revolving Gear Rivet.
32-Center Locking Spring.
33-Month Click Plate.
34-Day Click Plate.
35-Center Click Plate.
36-Detent Spring.
37-Leap Year Pawl Spring.
38-Passing Spring.
39-Leap Year Pawl.
40-Center Lift Loop.
41-Slide Loop (2 in each Movement).
42-Friction Springs (3 in each Movement).
43-Month Click Spring.
44-Day Click Spring.
45-Center Click Spring.
46-Day Click.
47-Month Click.
48-Center Click.
49-Day Locking Spring.
50-Month Locking Spring.
51-Center Lift.
52-Center Lift Pin.
53-Disk Hub Screw.
54-Connection Check Nut.
55-Slide Rivets (2 in each movement).
56-Locking Spring Screws (5 in each movement).
57-Standard Screws (6 in each Movement).
58-Connection Loop Screw.
59-Day & Month Spiral Spring (2 in each Movement).
60-Center Spiral Spring.
61-Connection Rod.
62-Day Roll.
63-Month Roll.

# 1901 SETH THOMAS CLOCK MOVEMENT DESCRIPTIONS

**No. 1.**—1 Day, Lever, Time, Small. 240 beats. Space required, 2⅞ inches high, 3⅛ inches wide, and 2 inches between back of case and glass. Suitable for use with dials less than 6 inches in diameter. Price includes key and hands. **$30.00**

**No. 2.**—1 Day, Lever, Time, Large. 240 beats. Space required, 4⅜ inches high, 5 inches wide, and 2½ inches between back of case and glass. Suitable for use with dials from 6 to 12 inches in diameter. Price includes key and hands. **$30.00**

**No. 10.**—8 Day, Lever, Time. 240 beats. Space required, 5 inches high, 6 inches wide, and 2⅛ inches between back of case and glass. Has two springs in barrels; is a substantial Movement; excellent for electrical purposes or for registering devices, having a surplus of power. For time-keeping, is not as good as a 1 Day Lever Movement. Price includes key and hands. **$45.00**

**No. 10 A.**—Same as No. 10, but has balance pivots jeweled. **$50.00**

**No. 10 B.**—Same as No. 10, but with four pairs of jewels. The escape wheel, pallets and balance are shortened and run in jewels, and the pallets are jeweled. **$65.00**

**No. 10 H.**—Same as No. 10 A, but has the centre staff ¼ inch longer and alarm dial fitted. Space required, 5 inches high, 6 inches wide, and 2¾ inches between back of case and glass. Suitable for electrical purposes. **$60.00**

**No. 22.**—8 Day, Lever, Time. 300 beats. Space required, 4⅛ inches high, 4⅛ inches wide, and 2⅛ inches between back of case and glass. Will fit in circular case having inside diameter of 5 inches. Has two springs in barrels; steel cut pinions—tempered, and chronometer balance. The escape wheel, pallets and balance run in jewels and the pallets are jeweled—which, with the impulse pin, makes nine jewels in all. Has an excellent regulator, which will admit of close adjustment. All parts polished, and the plates damaskeened. Movement should not be used for a dial exceeding 12 inches in diameter, and will then keep good time. Price includes key and hands. **$60.00**

**No. 30 A.**—1 Day, Pendulum, Time, Small, Spring. Recoil pallets. Space required, 3 inches high, 3⅛ inches wide, and 2⅛ inches between back of case and glass. Distance from center to lowest point of pendulum 5⅞ inches. Price includes key, hands, rod and pendulum ball. **$40.00**

**No. 33.**—1 Day, Pendulum, Strike, Large, Spring. Space required, 7 inches high, 5⅛ inches wide, and 2⅞ inches between back of case and glass. Distance from center to lowest point of pendulum, 8⅛ inches. Price includes key, hands, rod and pendulum ball. **$45.00**

**No. 35 C.**—Detached Alarm part, with fixed key. Runs about 20 seconds. Space required, 2⅛ inches high, 1⅛ inches wide, and 2⅛ inches from back to end of key. Can be applied to any one of the pendulum movements from No. 30 to No. 44 E inclusive; also No. 80, 82 A and 89 D. Price includes cup bell, brass alarm disk, latch and connecting wire. **$15.00**

**No. 40.**—8 Day, Pendulum, Small, Time, Spring. Space required, 4¼ inches high, 3 inches wide, and 2⅞ inches from back of case to glass. Distance from center to lowest point of pendulum, 6¾ inches. Center is of sufficient length for dial ⅛ inch thick. Specially adapted to scroll-saw and kindred work. Price includes key, hands, rod and pendulum ball. **$45.00**

**No. 40 A.**—Same as No. 40, but with longer rod. Distance from center to lowest point of pendulum, 10¼ inches. **$55.00**

**No. 41 A.**—8 Day, Pendulum, Time, Large, Spring. Space required, 6 inches high, 5⅛ inches wide, and 4 inches between back of case and glass. Distance between center and lowest point of pendulum, 13 inches. Price includes key, hands (not larger than 12 inches), rod and pendulum ball. **$55.00**

**No. 41 H.**—Same as No. 41 A, but with escapement on back and with longer rod of wood and 1 lb. ball. Distance from center to lowest point of pendulum, 15 inches. Space required for swing of pendulum, 8 inches. **$95.00**

**No. 42.**—8 Day, Pendulum, Strike, Small, Spring. Strikes hours and half hours. Space required 8 inches high, 4¾ inches wide, and 3⅛ inches between back of case and glass. Distance from center to lowest point of pendulum, 9 inches. Adapted for cup bell or cathedral bell. Suitable for scroll-saw and kindred work, having long center, permitting the use of wood dial. Price includes key, hands, rod and ball. **$75.00**

**No. 42 A.**—Same as No. 42, but with shorter rod. Distance from center to lowest point of pendulum, 6½ inches. **$80.00**

No. 1.

No. 2.

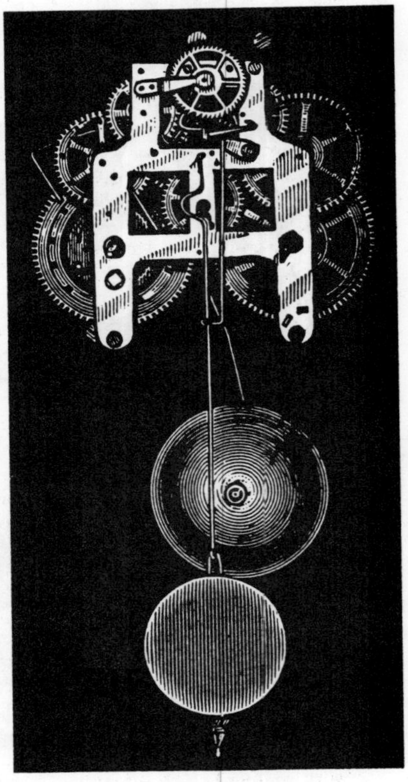

No. 33.

# 1901 SETH THOMAS CLOCK MOVEMENT DESCRIPTIONS

*To receive top prices, movements must be complete and original, and they must be in good running or repairable condition. Damaged, worn-out or missing parts will reduce prices accordingly.*

**No. 44 E.**—8 Day, Pendulum, Strike, Spring. Strikes the hours and half hours. Suitable for cathedral bell. Space required, 10 inches high, 6½ inches wide, and 3 inches between back of case and glass. Distance from center of movement to lowest point of pendulum, 13⅜ inches. Space required for swing of pendulum, 7 inches. The lower section of the pendulum rod is of wood, with a 10 ounce nickel plated ball. Price includes key, hands, rod and ball.

**$85.00**

**No. 48.**—8 Day, Pendulum, Strike, Spring, Visible Escapement. Has steel cut pinions, and strikes the hour and half hour on cathedral bell. Suitable for marble or porcelain cases. Fitted to sash for 5 inch opening only. Space required, from front of case outside to back of case inside, 4¼ inches. Distance from center to lowest point of pendulum, 5¼ inches. Price includes cathedral bell and stand, ornamented front sash, with beveled glass, porcelain dial, back sash, straps, face plate, key and pendulum.

**$90.00**

**No. 48 A.**—Similar to No. 48, but with plain dead-beat escapement, lantern pinions and shorter pendulum, fitted to sash for 4½ inch opening only. Space required from front of case outside to back of case inside, 4¼ inches. Distance from center to lowest point of pendulum, 4⅜ inches. Price includes cathedral bell and stand, ornamented front sash with beveled glass, porcelain dial, back sash, straps, face plate, key and pendulum.

**$80.00**

**No. 51.**—15 Day, Pendulum, Strike. A fine polished Movement, with springs in barrels and steel cut pinions. Strikes the hours and half hours on cathedral bell. Fitted to sashes for 4, 4½ and 5 inch openings. Space required from front of case outside to back of case inside, 4 inches. Distance from center to lowest point of pendulum, 4 inches. Price includes 3 inch cathedral bell and stand, ornamented front sash with beveled glass, porcelain dial, back sash, straps, face plate, key and pendulum.

**$90.00**

**No. 51 M.**—15 Day, Pendulum, Strike. A fine polished movement with springs in barrels and steel cut pinions. Strikes the hours and half hours on cup bell. Same measurements as No. 51. Front plate has holes fitted with screws so that it can be used in old style cases which had zinc plates fitted in the front for that purpose. This movement can be fitted with hammer for cathedral bell, if desired. Price includes key, pendulum, hands and cup bell, or cathedral bell and stand in place of the cup bell.

**$90.00**

**No. 61.**—8 Day, Weight, Time, 80 beats. Space required 4⅛ inches high, 4¾ inches wide, and 4⅛ inches between back of case and glass. Distance from center to lowest point of pendulum, 22½ inches. Space for swing of pendulum, 8 inches. Movement and pendulum are hung on an iron bracket. Has Graham dead-beat escapement, lantern pinions, maintaining power, wood rod, and 2 lb. brass ball. Price includes key, 12 inch hands, weight, rod and ball.

**$200.00**

**No. 61 A.**—60 beats. Same as No. 61, but with seconds pendulum and 6 lb. ball of 8 inches diameter. Distance between center and lowest point of pendulum 41¾ inches. Space for swing of pendulum, 12 inches. Price includes key, 12 or 14 inch hands, weight, rod and ball.

**$275.00**

**No. 68.**—8 Day, Weight, Time, 60 beats. Space required, 10¼ inches high, 8¼ inches wide, and 5⅞ inches between back of case and glass. Has polished plates and cut steel pinions, Graham pallets, maintaining power, and wood rod with 10 lb. ball 8 inches in diameter. The Movement rests on a heavy iron bracket, from which the pendulum is also suspended. Distance from center to lowest point of pendulum, 40½ inches. Space for swing of pendulum, 12 inches. Price includes key, weight, rod and ball. This Movement is made with and without dial feet, the latter used when it is intended to have dial fastened to Movement.

**$375.00**

**No. 69.**—8 Day, Weight, Time, 60 beats. Similar to No. 68, but with lacquered plates and lantern pinions with 6 lb. ball 8 inches in diameter. Space for swing of pendulum, 12 inches.

**$300.00**

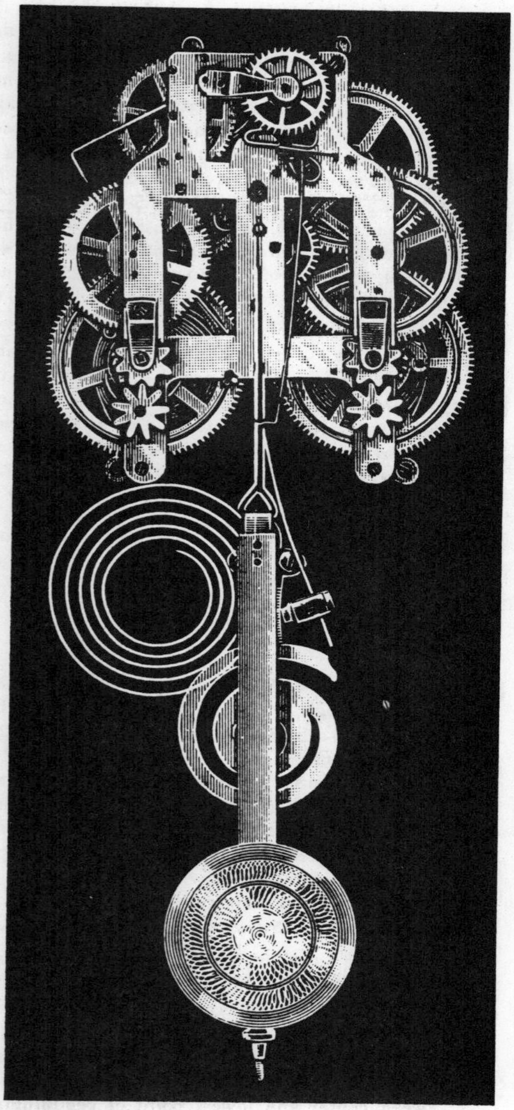

No. 44 E.

# 1901 SETH THOMAS CLOCK MOVEMENT DESCRIPTIONS

**No. 72 A.**—8 Day, Weight, striking hours only. A good Movement for hall clocks. Has Graham escapement and lantern pinions, also maintaining power. Plates are damaskeened. Movement rests on an iron bracket from which the pendulum is also suspended. Supplied with long brackets or short brackets. Space required, 10 inches high, 10 inches wide; and distance between back of case and glass with long bracket, 8¾ inches; with short bracket, 6¾ inches. Distance from centre to lowest point of pendulum, 41 inches. Distance required for drop of weight from center of dial to bottom of weights, 56 inches. Four posts project from front plate for support of dial. The rod is of wood with brass-covered ball weighing 4 lbs. and measuring 6 inches in diameter. Weights are brass shells 10 inches long and 2¼ inches in diameter, filled with lead, and weigh respectively, strike side, 11½ lbs.; time side, 5½ lbs. Space for swing of pendulum, 11½ inches; or, if a 4 inch ball be used, 9½ inches. Price includes rod, ball, weights, key, 12 inch hands, 6 inch cathedral bell and stand. When ordering, state if wanted with long bracket or with short bracket, also size of pendulum ball. Dials suitable, No. 72 A, No. 373 and No. 547. Can be fitted with seat board, 13 3-16 or 15 inches long, in place of bracket, if desired.

**$250.00 w/Bracket**

**No. 72 A**

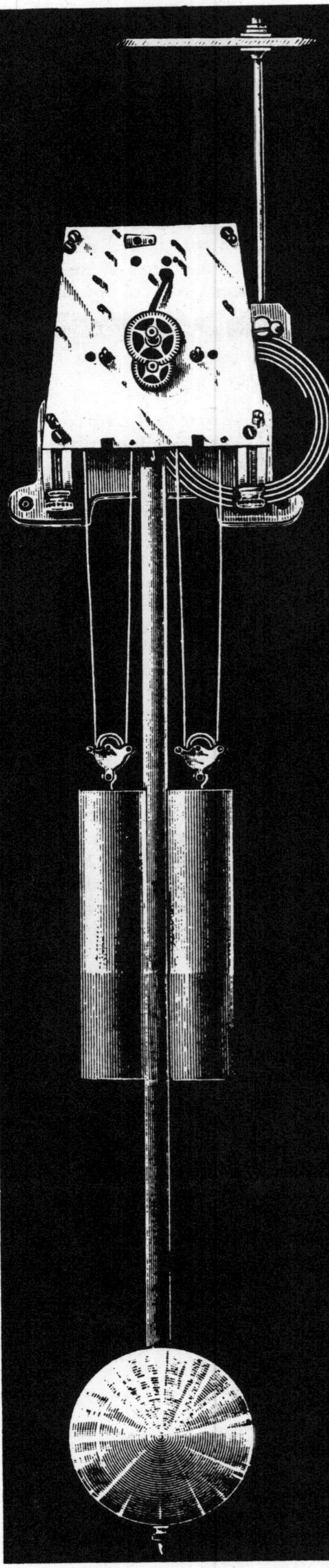

**No. 72 B.**—Same as No. 72 A, but fitted with attachments for moon. Dials suitable, No. 558, No. 507 and No. 514. Can be fitted with seat-board, 13 3-16 inches or 15 inches long, in place of bracket if desired.

**$250.00 w/Bracket**

**No. 72 C.**—Same as No. 72 A, but fitted with moon disk 8⅝ inches in diameter, showing phases of the moon, and without seconds. Dial suitable, No. 405. Can be fitted with seat-board 13 3-16 inches or 15 inches long, in place of bracket, if desired.

**$275.00 w/Bracket**

**No. 74 B**—8 Day, Weight, Quarter Strike, Westminster Chimes. Space required, 9½ inches high, 9 inches wide, and 8¾ inches between back of case and glass. Space above center for bells, 8½ inches. Width required for swing of pendulum, 11½ inches. Distance from center for drop of weights 58 inches. Distance from center to lowest point of pendulum 40½ inches. The Movement rests on a heavy iron bracket, from which the pendulum is also suspended. The plates are damaskeened. The chimes are struck on four cup bells, and the hours on cathedral bell. Has Graham escapement, lantern pinions, wood rod and maintaining power, without seconds. The ball is of zinc covered with brass, 6 inches in diameter, and weighs 4 lbs. In fitting these Movements in old cases with narrow bodies, the required space for swing of pendulum can be reduced to 9½ inches by using a 4 inch 2 lb. ball instead of the 6 inch 4 lb. ball. The weights are brass shells of a uniform length of 10 inches and diameter of 2¼ inches, filled with lead, and weigh respectively, counting them from left to right: 1st Strike, 11½ lbs.; 2nd Time, 5½ lbs.; 3d Chime, 15 lbs. The moon disk is 8¾ inches in diameter. Price includes the Movement with bells, moon attachment, weights, 10 inch hands, key, rod and ball. Dial suitable, No. 74 B. Can be fitted with seat-board 13 3-16 inches or 15 inches long, in place of bracket, if desired.

**$375.00 w/Bracket**

**No. 74 C.**—Same as No. 74 B, but without moon disk. Dials suitable, No. 74 C No. 374 and No. 562. Can be fitted with seat-board 13 3-16 inches or 15 inches long, in place of bracket, if desired.

**$350.00 w/Bracket & Bell**

**No. 74 D.**—Same as No. 74 C, but with attachments for moon. Dials suitable, No. 599, No. 523 and No. 536. Can be fitted with seat-board, 13 3-16 inches or 15 inches long, in place of bracket, if desired.

**$350.00 w/Bracket & Bell**

# 1901 SETH THOMAS CLOCK MOVEMENT DESCRIPTIONS

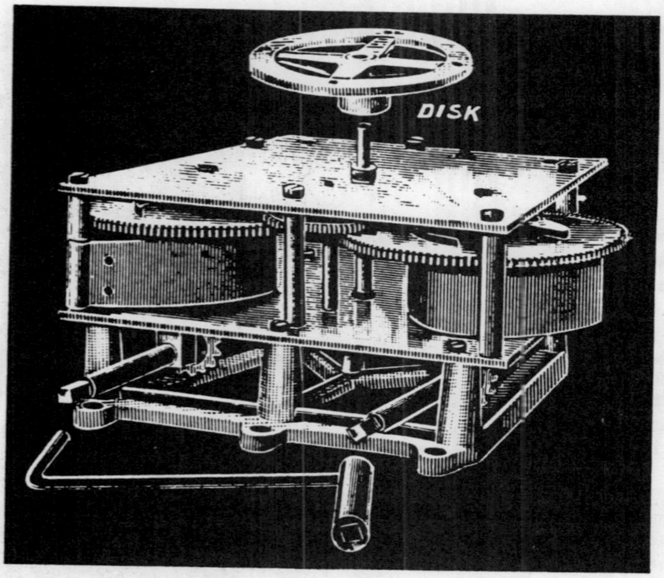


No. 75.

*To receive top prices, movements must be complete and original, and they must be in good running or repairable condition. Damaged, worn-out or missing parts will reduce prices accordingly.*

**No. 75.**—Motor Movement. An extra heavy train of six wheels, driven by two springs, each 1 5-16 inches wide, 18 feet long, 23 guage, ending with a fly. Intended for turning a table in show window, or for any other purpose for which a large spring motor is required. Table and contents should not weigh more than 50 lbs. well balanced. Will run 7 hours with one turn of table per minute. Size: 10 ¼ inches high, 10 inches wide, 5 ¼ inches deep. **$75.00**

**Disk.**—For use with No. 75 when required. A circular iron frame, 4 inches in diameter, for mounting on center shaft, to which a table may be secured. **$25.00**

**No. 80.**—1 Day, Strike, Weight. Movement on seat-board 6 ¾ inches long. Space required, 9 inches high, 5 inches wide, and 3¼ inches between back of case and glass. Required drop for weights, 24 inches. Distance from center to lowest point of pendulum, 14½ inches. Price includes seat-board. key, hands, rod and ball. Weights are extra; time side, 3 lbs. 4 oz; strike side, 3 lbs. **$50.00**

**No. 82 A.**—Not illustrated, but similar in appearance to No. 80. Large 8 Day, Strike, Weight. Movement on seat board 7 inches long Space required, 12 inches high, 7 inches wide, and 3¾ inches between back of case and glass. Required drop for weights, 28 inches. Distance from center to lowest point of pendulum, 19¼ inches. Price includes key, hands, rod and ball. Weights are extra; time side, 8 lbs. 10 oz.; strike side, 7 lbs. 10 oz. **$110.00**

**No. 84.**—8 Day, Weight, Quarter Strike, Westminster Chimes; with disk showing moon phases. Space required, 8 inches high, 8 inches wide, and 11¾ inches between back of case and glass. Space required for swing of pendulum 11¾ inches. Distance from center for drop of weights, 71 inches. Distance from center to lowest point of pendulum, 42 inches. The Movement rests on a seat board 15 inches long. The pendulum is suspended from a stud screwed to the back plate of Movement. The chimes are struck on four cathedral bells, and the hours on a large cathedral bell. Can be made silent or chime at will. Has Graham escapement, steel cut pinions tempered and maintaining power. The lower arms of the hour and quarter racks are provided with springs, which prevent any straining or bending of those parts in case the hands should be turned backward. The crutch wire is also provided with beat screws, by which the beat can be adjusted with precision. All of the parts are highly finished, and the plates are damaskeened. The rod is of soft steel. The ball is of zinc covered with brass, 5¾ inches in diameter, and weighs 3¼ lbs. The weights are brass shells filled with lead, and weigh respectively, counting them from left to right: 1st Hour, 20 lbs.; 2nd Time, 14¼ lbs.; 3d chime, 32¼ lbs. Price includes the movement with all bells with their stands, seat-boards, weights, key, hands, rod and ball, also No. 84 H Dial and moon disk. This Movement is equal in every detail to the finest of English makes, and out-classes those of German manufacture. **$700.00 Complete W/Dial & Gongs**

# 1901 SETH THOMAS CLOCK MOVEMENT DESCRIPTIONS

**No. 85 A.**—8 Day, Pendulum, Strike, Spring, 60 beats. Is as large as and similar in appearance to, Movement No. 86. Strikes hours only. Suitable for cathedral bell. Space required, 12 inches high, 7¼ inches wide, and 4 inches between back of case and glass. Distance from center to lowest point of pendulum, 39¼ inches. Space required for swing of pendulum, 10 inches. Graham pallets and lantern pinions. Movement and pendulum are hung on an iron bracket. Can supply suitable Dial, like No. 373 and No. 547. Price includes key, 12 inch hands, wood rod and 2 lb. ball. **$150.00**

**No. 85 E.**—80 beats. Same as No. 85 A, but with shorter center and shorter rod. Distance from center to lowest point of pendulum, 22 inches. Space required between back of case and glass, 3¾ inches. Space required for swing of pendulum, 8 inches. **$150.00**

**No. 86.**—15 Day, Pendulum, Time, Duplex, Spring, 80 beats. Space required, 8 inches high, 7 inches wide, and 4 inches between back of case and glass. Distance from center to lowest point of pendulum, 21⅛ inches. Space required for swing of pendulum, 8 inches. Movement and rod are hung on an iron bracket. Has Graham pallets and lantern pinions. Price includes key, 12 inch hands, rod and ball. **$100.00**

**No. 86 C.**—15 Day, Pendulum, Time, 60 beats (seconds). Same as No. 86, but with longer rod and heavier ball. Distance from center to lowest point of pendulum, 39 inches. Space required for swing of pendulum, 10 inches. **$150.00**

No. 86 F.

**No. 86 F.**—15 Day, Pendulum, Time, 120 beats (half seconds). Same as No. 86, but with shorter rod and longer center. Space required, 8 inches high, 7 inches wide, and 4⅛ inches between back of case and glass. Distance from center to lowest point of pendulum, 9 inches. Space required for swing of pendulum, 6 inches. **$75.00**

**No. 86 G.**—15 Day, Pendulum, Time, 100 beats. Same as No. 86 F, but with longer pendulum. Distance from center to lowest point of pendulum, 13½ inches. Space required for swing of pendulum 7 inches. **$75.00**

**No. 89 C.**—8 Day, Pendulum, Strike, Spring. Strikes the hours and half-hours. Suitable for cathedral bell. The brackets for securing Movement to case are attached to the front plate. The escapement and pendulum are on the back of the Movement. Space required, 8 inches high, 6 inches wide, and 4¾ inches between back of case and glass. Distance from center of Movement to lowest point of pendulum, 5 inches. Price includes key, hands, rod and ball. **$70.00**

**No. 89 D.**—Same as No. 89 C, but with longer rod. The escapement and pendulum are on the front of Movement. The brackets are on the back plate. Space required, 8 inches high, 6 inches wide, and 2¾ inches between back of case and glass. Distance from center to lowest point of pendulum, 9¾ inches. Price includes key, hands, rod and ball. **$80.00**

No. 89 D.

**No. 98.**—Turn-table Movement. A heavy train of four wheels driven by a spring 1 inch wide, 10 feet long, 25 gauge, ending with a fly. Intended for turning a small table in show windows. Table and contents should not weigh more than 8 lbs., well balanced. Will run 10 hours with one turn of table per minute. Size, 6½ inches high, 4¾ wide, and 3 inches deep. **$60.00**

**Zinc and Steel Pendulum.**—The compensation pendulum differs from the usual forms, and is constructed on scientific principles. The central steel rod has the rating nut on the lower end at the bottom of the pendulum. The lower end of the zinc tube is supported by a brass collar which rests on the rating nut; to the upper end is secured a brass cross piece from which the two outside steel rods are suspended. To the lower end of these rods the bob is secured. The bob being suspended from its center to the steel rod, leaves it practically out of the compensation, thus avoiding the irregularity caused by the changes in temperature, which act more slowly upon a heavy bob than on the lighter tube and rod. The pendulum measures from the top of suspension to its extreme end, 47¼ inches and weighs 16 lbs. It can be used with movements 68, 69, 72 A, 72 B, 74 C, and 74 D. Space required for swing of pendulum, 11 inches. **$185.00**

When making a Pendulum, bear in mind that while the weight of the Ball may be varied with impunity, the distance from the point of suspension of the pendulum to the center of gravity in the Ball is fixed, and cannot be changed.

# Ca. 1904 SETH THOMAS CLOCK MOVEMENT DESCRIPTIONS

**No. 1.**—1 Day, Lever, Time, Small. 240 beats. Space required, 2⅞ inches high, 3½ inches wide, and 2 inches between back of case and glass. Suitable for use with dials less than 6 inches in diameter. Price includes key and hands. $30.00

No. 1.

**No. 2.**—1 Day, Lever, Time, Large. 240 beats. Space required, 4⅜ inches high, 5 inches wide, and 2½ inches between back of case and glass. Suitable for use with dials from 6 to 12 inches in diameter. Price includes key and hands.

**No. 10.**—8 Day, Lever, Time. 240 beats. Space required, 5 inches high, 6 inches wide, and 2½ inches between back of case and glass. Has two springs in barrels; is a substantial Movement; excellent for electrical purposes or for registering devices, having a surplus of power. For time-keeping, is not as good as a 1 Day Lever Movement. Price includes key and hands. $45.00

**No. 10 A.**—Same as No. 10, but has balance run in jewels. $50.00

**No. 10 B.**—Same as No. 10, but with four pairs of jewels. The escape wheel, pallets and balance are shortened and run in jewels, and the pallets are jeweled. $65.00

**No. 10 H.**—Same as No. 10 A, but has the center staff ¼ inch longer and alarm dial fitted. Space required, 5 inches high, 6 inches wide and 2¾ inches between back of case and glass. Suitable for electrical purposes. $60.00

No. 2.

**No. 22.**—8 Day, Lever, Time. 300 beats. Space required, 4½ inches high, 4⅛ inches wide, and 2½ inches between back of case and glass. Will fit in circular case having inside diameter of 5 inches. Has two springs in barrels; steel cut pinions—tempered, and chronometer balance. The escape wheel, pallets and balance run in jewels and the pallets are jeweled—which, with the impulse pin, makes eleven jewels in all. Has an excellent regulator, which will admit of close adjustment. All parts polished, and the plates damaskeened. Movement should not be used for a dial exceeding 12 inches in diameter, and will then keep good time. Price includes key and hands. $60.00

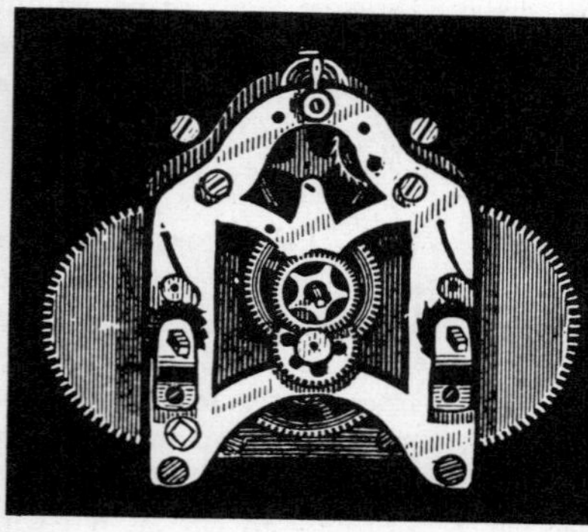

No. 10.

**No. 30 A.**—1 Day, Pendulum, Time, Small, Spring. Recoil pallets. Space required, 3 inches high, 3½ inches wide, and 2½ inches between back of case and glass. Distance from center to lowest point of pendulum 5⅞ inches. Price includes key, hands, rod and pendulum ball. $40.00

**No. 33.**—1 Day, Pendulum, Strike, Large, Spring. Space required, 7 inches high, 5½ inches wide, 2⅞ inches between back of case and glass. Distance from center to lowest point of pendulum, 8½ inches. Price includes key, hands, rod and pendulum ball. $45.00

**No. 35 C.**—Detached Alarm part, with fixed key. Runs about 20 seconds. Space required, 2½ inches high, 1½ inches wide and 2½ inches from back to end of key. Can be applied to any one of the pendulum movements from No. 30 to No. 44 E inclusive; also No. 80, 82 A and 89 D. Price includes cup bell, brass alarm disk, latch and connecting wire. $15.00

No. 22.

**No. 40.**—8 Day, Pendulum, Small, Time, Spring. Space required, 4¼ inches high, 3 inches wide and 2⅞ inches from back of case to glass. Distance from center to lowest point of pendulum, 6¾ inches. Center is of sufficient length for dial ½ inch thick. Specially adapted to scroll-saw and kindred work. Price includes key, hands, rod and pendulum ball. $45.00

**No. 40 A.**—Same as No. 40, but with longer rod. Distance from center to lowest point of pendulum, 10¼ inches. $55.00

# Ca. 1904 SETH THOMAS CLOCK MOVEMENT DESCRIPTIONS

**No. 41 A.**—8 Day, Pendulum, Time, Large, Spring. Space required, 6 inches high, 5½ inches wide, and 4 inches between back of case and glass. Distance between center and lowest point of pendulum, 13 inches. Price includes key, hands (not larger than 12 inches), rod and pendulum ball. $55.00

**No. 41 M.**—Same as No. 41 A, but with shorter rod. Distance from center to lowest point of pendulum, 7 inches. $75.00

**No. 41 H.**—Same as No. 41 A, but with escapement on back and with longer rod of wood and 1 lb. ball. Distance from center to lowest point of pendulum, 15 inches. Space required for swing of pendulum, 8 inches. $95.00

No. 41 A.

**No. 42.**—8 Day, Pendulum, Strike, Small, Spring. Strikes hours and half hours. Space required, 8 inches high, 4¾ inches wide and 3½ inches between back of case and glass. Distance from center to lowest point of pendulum, 9 inches. Adapted for cup bell or cathedral bell. Suitable for scroll-saw and kindred work, having long center, permitting the use of wood dial. Price includes key, hands, rod and ball. $75.00

**No. 42 A.**—Same as No. 42, but with shorter rod. Distance from center to lowest point of pendulum, 6½ inches. $80.00

**No. 44 E.**—8 Day, Pendulum, Strike, Spring. Strikes the hours and half hours. Suitable for cathedral bell. Space required, 10 inches high, 6½ inches wide, and 3 inches between back of case and glass. Distance from center of movement to lowest point of pendulum, 13⅜ inches. Space required for swing of pendulum, 7 inches. The lower section of the pendulum rod is of wood, with a 10-ounce ball. Price includes key, hands, rod and ball. $85.00

**No. 48.**—8 Day, Pendulum, Strike, Spring, Visible Escapement. Has steel cut pinions, and strikes the hour and half hour on cathedral bell. Suitable for marble or porcelain cases. Fitted to sash for 5-inch opening only. Space required, from front of case outside to back of case inside, 4¼ inches. Distance from center to lowest point of pendulum, 5¼ inches. Price includes cathedral bell and stand, ornamented front sash, with beveled glass, porcelain dial, back sash, straps, face plate, key and pendulum. $90.00

**No. 48 A.**—Similar to No. 48, but with plain dead-beat escapement, lantern pinions and shorter pendulum, fitted to sash for 4½-inch opening only. Space required from front of case outside to back of case inside, 4¼ inches. Distance from center to lowest point of pendulum, 4⅜ inches. Price includes cathedral bell and stand, ornamented front sash with beveled glass, porcelain dial, back sash, straps, face plate, key and pendulum. $80.00

**No. 51.**—15 Day, Pendulum, Strike. A fine polished Movement, with springs in barrels and steel cut pinions. Strikes the hours and half hours on cathedral bell. Fitted to sashes for 4, 4½ and 5-inch openings. Space required from front of case outside to back of case inside, 4 inches. Distance from center to lowest point of pendulum, 4 inches. Price includes 3-inch cathedral bell and stand, ornamented front sash with beveled glass, porcelain dial, back sash, straps, face plate, key and pendulum. $90.00

**No. 51 M.**—15 Day, Pendulum, Strike. A fine polished movement with springs in barrels and steel cut pinions. Strikes the hours and half hours on cup bell. Same measurements as No. 51. Front plate has holes fitted with screws so that it can be used in old style cases which had zinc plates fitted in the front for that purpose. This movement can be fitted with hammer for cathedral bell, if desired. Price includes key, pendulum, hands and cup bell, or cathedral bell and stand in place of the cup bell. $90.00

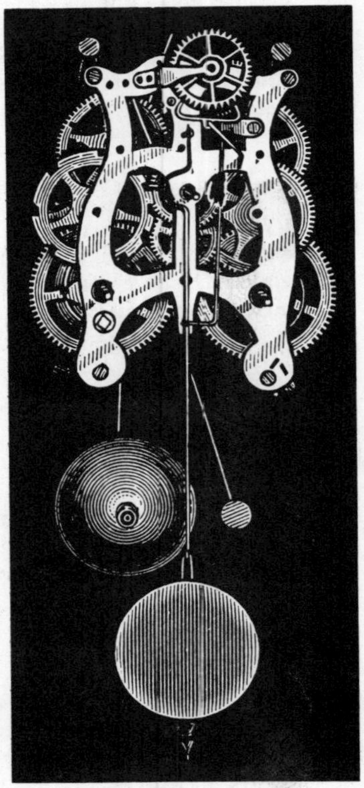

No. 42.

# Ca. 1904 SETH THOMAS CLOCK MOVEMENT DESCRIPTIONS

**No. 61.**—8 Day, Weight, Time, 80 beats. Space required 4½ inches high, 4¾ inches wide, and 4½ inches between back of case and glass. Distance from center to lowest point of pendulum, 22½ inches. Space for swing of pendulum, 8 inches. Movement and pendulum are hung on an iron bracket. Has Graham dead-beat escapement, lantern pinions, maintaining power, wood rod and 2-lb. brass ball. Price includes key, 12-inch hands, weight, rod and ball. $200.00

**No. 61 A.**—60 beats. Same as No. 61, but with seconds pendulum and 6-lb. ball of 8 inches diameter. Distance between center and lowest point of pendulum 41¾ inches. Space for swing of pendulum, 12 inches. Price includes key, 12 or 14-inch hands, weight, rod and ball. $275.00

**No. 68.**—8 Day, Weight, Time, 60 beats. Space required, 10¼ inches high, 8¼ inches wide and 5⅞ inches between back of case and glass. Has polished plates and cut-steel pinions, Graham pallets, maintaining power and wood rod with 10-lb. ball, 8 inches in diameter. The Movement rests on a heavy iron bracket, from which the pendulum is also suspended. Distance from center to lowest point of pendulum, 40½ inches. Space for swing of pendulum, 12 inches. Price includes key, weight, rod and ball. This Movement is made with and without dial feet, the latter used when it is intended to have dial fastened to Movement. $375.00

**No. 69.**—8 Day, Weight, Time, 60 beats. Similar to No. 68, but with lacquered plates and lantern pinions with 6-lb. ball 8 inches in diameter. Space for swing of pendulum, 12 inches. $300.00

**No. 72 A.**—8 Day, Weight, striking hours and half hours. A good Movement for hall clocks. Has Graham escapement and lantern pinions, also maintaining power. Plates are damaskeened. Movement rests on an iron bracket from which the pendulum is also suspended. Supplied with long brackets or short brackets. Space required, 10 inches high, 10 inches wide; and distance between back of case and glass with long bracket, 8¾ inches; with short bracket, 6¾ inches. Distance from center to lowest point of pendulum, 41 inches. Distance required for drop of weights from center of dial to bottom of weights, 56 inches. Four posts project from front plate for support of dial. The rod is of wood with brass-covered ball weighing 4 lbs. and measuring 6 inches in diameter. Weights are brass shells, 10 inches long and 2¼ inches in diameter, filled with lead, and weigh, respectively, strike side, 11½ lbs.; time side, 5½ lbs. Space for swing of pendulum, 11½ inches; or, if a 4½-inch ball be used, 9½ inches. Price includes rod, ball, weights, key, 12-inch hands, 6-inch cathedral bell and stand. When ordering, state if wanted with long bracket or with short bracket, also size of pendulum ball. Dials suitable, No. 366, No. 373 and No. 435. Can be fitted with seat-board, 13 3-16 or 15 inches long, in place of bracket, if desired. $250.00

**No. 72 B.**—Same as No. 72 A, but fitted with attachments for moon. Dials suitable, No. 437, No. 441 and No. 439. Can be fitted with seat-board, 13 3-16 or 15 inches long, in place of bracket, if desired. $250.00

**No. 72 C.**—Same as No. 72 A, but fitted with moon disk 8⅝ inches in diameter, showing phases of the moon, and without seconds. Dial suitable, No. 405. Can be fitted with seat-board 13 3-16 or 15 inches long, in place of bracket, if desired. $275.00

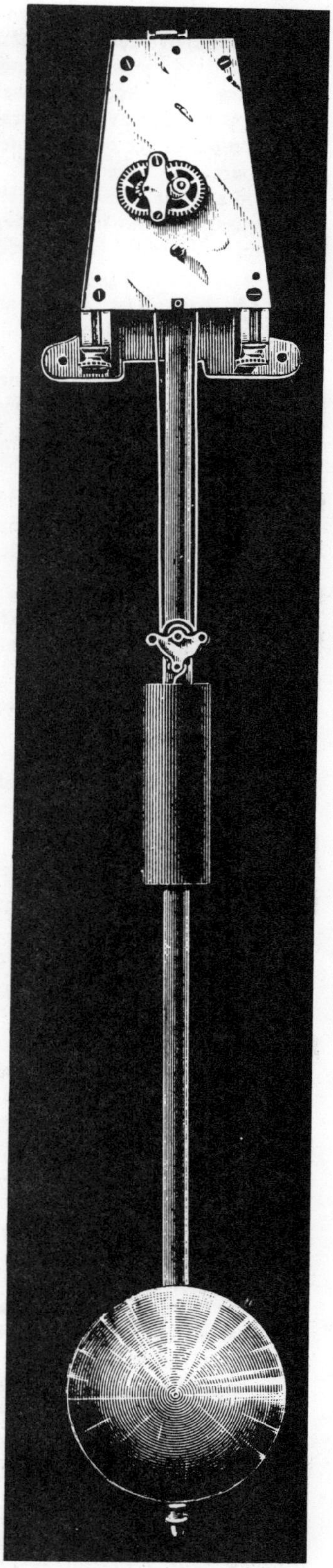

**No. 68.**

# Ca. 1904 SETH THOMAS CLOCK MOVEMENT DESCRIPTIONS

**No. 74 B.**—8 Day, Weight, Quarter Strike, Westminster Chimes. Space required, 9½ inches high, 9 inches wide and 8¾ inches between back of case and glass. Space above center for bells, 8½ inches. Width required for swing of pendulum, 11½ inches. Distance from center for drop of weights 58 inches. Distance from center to lowest point of pendulum 40½ inches. The Movement rests on a heavy iron bracket, from which the pendulum is also suspended. The plates are damaskeened. The chimes are struck on four cup bells and the hours on cathedral bell. Has Graham escapement, lantern pinions, wood rod and maintaining power, without seconds. The ball is of zinc covered with brass, 6 inches in diameter and weighs 4 lbs. In fitting these Movements in old cases with narrow bodies, the required space for swing of pendulum can be reduced to 9½ inches by using a 4½-inch 2-lb. ball instead of the 6-inch 4-lb. ball. The weights are brass shells of a uniform length of 10 inches and diameter of 2¼ inches, filled with lead, and weigh, respectively, counting them from left to right: 1st Strike, 11½ lbs.; 2d Time, 5½ lbs.; 3d Chime, 15 lbs. The moon disk is 8¾ inches in diameter. Dial suitable, No. 434. Price includes the Movement with bells, moon attachment, weights, 10-inch hands, key, rod and ball. Can be fitted with seat-board 13-3-16 or 15 inches long, in place of bracket, if desired. $375.00

**No. 74 C.**—Same as No. 74 B, but without moon disk. Dials suitable, No. 266, No. 374 and No. 436. Can be fitted with seat-board 13 3-16 or 15 inches long, in place of brackets, if desired. $350.00

**No. 74 D.**—Same as No. 74 C, but with attachments for moon. Dials suitable, No. 438, No. 442 and No. 440. Can be fitted with seat-board, 13 3-16 or 15 inches long, in place of bracket, if desired. $350.00

**No. 75.**—Motor Movement. An extra heavy train of six wheels, driven by two springs, each 1 5-16 inches wide, 18 feet long, 23 gauge, ending with a fly. Intended for turning a table in show window, or for any other purpose for which a large spring motor is required. Table and contents should not weigh more than 50 lbs. well balanced. Will run 7 hours with one turn of table per minute. Size: 10½ inches high, 10 inches wide, 5½ inches deep. $75.00

**Disk.**—For use with No. 75 when required. A circular iron frame, 4 inches in diameter, for mounting on center shaft, to which a table may be secured. $25.00

**No. 80.**—1 Day, Strike, Weight. Movement on seat-board 6½ inches long. Space required, 9 inches high, 5 inches wide and 3½ inches between back of case and glass. Required drop for weights, 24 inches. Distance from center to lowest point of pendulum, 14½ inches. Price includes seat-board, key, hands, rod and ball. Weights are extra; time side, 3 lbs. 4 oz.; strike side, 3 lbs. $50.00

**No. 82 A.**—Not illustrated, but similar in appearance to No. 80. Large 8 Day, Strike, Weight. Movement on seat-board 7 inches long. Space required, 12 inches high, 7 inches wide and 3½ inches between back of case and glass. Required drop for weights, 28 inches. Distance from center to lowest point of pendulum, 19½ inches. Price includes key, hands, rod and ball. Weights are extra; time side, 8 lbs. 10 oz.; strike side, 7 lbs. 10 oz. $110.00

**Zinc and Steel Pendulum.**—The compensation pendulum differs from the usual forms, and is constructed on scientific principles. The central steel rod has the rating nut on the lower end at the bottom of the pendulum. The lower end of the zinc tube is supported by a brass collar which rests on the rating nut; to the upper end is secured a brass cross piece from which the two outside steel rods are suspended. To the lower end of these rods the bob is secured. The bob being suspended from its center to the steel rod, leaves it practically out of the compensation, thus avoiding the irregularity caused by the changes in temperature, which act more slowly upon a heavy bob than on the lighter tube and rod. The pendulum measures from the top of suspension to its extreme end, 47¼ inches and weighs 16 lbs. It can be used with movements 68, 69, 72 A, 72 B, 74 C and 74 D. Space required for swing of pendulum, 11 inches. $185.00

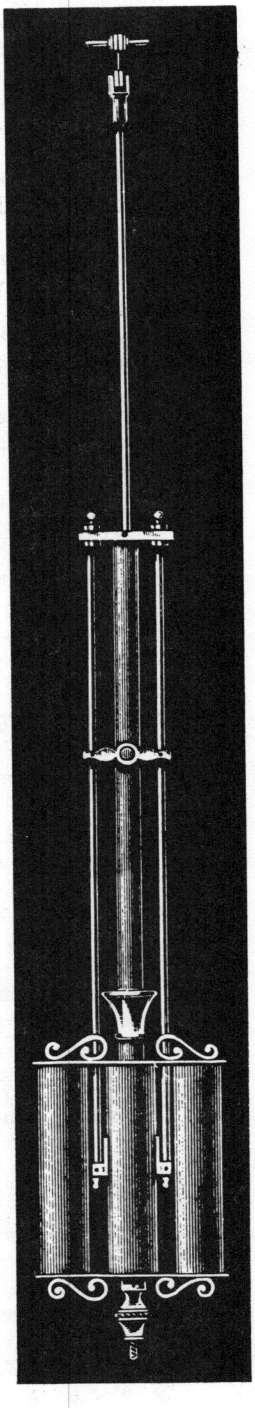

Zinc and Steel Pendulum.

When making a Pendulum, bear in mind that while the weight of the Ball may be varied with impunity, the distance from the point of suspension of the pendulum to the center of gravity in the Ball is fixed and cannot be changed.

# Ca. 1904 SETH THOMAS CLOCK MOVEMENT DESCRIPTIONS

**No. 84.**—8 Day, Weight, Quarter Strike, Westminster Chimes, with disk showing moon phases. Space required, 8 inches high, 8 inches wide and 11¾ inches between back of case and glass. Space required for swing of pendulum 11½ inches. Distance from center for drop of weights, 71 inches. Distance from center to lowest point of pendulum, 42 inches. The Movement rests on a seat-board 15 inches long. The pendulum is suspended from a stud screwed to the back plate of Movement. The chimes are struck on four cathedral bells, and the hours on a large cathedral bell. Can be made silent or chime at will. Has Graham escapement, steel cut pinions tempered and maintaining power. The lower arms of the hour and quarter racks are provided with springs, which prevent any straining or bending of those parts in case the hands should be turned backward. The crutch wire is also provided with beat screws, by which the beat can be adjusted with precision. All of the parts are highly finished, and the plates are damaskeened. The rod is of soft steel. The ball is of zinc covered with brass, 5¾ inches in diameter, and weighs 3½ lbs. The weights are brass shells filled with lead, and weigh, respectively, counting them from left to right: 1st Hour, 20 lbs.; 2d Time, 14½ lbs.; 3d Chime, 32½ lbs. Price includes the movement with all bells, with their stands, seat-boards, weights, key, hands, rod and ball, also No. 443 dial and moon disk. This Movement is equal in every detail to the finest of English makes, and outclasses those of German manufacture.

$700.00 Complete
W/Dial & Gongs

**Dial No. 443.**
Suitable for No. 84 movement only.
$200.00

**No. 85 A.**—8 Day, Pendulum, Strike, Spring, 60 beats. Is as large as, and similar in appearance to, Movement No. 86. Strikes hours only. Suitable for cathedral bell. Space required, 12 inches high, 7½ inches wide and 4⅜ inches between back of case and glass. Distance from center to lowest point of pendulum, 39½ inches. Space required for swing of pendulum, 10 inches. Graham pallets and lantern pinions. Movement and pendulum are hung on an iron bracket. Can supply suitable Dial, like No. 373 and No. 435. Price includes key, 12-inch hands, wood rod and 2-lb. ball. $150.00

**No. 85 E.**—80 beats. Same as No. 85 A, but with shorter center and shorter rod. Distance from center to lowest point of pendulum, 22 inches. Space required between back of case and glass, 4⅛ inches. Space required for swing of pendulum, 8 inches. $150.00

**No. 86.**—15 Day, Pendulum, Time, Duplex, Spring, 80 beats. Space required, 8 inches high, 7 inches wide and 4 inches between back of case and glass. Distance from center to lowest point of pendulum, 21½ inches. Space required for swing of pendulum, 8 inches. Movement and rod are hung on an iron bracket. Has Graham pallets and lantern pinions. Price includes key, 12-inch hands, rod and ball. $150.00

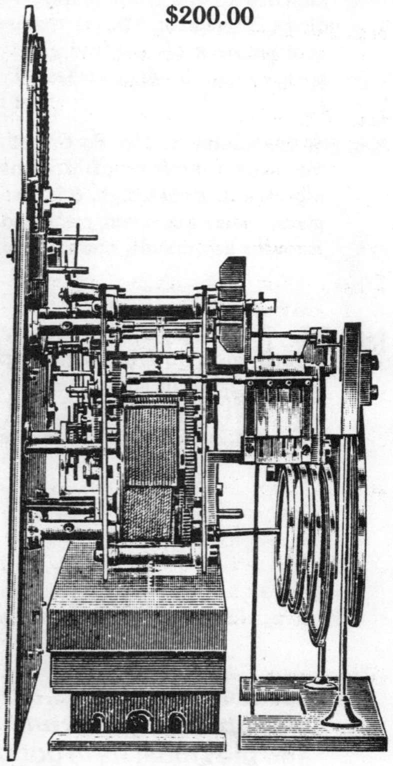

**No. 84.**

**No. 86 C.**—15 Day, Pendulum, Time, 60 beats (seconds). Same as No. 86, but with longer rod and heavier ball. Distance from center to lowest point of pendulum, 39 inches. Space required for swing of pendulum, 10 inches. $150.00

**No. 86 F.**—15 Day, Pendulum, Time, 120 beats (half seconds.) Same as No. 86, but with shorter rod and longer center. Space required, 8 inches high, 7 inches wide and 4½ inches between back of case and glass. Distance from center to lowest point of pendulum, 9 inches. Space required for swing of pendulum, 6 inches. $75.00

**No. 86 G**—15 Day, Pendulum, Time, 100 beats. Same as No. 86 F, but with longer pendulum. Distance from center to lowest point of pendulum, 13½ inches. Space required for swing of pendulum, 7 inches. $75.00

# Ca. 1904 SETH THOMAS CLOCK MOVEMENT DESCRIPTIONS

**No. 87.**—8 Day, Weight, Chain Wind Movement. Strikes hours and half hours. Has Graham escapement and lantern pinions. Suitable for hall clocks and other places where a chain wind and long pendulum are desired. Movement rests on a 10-inch seat-board, is 7 inches wide and 7½ inches high, exclusive of the hammer. Depth required between back of case and glass, 5½ inches. Space for cathedral bell, as shown in cut, must be at least 14 inches wide and 10 inches high, measured from center of Movement; but if 4⅜-inch cup bell is used 7 inches in width only are required. The pendulum rod is of wood, with ornamental brass-covered ball 5 inches in diameter, and is suspended from a stud on back of movement. Distance from center of Movement to lowest point of pendulum, 39½ inches. Width of case required for weights and swing of pendulum at least 9 inches. The weights are brass shells filled with lead, and weigh, respectively, strike side, 5¼ lbs.; time side, 3½ lbs.; they require a space of 57½ inches for drop, measured from center of Movement to bottom of weights. Price includes rod, ball, weights, 12-inch hands, 5-inch cathedral bell and spiral stand, or 4⅜-inch cast cup bell, if desired. Movements will be sent with cathedral bell unless otherwise ordered.

$250.00 Complete with Pendulum & Weights.

**No. 89 C.**—8 Day, Pendulum, Strike, Spring. Strikes the hours on cathedral bell and half hours on cup bell. The brackets for securing Movement to case are attached to the front plate. The escapement and pendulum are on the back of the Movement. Space required, 8 inches high, 6 inches wide and 4¾ inches between back of case and glass. Distance from center of Movement to lowest point of pendulum, 5 inches. Price includes key, hands, rod, ball, cup bell and small cathedral bell, with short iron stand. **$70.00**

**No. 89 D.**—Same as No. 89 C, but with longer rod. The escapement and pendulum are on the front of Movement. The brackets are on the back plate. Space required, 8 inches high, 6 inches wide and 2¾ inches between back of case and glass. Distance from center to lowest point of pendulum, 9¾ inches. Price includes key, hands, rod and ball. **$80.00**

**No. 98.**—Turn-table Movement. A heavy train of four wheels driven by a spring 1 inch wide, 10 feet long, 25 gauge, ending with a fly. Intended for turning a small table in show windows. Table and contents should not weigh more than 8 lbs., well balanced. Will run 10 hours with one turn of table per minute. Size, 6½ inches high, 4¾ inches wide and 3 inches deep. **$60.00**

*To receive top prices, movements must be complete and original, and they must be in good running or repairable condition. Damaged, worn-out or missing parts will reduce prices accordingly.*

No. 98.

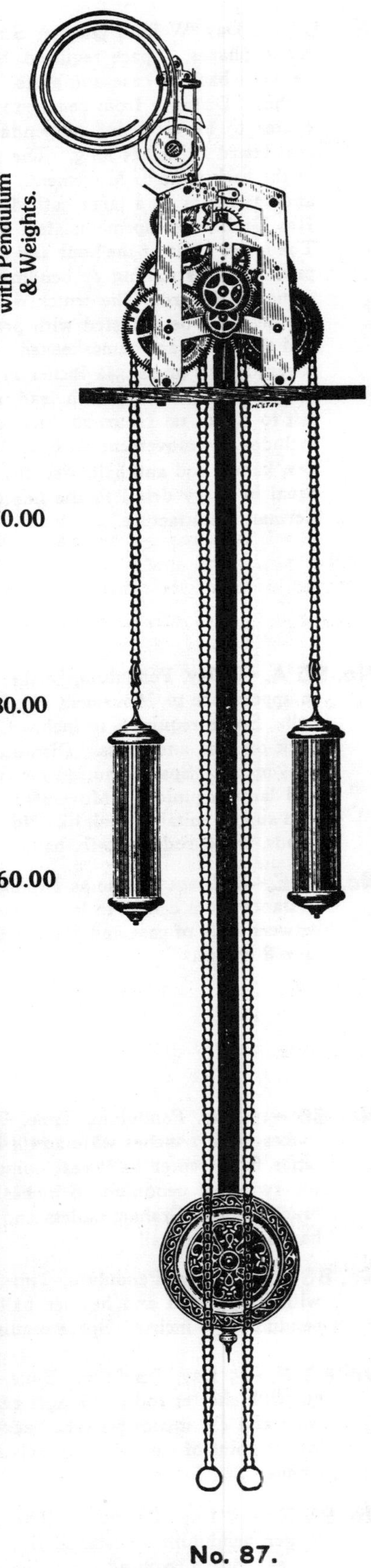

No. 87.

# Ca. 1904 SETH THOMAS CLOCK MOVEMENT DESCRIPTIONS

## DESCRIPTION OF DIALS

The Dials enumerated in our price list by size in inches, 6 inches to 24 inches, inclusive, are the kind known to the trade as "Zinc" Dials. The sizes 6 inches to 8 inches, inclusive, except weight Dials, are usually made of paper—zinc backed, and the larger sizes, 10 inches to 24 inches, inclusive, also weight Dials, are of zinc, painted.

When Dials are made Roman or Arabic it is the rule to send Roman unless Arabic is specified.

**Dial No. 266**
$40.00

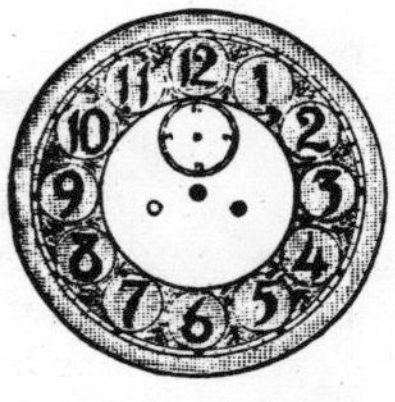

**Dial No. 366**
$40.00

The following dials only can be supplied, either Roman or Arabic:

| | | |
|---|---|---|
| 5-inch | Dials, | with ornamental sash and glass, for Movement No. 89. |
| 12 " | " | with rim, for Movements No. 61, No. 68 and No. 69. |
| 14 " | " | with rim, for Movements No. 68 and No. 69. |
| 14 " | " | flat, for Movement No. 86. |
| 18 " | " | flat, for Movement No. 86. |
| 24 " | " | flat, without seconds for Movement No. 86. |

The largest sashes and glasses made are suitable for 12-inch Dials. Where larger Dials are used the door of the clock forms the sash—a metal ring being secured inside to give a finished appearance.

**Dial No. 373**
$75.00

**Dial No. 374**
$75.00

**Dial No. 266.**—Suitable for Movement No. 74 C only. Made of brass, silvered and polished. Black Arabic numerals. Diameter, 12½ inches; minute circle, 11¼ inches.

**Dial No. 366.**—Suitable for Movement No. 72 A only. Made of brass, silvered. The center section is polished, and the outer section dead finish. Black Arabic numerals, ornamental lines being in gilt. Total diameter, 12½ inches; minute circle, 11¼ inches.

**Dial No. 373.**—Suitable for Movement No. 72 A only. Made of brass, silvered. The center section is polished, and the outer section dead finish. Black Arabic numerals, trimmed with gilt. Ornamental lines are gilt. 15⅝ inches high, 11⅜ inches wide. Minute circle, 9⅝ inches.

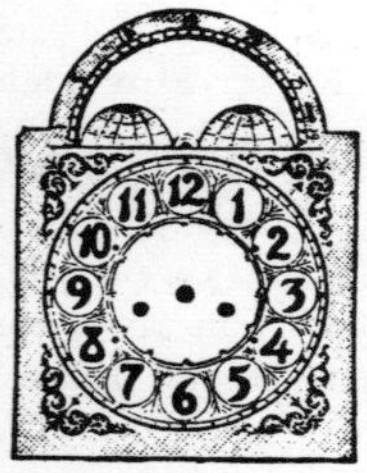

**Dial No. 405**
$75.00

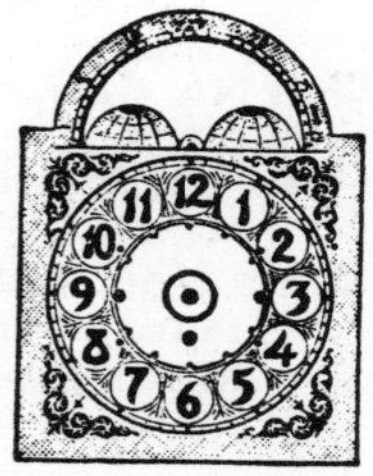

**Dial No. 434**
$75.00

**Dials Similar** to No. 373 can be supplied at price of No. 373 Dial to fit any other movements, provided no seconds circle be required.

**Dial No. 374.**—Suitable for movement No. 74 C only. Description same as Dial No. 373.

**Dial No. 405.**—Suitable for movement No. 72 C only. In every other respect like Dial No. 434.

**Dial No. 434.**—Suitable for movement No. 74 B only. Made of brass, silvered. The center section being polished, and the remainder of Dial dead finish. Black Arabic numerals trimmed with gilt. Ornamental lines are gilt. 15⅝ inches high, 11⅜ inches wide, minute circle, 9⅝ inches.

**Dial No. 435.**—Suitable for Movement No. 72 A only.

**Dials similar to Dial No. 435** can be supplied at price of No. 435, to fit any other movement.

**Dial No. 436.**—Suitable for Movement No. 74 C only.

**Dials Nos. 435 and 436** are like Dials Nos. 437 and 438, except that the arched section is solid—showing a mock moon.

**Dial No. 437.**—Suitable for Movement No. 72 B only.

**Dial No. 438.**—Suitable for Movement No. 74 D only.

**Dials Nos. 437 and 438** are made of heavy iron, painted cream white, with floral corners in gilt and color. Lithographed moon disk. 18⅞ inches high, 13⅜ inches wide, 12⅝ inch minute circle.

**Dial No. 435**
**Dial No. 436** $125.00
**Dial No. 437**
**Dial No. 438** $150.00

# Ca. 1904 SETH THOMAS CLOCK MOVEMENT DESCRIPTIONS

## DESCRIPTION OF DIALS

**Dial No. 439**
**Dial No. 440** $175.00

**Dial No. 441**
**Dial No. 442** $185.00

Dial No. 443.
**$200.00**

**Dial No. 439.**—Suitable for Movement No. 72 B only.

**Dial No. 440.**—Suitable for Movement No 74 D only.

**Dials Nos. 439 and 440** are made of heavy brass, with gilt raised numerals on raised silver circle. Engraved center and cast corners. Lithographed moon. Well finished throughout. 19 inches high, 13¼ inches wide, 12-inch minute circle.

**Dial No. 441.**—Suitable for Movement No. 72 B only.

**Dial No. 442.**—Suitable for Movement No. 74 D only.

**Dials Nos. 441 and 442** are made of heavy brass, with gilt raised numerals on raised silvered circle; gilt ornamental pierced center and corners. Hand painted moon disk. Well finished throughout. 19 inches high, 13¼ inches wide, 12 inch minute circle.

**Dial No. 443.**—Suitable for No. 84 movement only. Made of heavy brass, with gilt raised numerals on raised silver circle; gilt ornamental pierced center and corners. Fine painted moon disk. Handsomely finished throughout. 19 inches high, 13¼ inches wide, 12 inch minute circle. Sold only with movement 84, and included in its price.

**All Movements made by the Seth Thomas Clock Co. are stamped with their Trade Mark, thus:**

ST

# Ca. 1904 SETH THOMAS CLOCK MOVEMENT DESCRIPTIONS

The earliest movement of which any considerable number were sent out was that of the rotary winding from a three-pole motor, as shown in Fig. 137. Each of these magnet spools is of two ohms, with twelve ohms resistance, placed in parallel with the winding of each set of magnet spools, thus making a total of nine spools for the three-pole motor.

On the front end of the armature drum arbor is a commutator having six points, corresponding to the six armatures in the drum. There are three magnets marked O, P and X; each magnet has its own brush marked O′, P′ and X′. When an armature approaches a magnet (see Fig. 137) the brush makes contact with a point of the commutator, and remains in contact until the magnet has done its work and the next magnet has come into action. When properly adjusted the brush O′ will make contact when armatures 1 and 2 are in the position shown, with No. 2 a little nearer the core of the magnet than No. 1; and it will break contact when the armature has advanced into the position shown by armature No. 3, the front edge of the armature being about one-sixteenth of an inch from the corner of the core, armature No. 4 being entirely out of circuit, as brush X′ is not touching the commutator.

The back stop spring, S, Fig. 137, must be adjusted so that the brush O′ is in full contact with a point of the commutator when the motor is at rest, with a tooth of the ratch touching the end of the spring, S.

Sometimes the back stop spring, S, becomes broken or bent. When this occurs it is usually from overwinding. It must be repaired by a new spring, or by straightening the old one by burnishing with a screwdriver. Set the spring so that it will catch about half way down the last tooth.

Having explained the action of the motor we come now to the means of temporarily closing the circuit and keeping it closed until such time as the spring is wound a sufficient amount to run the clock for one hour; as the spring is on the center arbor this requires one complete turn.

This is the distinguishing feature of this system of clocks and is not possessed by any of the others. It varies in construction in the various movements, but in all its forms it maintains the essential properties of holding the current on to the circuit until such time as the spring has been wound a sufficient quantity, when it is again forcibly broken by the action of the clock. This is termed the "knock away," and exists in all of these movements.

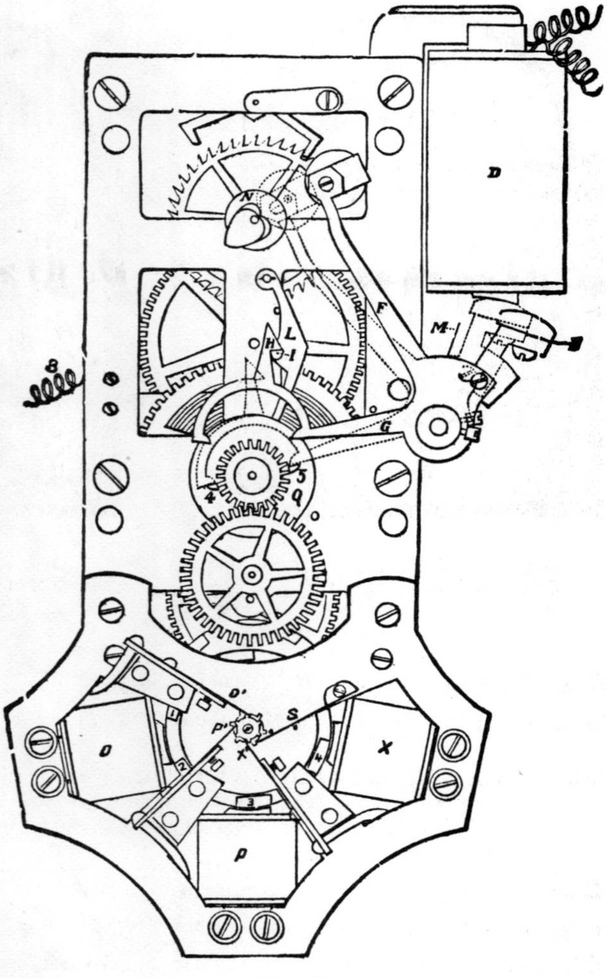


Fig. 137

**$140.00**

*Reprinted from the Modern Clock*
*by Ward L. Goodrich*

*To receive top prices, movements must be complete and original, and they must be in good running or repairable condition. Damaged, worn-out or missing parts will reduce prices accordingly.*

# 1924 SETH THOMAS CLOCK MOVEMENT

SELF WINDING
MOVEMENT

$120.00

# 1922 SETH THOMAS CLOCK MOVEMENT DESCRIPTIONS

$50.00

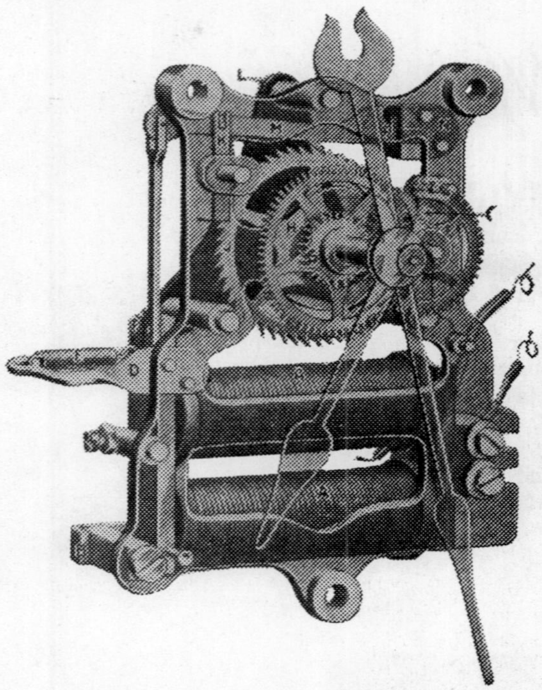

## Secondary Movement

A Magnet
B Armature Bracket
C Armature Adjusting Screw
D Armature Spring Bracket
E Armature Spring
F Armature
G, H, I } Dial Works
J Center Wheel
K Locking Lever
L Back Plate
M Actuating Lever
N Actuating Lever Bracket
O Locking Pawl
P Locking Angle

$125.00

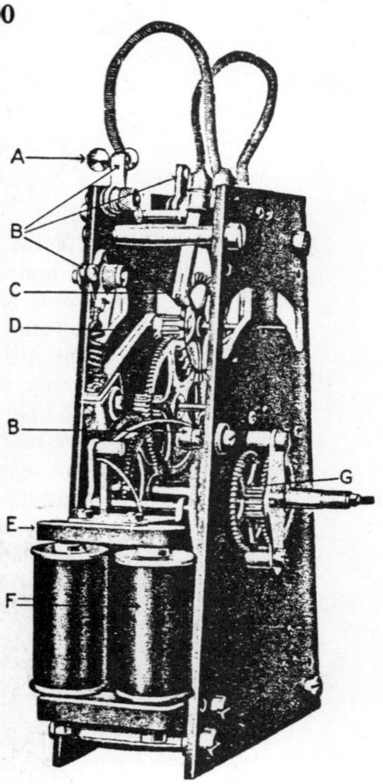


**Self-Winding Master Movement**

RIGHT SIDE

A. Micrometer beat adjuster
B. Terminal connectors
C. Jeweled pallets
D. Bell contact
E. Armature
F. Winding magnets
G. Dial works

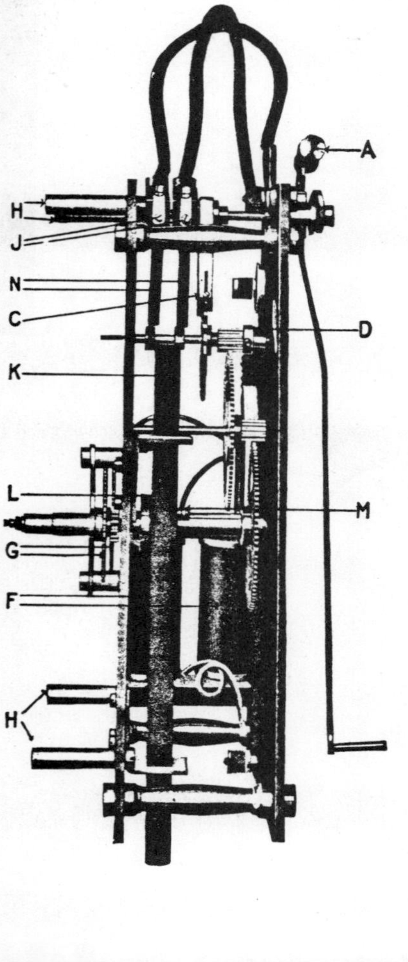


**Self-Winding Master Movement**

LEFT SIDE

A. Micrometer beat adjuster
C. Jeweled pallets
D. Bell contact
F. Winding magnets
G. Dial works
H. Dial feet
J. Contact adjusting screws
K. Weight
L. Maintaining power
M. Winding contact
N. Secondary and program contacts

*To receive top prices, movements must be complete and original, and they must be in good running or repairable condition. Damaged, worn-out or missing parts will reduce prices accordingly.*

# 1928 SETH THOMAS CLOCK MOVEMENT

## Electric Clock Movement

IN the Seth Thomas Electric Clock a General Electric motor winds the mainspring until the counter torque (which acts downward on the motor shaft) is sufficient to depress the shaft, causing the rotor to strike the stationary brake. This causes the motor speed to reduce and automatically hold at a value that winds the spring at the same rate that the spring is run down in driving the escapement.

The motor thrust bearing constantly weighs the mainspring tension and maintains the latter at normal value through the action of the brake. The adjustable thrust bearing is set at the factory at proper value to produce normal mainspring tension.

A complete, standard, positive, electric motor unit applied to a standard clock train. All units independent and replaceable. Timekeeping accuracy independent of variation in voltage and frequency of supply current. Depth 2¾ inches. Diameter 2¾ inches.

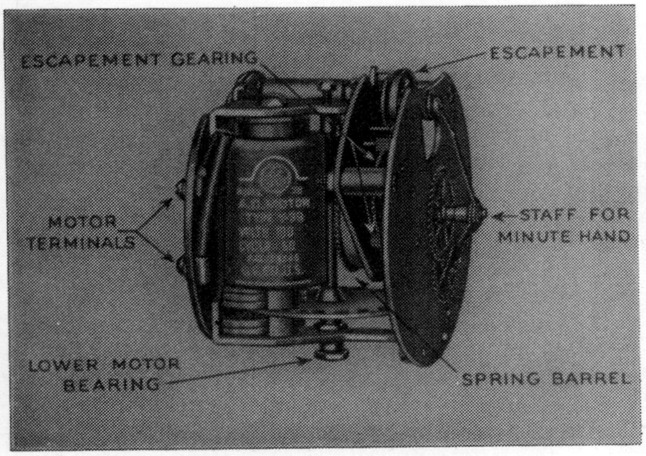


*Seth Thomas Electric Clock movement, showing application of Type Y-39 GE Motor.*

**$10.00**

# 1907 SETH THOMAS CLOCK MOVEMENT DESCRIPTIONS

**No. 0.**—1 Day, lever, time, small, for glass case. Discontinued. Crown 15, center 19/32". Length 2 1/2", width 2 3/4", depth 1 11/16". Mainspring No. 70, 4' x 3/8" x .013. $35.00

**No. 1.**—1 Day, lever, time, small, Banner, front wind. Crown 15, center 25/32". Length 2 3/4", width 3 1/2", depth 1 11/16". Mainspring No. 15, 5' x 3/8" x .016. $30.00

No. 1.

**No. 1 A.**—1 Day, lever, time, small, Plaque, back wind. Crown 15, center 7/8". Length 2 3/4", width 3 1/2", depth 2 3/16". Mainspring No. 15, 5' x 3/8" x .016. $30.00

**No. 2.**—1 Day, lever, time, large. Crown 15, center 7/8". Length 4 1/2", width 4 1/2", depth 2 3/8". Mainspring No. 54, 6' x 5/8" x .018. $30.00

**No. 2 A.**—1 Day, lever, time, large, side wind, Navy Lever. Crown 15, center 7/8". Length 4 3/16", width 4 1/2", depth 2 3/8". Mainspring No. 54, 6' x 5/8" x .018. $60.00

**No. 2 B.**—1 Day, lever, time, large, alarm. Crown 15, center 7/8". Length 5", width 4 1/2", depth 2 3/8". Time mainspring No. 54, 6' x 5/8" x .018. Alarm mainspring No. 80, 2 1/2' x 1/4" x .016. $30.00

**No. 2 C.**—1 Day, lever, time, sweep seconds. Discontinued. Crown 15, center 11/16". Length 4 1/2", width 4 1/2", depth 2 1/2". Mainspring No. 54, 6' x 5/8" x .018. $35.00

No. 2.

**No. 2 D.**
**No. 2 E.**
**No. 2 F.** No information at this time.
**No. 2 G.**
**No. 2 H.**

**No. 2 I.**—1 Day, lever, time, large, Schultz Watchman. Crown 15, center 1". Length 4 1/2", width 4 1/2", depth 2 1/2". Mainspring No. 54, 6' x 5/8" x .018. $25.00

**No. 3.**—1 Day, lever, time, Long Alarm, Owl Alarm. Crown 15, center 5/8". Length 8", width 4 1/2"' depth 2". Time mainspring No. 66, 5' x 3/8" x .016. Alarm mainspring No. 138, 7' x 1/2" x .018. $30.00

**No. 3 A.**—1 Day, lever, time, new Long Alarm and Owl Alarm. Crown 15, center 5/8". Length 6 1/8", width 4 3/4", depth 2". Time mainspring No 66, 5' x 3/8" x .016. Alarm mainspring No. 138, 7' x 1/2" x .018. $30.00

**No. 4.**—1 Day, lever, strike. Crown 15, center 7/8". Length 4 1/2", width 4 13/16", depth 2 7/16". Time mainspring No. 54, 6' x 5/8" x .018. Strike mainspring No. 75, 4' x 5/16" x .016. $35.00

**No. 4 A.**—1 Day, lever, strike, alarm. Crown 15, center 7/8". Length 5", width 4 13/16", depth 2 7/16. Time mainspring No. 54, 6' x 5/8" x .018. Strike mainspring No. 75, 4' x 5/16" x .016. Alarm mainspring No. 80. 2 1/2' x 1/4" x .016. $35.00

**No. 4 B.**—1 Day, lever, strike, Ship's Bell. Crown 15, center 7/8". Length 4 1/2", width 4 13/16", depth 2 7/16". Time mainspring No. 54, 6' x 5/8" x .018. Strike mainspring No. 75, 4' x 5/16" x .016. $80.00

**No. 4 C.**—1 Day, lever, strike, sweep seconds. Discontinued. Crown 15, center 11/16". Length 4 7/8", width 4 13/16", depth 2 1/4". Time mainspring No. 54, 6' x 5/8" x .018. Strike mainspring No. 75, 4' x 5/16" x .016. $45.00

**No. 5.** No information at this time.

**No. 6.**—1 Day, lever, time, alarm, Dix Alarm. Crown 15, center 5/8". Length 5", width, 4" depth 2". Time mainspring No. 66, 5' x 3/8" x .016. Alarm mainspring No. 89, 5' x 7/16" x .016. $30.00

**No. 7.**—1 Day, lever, strike, New Ship's Bell. Crown 15, center 7/8". Length 7 3/4", width 5 1/2", depth 2 1/4". Time mainspring No. 54, 6' x 5/8" x .018. Strike mainspring No. 73, 6' x 5/16" x .016. $80.00

# 1907 SETH THOMAS CLOCK MOVEMENT DESCRIPTIONS

**No. 7 A.**—1 Day, lever, strike, wardroom, bell on back of movement. Crown 15, center 7/8". Length 5 3/8" with bell, width 5 1/2", depth 3 7/8". Time mainspring No. 54, 6' x 5/8" x .018. Strike mainspring No. 73, 6' x 5/16" x .016. $75.00

No. 10.

**No. 8.**— No information at this time.

**No. 9.**—8 Day, lever, time, Carriage No. 1, steel cut pinions, polished, watch escapement. Center 9/16". Length 3 5/8", width 2 3/8", depth 2 3/16". Mainspring No. 144, 50" x 3/4" x .011. $65.00

**No. 10.**—8 Day, lever, time, double spring. Crown 15, center 27/32". Length 4 15/16", width 5 3/4", depth 2 7/16". Mainspring No. 47, 9' x 5/8" x .022. $45.00

**No. 10 A.**—8 Day, lever, time, double spring, jeweled balance. Crown 15, center 27/32". Length 4 15/16", width 5 3/4", depth 2 7/16". Mainspring No. 47, 9' x 5/8" x .022. $50.00

**No. 10 B.**—8 Day, lever, time, double spring, 4 pairs jewels. Crown 15, center 27/32". Length 4 15/16", width 5 3/4", depth 2 7/16". Mainspring No. 47, 9' x 5/8" x .022. $65.00

**No. 10 C.**
**No. 10 D.** No information at this time.
**No. 10 E.**

**No. 10 F.**—8 Day, lever, time, back wind, double spring. Crown 15, center 27/32". Length 4 15/16", width 5 3/4", depth 2 13/16". Mainspring No. 47, 9' x 5/8" x .022. $45.00

**No. 10 G.** No information at this time.

**No. 10 H.**—8 day, lever, time, double spring, jeweled balance. Crown 15, center 1 5/16". Length 4 15/16", width 5 3/4", depth 2 11/16". Mainspring No. 47, 9' x 5/8" x .022. $50.00

**No. 11.**—8 Day, lever, strike. Crown 15, center 7/8". Length 6 3/16", width 5 3/4", depth 2 9/16". Time mainspring No. 52, 8' x 5/8" x .018. Strike mainspring No. 10, 8' x 7/8" x .020. $45.00

**No. 11 A.**—8 Day, lever, strike, jeweled balance. Crown 15, center 7/8". Length 6 3/16", width 5 3/4", depth 2 9/16". Time mainspring No. 52, 8' x 5/8" x .018. Strike mainspring No. 10, 8' x 7/8" x .022. $50.00

**No. 12.**—Metronome, without bell. Mainspring No. 140, 2 1/2' x 9/32" x .0075. $15.00

**No. 13.**—Metronome, with bell. Mainspring No. 140, 2 1/2' x 9/32" x .0075. $25.00

**No. 14.**
**No. 15.**
**No. 16.**
**No. 17.** No information at this time.
**No. 18.**
**No. 19.**

*To receive top prices, movements must be complete and original, and they must be in good running or repairable condition. Damaged, worn-out or missing parts will reduce prices accordingly.*

# 1907 SETH THOMAS CLOCK MOVEMENT DESCRIPTIONS

**No. 20.**—1 Day, lever, time, for Nos. 5003 and 5004 cases. Discontinued. Round movement, polished, steel cut pinions. Crown 15, center 11/16". Length 3 3/8", width 3 3/8", depth 1 13/16". Mainspring No. 56, 4' x 9/16" x .013. $30.00

**No. 20 A.**—1 Day, lever, time, for side wind, for No. 5006 case. Discontinued. Round movement, polished, steel cut pinions. Crown 15, center 11/16". Length 3 3/8", width 4", depth 2 1/8". Mainspring No. 56, 4' x 9/16" x .013. $45.00

**No. 20 B.**—1 Day, lever, time, for Nos. 5003 and 5004 cases. Round movement, polished, steel cut pinions, watch escapement. Center 11/16". Length 3 3/8", width 3 3/8", depth 1 13/16". Mainspring No. 56, 4' x 9/16" x .013. $45.00

**No. 20 C.**— 1 Day, lever, time, for No. 5006 case. Round movement, polished, steel cut pinions, watch escapement. Center 11/16". Length 3 3/8", width 4", depth 2 1/8". Mainspring No. 56, 4' x 9/16" x .013. $45.00

**No. 21.**—8 Day, Chronometer lever, for No. 5005 case. Discontinued. Crown 15, center 13/16". Length 5 3/16", width 3 1/4", depth 2 9/16". $80.00

**No. 22.**—8 Day, fine, lever, time, double spring, watch escapement, for No. 1 Fine Lever case. Center 13/16". Length 4 5/8", width 4 1/8", depth 2 1/4". Mainspring No. 141, 7' x 3/4" x .0115. $60.00

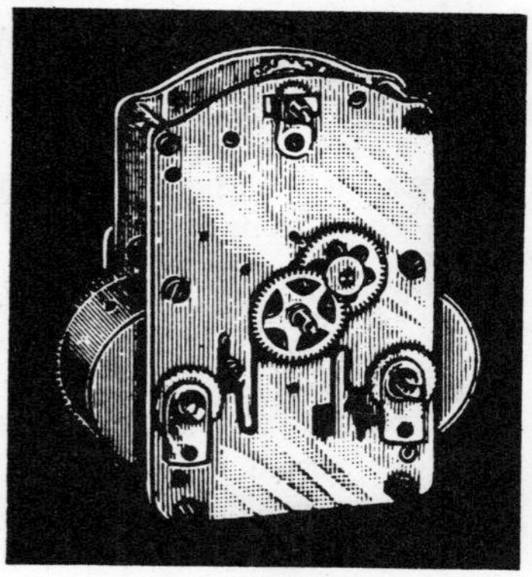

No. 22.

**No. 23.**
**No. 24.**
**No. 25.**
**No. 26.** No information at this time.
**No. 27.**
**No. 28.**
**No. 29.**

*To receive top prices, movements must be complete and original, and they must be in good running or repairable condition. Damaged, worn-out or missing parts will reduce prices accordingly.*

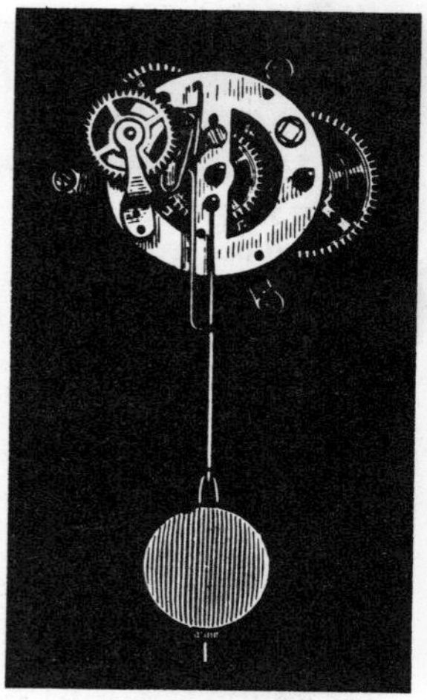

No. 30 A.

**No. 30.**—1 Day, small pendulum, time, spring, round movement. Pendulum, from center of movement to end of rating screw, 5 5/8", with No. 16 ball; diameter of ball 1 3/8". Crown 40, recoil center 15/16". Length 3 1/2", width 3", depth 2 9/16". Mainspring No. 74, 5' x 5/16" x .015. $35.00

**No. 30 A.**—1 Day, small pendulum, time, spring, Bradley. Pendulum, from center of movement to end of rating screw, 5 3/4", with No. 16 ball; diameter of ball 1 3/8". Crown 42, recoil, center 15/16". Length 3 1/16", width 3 9/16", depth 2 9/16". Mainspring No. 74, 5' x 5/16" x .015. $40.00

**No. 31.**—1 Day, small, pendulum, time, alarm, spring. Pendulum, from center of movement to end of rating screw, 5 3/8", with No. 16 ball; diameter of ball 1 3/8". Crown 40, recoil, center 15/16". Length 3 3/4", width 3 3/8", depth 2 9/16". Time mainspring No. 74, 5' x 5/16" x .015. Alarm mainspring No. 76, 2 1/2' x 5/16" x .013. $35.00

**No. 32.**—1 Day, small, pendulum, strike, spring. Discontinued. Crown 48, recoil, center 7/8". Length 4 3/4", width 4 3/4", depth 2 9/16". Time mainspring No. 77, 4 1/2' x 5/16" x .012. Strike mainspring No. 77. $35.00

**No. 32 A.**—1 Day, small, pendulum, strike, spring. Discontinued. Crown 50, recoil, center 1 1/4". Length 4 3/4", width 4 3/4", depth 2 7/8". Time mainspring No. 77, 4 1/2' x 5/16" x .012. Strike mainspring No. 77. $35.00

# 1907 SETH THOMAS CLOCK MOVEMENT DESCRIPTIONS

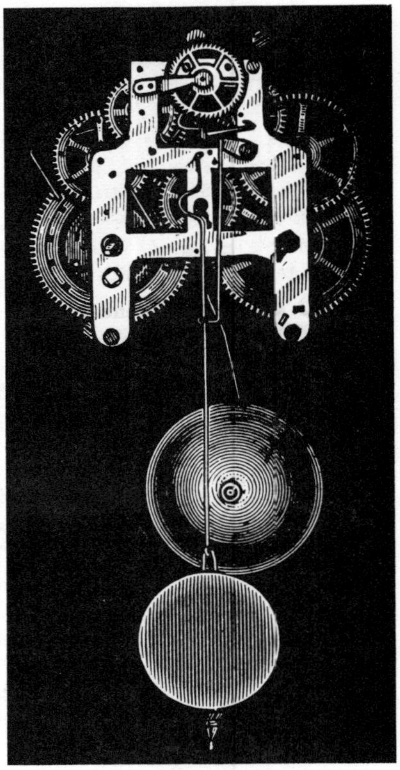

No. 33.

**No. 33.**—1 Day, large, pendulum, strike, spring. Pendulum, from center of movement to end of rating screw, 8 3/4", with No. 14 ball; diameter of ball 1 5/8". Crown 46, recoil, center 1 1/8". Length 5 9/16", width 5 1/8", depth 2 13/16". Time mainspring No. 69, 5' x 3/8" x .013. Strike mainspring No. 69. $45.00

**No. 34.**—8 Day, large, pendulum, half-hour strike, spring, F. E., W. B. Pendulum, from center of movement to end of rating screw, 9 1/2", with No. 36 ball; diameter of ball 2 7/16". Crown 40, recoil, center 1". Length 6 3/16", width 5 15/16", depth 2 13/16". Time mainspring No. 41, 10' x 11/16" x .018. Strike mainspring No. 41. $60.00

**No. 34 A.**—8 Day, large, pendulum, half hour strike, spring, B. E., G. B. Pendulum, from center of movement to end of rating screw, 5 3/4", with No. 46 ball; diameter of ball 1 11/16". Crown 50, recoil, center 1 3/4". Length 6 3/16", width 5 15/16", depth 3 15/16". Time mainspring No. 41, 10' x 11/16" x .018. Strike mainspring No. 41. $60.00

**No. 34 B.**—8 Day, large, pendulum, spring, Devotional, B. E., G. B. Pendulum. from center of movement to end of rating screw, 8 7/8", with No. 12 ball; diameter of ball 2 7/32". Crown 40, recoil, center 1". Length 6 3/16", width 5 15/16", depth 3 1/4". Time mainspring No. 41, 10' x 11/16" x .018, Strike mainspring No. 41. $60.00

**No. 34 C.**—8 Day, large, pendulum, half-hour strike, spring, B. E., G. B. Pendulum, from center of movement to end of rating screw, 5 3/4", with No 46 ball; diameter of ball 1 11/16". Crown 50, recoil, center 1". Length 6 3/16", width 5 15/16", depth 3 1/4". Time mainspring No. 41, 10' x 11/16" x .018. Strike mainspring No. 41. $60.00

**No. 34 D.**—8 Day, large, pendulum, half-hour strike, spring, F. E., G. B. Pendulum, from center of movement to end of rating screw, 9 1/2", with No. 36 ball; diameter of ball 2 7/16". Crown 40, recoil, center 1". Length 6 3/16", width 5 15/16"; depth 2 13/16". Time mainspring No. 41, 10' x 11/16" x .018. Strike mainspring No. 41. $60.00

**No. 34 E.**—8 Day, large, pendulum, spring, new Angelus. Pendulum, from center of movement to end of rating screw, 9 5/16", with No. 12 ball; diameter of ball 2 7/32". Crown 40, recoil, center 1". Length 6 5/16", width 5 15/16", depth 3 7/8". Time mainspring No. 41, 10' x 11/16" x .018. Strike mainspring No. 41. $75.00

**No. 34 F.**—8 Day, large pendulum, half-hour strike, spring, Summit, F. E., W. B. Pendulum, from center of movement to end of rating screw, 8 3/8", with No. 12 ball; diameter of ball 2 7/32". Crown 43, recoil, center, 1", Length 6 3/16", width 5 15/16", depth 2 13/16". Time mainspring No. 41 10' x 11/16" x .018. Strike mainspring No. 41. $65.00

**No. 34 G.**—8 Day, large, pendulum, half-hour strike, spring, Drummer, F. E, G. B. Brackets on front plate. Pendulum, from center of movement to end of rating screw, 6 3/4", with No. 12 ball; diameter of ball 2 7/32". Crown 53, recoil, center 1". Length 6 3/16", width 5 15/16", depth 3 1/4." Time mainspring No. 41, 10' x 11/16" x .018. Strike mainspring No. 41. $65.00

**No. 34 H.**—8 Day, large, pendulum, half-hour strike, spring, F. E., W. B., Dixie Calendar. No. 44 W now used in place of this. Pendulum, from center of movement to end of rating screw, 9 1/2", with No. 36 ball; diameter of ball 2 7/16". Crown 40, recoil, center 1 3/4". Length 6 3/16", width 5 15/16", depth 2 13/16". Time mainspring No. 41, 10' x 11/16" x. 018. Strike mainspring No. 41. $65.00

**No. 34 I.**—8 Day, large, pendulum, half-hour strike, spring, F. E., G. B., with Jackson Drop Calendar attachment, Victoria. Discontinued. Pendulum. from center of movement to end of rating screw, 9 1/2", with No. 36 ball; diameter of ball 2 7/16". Crown 40, recoil, center 1 3/4". Length 6 3/16", width 5 15/16", depth 3 9/16". Time mainspring No. 41, 10' x 11/16" x .018, Strike mainspring No. 41. $100.00

**No. 34 J.**—8 Day, large, pendulum, half-hour strike, spring, Panama. Pendulum, from center of movement to end of rating screw, 18", with No. 33 ball; diameter of ball 2 11/16". Rod in two sections, lower part wood. Crown 33, recoil, center 1". Length 6 3/16", width 5 15/16", depth 2 13/16". Time mainspring No. 41, 10' x 11/16" x .018. Strike mainspring No. 41. $60.00

# 1907 SETH THOMAS CLOCK MOVEMENT DESCRIPTIONS

**No. 35.**—Detached Alarm, No. 2. Length 2 5/16", width 2", depth 1 7/8". Mainspring No. 76, 2 1/2′ x 5/16" x .013. $15.00

**No. 35 A.**—Detached Alarm, No. 3. Length 2 5/16", width 2", depth 1 7/8", Mainspring No. 76, 2 1/2′ x 5/16" x .013. $15.00

**No. 35 B.**—Detached Alarm, No. 4. Length 2 5/16", width 2", depth 2 1/16". Mainspring No. 76, 2 1/2′ x 5/16" x .013. $15.00

**No. 35 C.**—Detached Alarm, No. 6, new style. Length 2 11/16", width 1 1/4", depth 2 11/16". Mainspring No. 130, 27" x 1/4" x .012. $15.00

**No. 36.**—Small Calendar, white rolls. Center 1/4". Length 3 1/2", width 7 7/8", depth 3 3/16". $200.00

**No. 36 A.**—Small Calendar, black rolls. Center 1/4". Length 3 1/2", width 7 7/8", depth 3 3/16". $250.00

**No. 36 B.**—Small Calendar, Spanish rolls. Center 1/4". Length 3 1/2", width 7 7/8", depth 3 3/16". $175.00

**No. 36 C.**—Small Calendar, Portuguese rolls. Center 1/4". Length 3 1/2", width 7 7/8", depth 3 3/16". $175.00

**No. 36 D.**—Small Calendar, German rolls. Center 1/4" Length 3 1/2", width 7 7/8", depth 3 3/16". $175.00

**No. 37.**—Large Calendar. Center 5/16". Length 4 1/2", width 12 1/2", depth 4 3/8". $300.00

**No. 37 A.**—Large Calendar, Spanish rolls. Center 5/16". Length 4 1/2", width 12 1/2", depth 4 3/8". $250.00

**No. 37 B.**—Large Calendar, Portuguese rolls. Center 5/16". Length 4 1/2" width 12 1/2", depth 4 3/8". $250.00

**No. 37 C.**—Large Calendar, German rolls. Center 5/16". Length 4 1/2", width 12 1/2", depth 4 3/8". $250.00

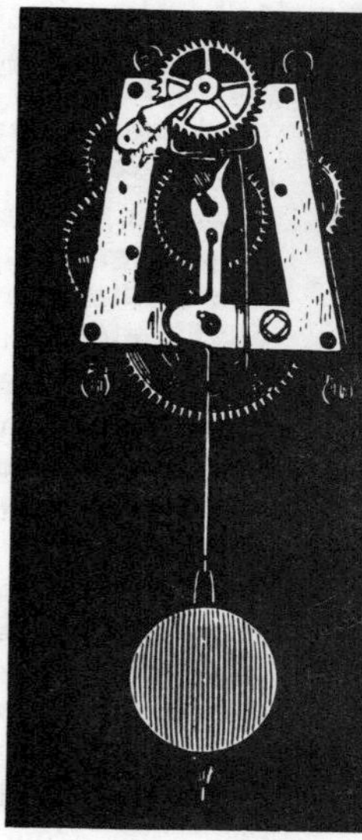

No. 40.

**No. 38.**
**No. 39.** No information at this time.

**No. 40.**—8 Day, small, pendulum, time, spring, N. Y. Store. Pendulum, from center of movement to end of rating screw, 6 5/8", with No. 14 ball; diameter of ball 1 5/8". Crown 38, recoil, center 1 7/16". Length 4 1/8", width 3 1/8", depth 1 13/16". Mainspring No. 149, 8′ x 1/2" x .016. $45.00

**No. 40 A.**—8 Day, small, pendulum, time, spring, N. Y. Store. Pendulum, from center of movement to end of rating screw, 10 1/16", with No. 14 ball; diameter of ball 1 5/8". Crown 29, recoil, center 1 7/16". Length 4 1/8", width 3 1/8", depth 1 13/16". Mainspring No. 149, 8′ x 1/2" x .016. $55.00

**No. 40 B.**—8 Day, small, pendulum, time, spring, J. W. Greenwood. Pendulum, from center of movement to end of rating screw, 5 7/16", with No. 14 ball; diameter of ball 1 5/8". Crown 44, recoil, center 2 1/4". Length 4 1/8", width 3 1/8", depth 3 1/2". Mainspring No. 149, 8′ x 1/2" x .016. $55.00

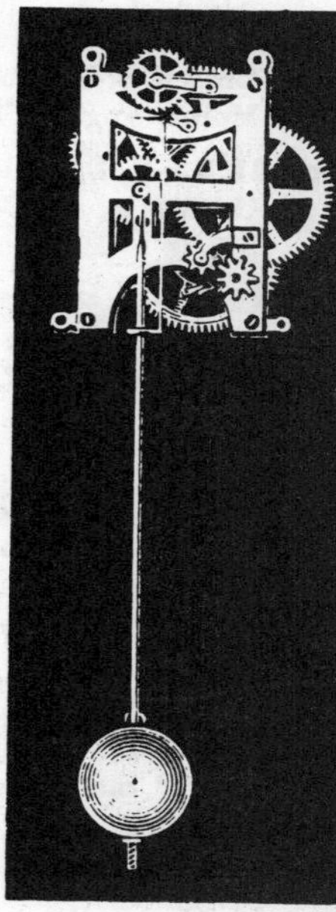

No. 41 A.

**No. 41.**—8 Day, large, pendulum, time, spring, No. 1 Office Time. Pendulum, from center of movement to end of rating screw, 13", with No. 12 ball; diameter of ball 2 7/32". Crown 38, recoil, center 1". Length 5 5/8", width 5 7/8", depth 2 3/4". Mainspring No. 131, 9′ x 11/16" x .015. $50.00

**No. 41 A.**—8 Day, large, pendulum, time, spring, 12 in. Drop Octagon Time. Pendulum, from center of movement to end of rating screw, 13", with No. 12 ball; diameter of ball 2 7/32". Crown 38, dead-beat, center 2". Length 5 5/8", width 5 7/8", depth 3 3/4". Mainspring No. 131, 9′ x 11/16" x .015. $55.00

**No. 41 B.**—8 Day, large, pendulum, time, spring, 10 in. Drop Octagon Time. Pendulum, from center of movement to end of rating screw, 11 7/8", with No. 12 ball; diameter of ball 2 7/32". Crown 40, dead-beat, center 2". Length 5 5/8", width 5 7/8", depth 3 3/4". Mainspring No. 131, 9′ x 11/16" x .015. $50.00

# 1907 SETH THOMAS CLOCK MOVEMENT DESCRIPTIONS

**No. 41 C.**—8 Day, large, pendulum, time, spring, Globe. Pendulum, from center of movement to end of rating screw, 17 7/8", with No. 22 ball; diameter of ball 3 11/16". Rod in two sections, lower part wood. Crown 32, dead-beat, center 2". Length 5 5/8", width 5 7/8", depth 3 3/4". Mainspring No. 131, 9' x 11/16" x .015. — $70.00

**No. 41 D.**—8 Day, large, pendulum, time, spring, English Office Time. Pendulum, from center of movement to end of rating screw, 13", with No. 12 ball; diameter of ball 2 7/32". Crown 38, recoil, center 2 1/4". Length 5 5/8", width 5 7/8", depth 4". Mainspring No. 131, 9' x 11/16" x .015. — $50.00

**No. 41 E.**—8 Day, large, pendulum, time, spring, English Office Time. Pendulum, from center of movement to end of rating screw, 7 5/8", with No. 12 ball; diameter of ball 2 7/32". Crown 53, recoil, center 2 1/4". Length 5 5/8", width 5 7/8", depth 4". Mainspring No. 131, 9' x 11/16" x .015. — $50.00

**No. 41 F.**—8 Day, large, pendulum, time, spring, English Office Time. Pendulum, from center of movement to end of rating screw, 5 13/16", with No. 12 ball; diameter of ball 2 7/32". Crown 55, recoil, center 2 1/4". Length 5 5/8", width 5 7/8", depth 4". Mainspring No. 131, 9' x 11/16" x .015. — $65.00 Unusual

**No. 41 G.**—8 Day, large, pendulum, time, spring, Ithica Drop Calendar. Pendulum, from center of movement to end of rating screw, 19 1/2", with special ball. Crown 31, recoil, center 2". Length 5 5/8", width 5 7/8", depth 4". Mainspring No. 131, 9' x 11/16" x .015. — $90.00

**No. 41 H.**—8 Day, large, pendulum, time, spring, Philadelphia and Queen Anne, B. E. Pendulum, from center of movement to end of rating screw, 15 7/8", with No. 22 ball; diameter of ball 3 11/16". Wood rod. Crown 30, dead-beat, center 1". Length 5 5/8", width 5 7/8", depth 3 5/16". Mainspring No. 131, 9' x 11/16" x .015. — $95.00

**No. 41 I.**—8 Day, large, pendulum, time, spring, Herald. Discontinued. Crown 44, dead-beat, center 1". Length 5 5/8", width 5 7/8", depth 3 5/16". Mainspring No. 131, 9' x 11/16" x .015. — $75.00

**No. 41 J.**—8 Day, large, pendulum, time, spring, change in train, 3rd wheel 90 teeth, English, James Stockwell. Pendulum, from center of movement to end of rating screw, 5 7/8", with No. 12 ball; diameter of ball 2 7/32". Crown 51, recoil, center 2 1/4". Length 5 5/8", width 5 7/8", depth 4". Main spring No. 131, 9' x 11/16" x .015. — $75.00

**No. 41 K.**—8 Day, large, pendulum, time, spring, Drop Octagon Calendar. Pendulum, from center of movement to end of rating screw, 11 7/8", with No. 12 ball; diameter of ball 2 7/32". Crown 38, dead-beat, center 2". Length 5 5/8", width 5 7/8", depth 3 3/4". Mainspring No. 131, 9' x 11/16" x .015. — $85.00

11 7/8 - 38

**No. 41 L.**—8 Day, large, pendulum, time, spring, with Calendar attachments, Nos. 6 and 7 Office Calendars. Pendulum, from center of movement to end of rating screw, 9 1/2", with No. 12 ball; diameter of ball 2 7/32". Crown 46, recoil, center 1 1/4". Length 5 5/8", width 5 7/8", depth 3 5/16". Mainspring No. 131, 9' x 11/16" x .015. — $100.00

**No 41 M.**—8 Day, large, pendulum, time, spring, Chatham. Pendulum, from center of movement to end of rating screw, 6 3/4", with No. 12 ball; diameter of ball 2 7/32". Crown 50, recoil, center 2 1/4". Length 5 5/8", width 5 7/8", depth 4". Mainspring No. 131, 9' x 11/16" x .015. — $75.00

**No. 41 N.**—8 Day, large, pendulum, time, spring, change in train, 3rd wheel 76 teeth, 6-leaf escape pinion, larger than regular, Juno. Pendulum, from center of movement to end of rating screw, 21 1/2", with No. 22 ball; diameter of ball 3 11/16". Wood rod. Crown 32, dead-beat, center 2". Length 5 5/8", width 5 7/8", depth 3 3/4". Mainspring No. 131, 9' x 11/16" x .015. — $75.00

**No. 41 O.**—8 Day, large, pendulum, time, spring, change in train, 3rd wheel 76 teeth, 6-leaf escape pinion, larger than for regular, Drop Calendar attachments, Juno. Pendulum, from center of movement to end of rating screw, 21 1/2", with No. 22 ball; diameter of ball 3 11/16". Wood rod. Crown 32, dead-beat, center 2". Length 5 5/8", width 5 7/8", depth 3 3/4". Mainspring No. 131, 9' x 11/16" x .015. — $85.00

# 1907 SETH THOMAS CLOCK MOVEMENT DESCRIPTIONS

**No. 41 P.**—8 Day, large, pendulum, time, spring, Nos 4 and 5 Office Time Pendulum, from center of movement to end of rating screw, 9 1/2", with No. 12 ball; diameter of ball 2 7/32". Crown 46, recoil, center 1". Length 5 5/8", width 5 7/8", depth 3 3/4". Mainspring No. 131, 9' x 11/16" x .015. — $75.00

**No. 42.**—8 Day, small, pendulum, half-hour strike, spring. Pendulum, from center of movement to end of rating screw, 9 1/2", with No. 12 ball; diameter of ball 2 7/32". Crown 40, recoil, center 1 1/2". Length 5 7/8", width 4 13/16", depth 3 3/8". Time mainspring No. 25, 6' x 3/4" x .020. Strike mainspring No. 25. — $75.00

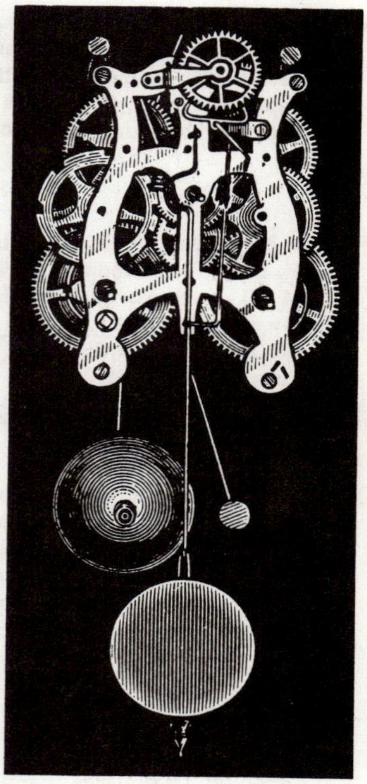

No. 42.

**No. 42 A.**—8 Day, small, pendulum, half-hour strike, spring. Pendulum, from center of movement to end of rating screw, 7", with No. 12 ball; diameter of ball 2 7/32". Crown 49, recoil, center 1 1/2". Length 5 7/8", width 4 13/16", depth 3 3/8". Time mainspring No. 25, 6' x 3/4" x .020. Strike mainspring No. 25. — $80.00

**No. 42 B.**—8 Day, small, pendulum, half-hour strike, spring. Pendulum, from center of movement to end of rating screw, 6 3/8", with No. 12 ball; diameter of ball 2 7/32". Crown 52, recoil, center 1 1/2". Length 5 7/8", width 4 13/16", depth 3 3/8". Time mainspring No. 25, 6' x 3/4" x .020. Strike mainspring No. 25. — $75.00

**No. 42 C.**—8 Day, small, pendulum, half-hour strike, spring. Pendulum, from center of movement to end of rating screw, 9 3/4", with No. 12 ball; diameter of ball 2 7/32". Crown 33, recoil, center 1 1/2". Length 5 7/8", width 4 13/16", depth 3 3/8". Time mainspring No. 25, 6' x 3/4" x .020. Strike mainspring No. 25. — $75.00

**No. 43.**—8 Day, large, pendulum, strike, spring, Harp Frame, V. P. Pendulum, from center of movement to end of rating screw, 9 1/4", with No. 12 ball; diameter of ball 2 7/32". Crown 46, recoil, center 1". Length 6 13/16", width 6 3/8", depth 2 13/16". Time mainspring No. 41, 10' x 11/16" x .018. Strike mainspring No. 41. — $75.00

**No. 43 A.**—8 Day, large, pendulum, strike, spring, Harp Frame, slow strike, V. P. Pendulum, from center of movement to end of rating screw, 9 1/4", with No. 12 ball; diameter of ball 2 7/32". Crown 46, recoil, center 1". Length 6 13/16", width 6 3/8", depth 2 13/16". Time mainspring No. 41, 10' x 11/16" x .018. Strike mainspring No. 41. — $75.00

**No. 43 B.**—8 Day, large, pendulum, strike, spring, Harp Frame, V. P., G. B., Toledo. Pendulum, from center of movement to end of rating screw, 8 3/16", with No. 12 ball; diameter of ball 2 7/32". Crown 50, recoil, center 1 1/2". Length 6 13/16", width 6 3/8", depth 3 5/13". Time mainspring No. 41, 10' x 11/16" x .018. Strike mainspring No. 41. — $75.00

**No. 43 C.**—8 Day, large, pendulum, strike, spring, Harp Frame, V. P., W. B. Pendulum, from center of movement to end of rating screw, 8 3/16", with No. 12 ball; diameter of ball 2 7/32". Crown 50, recoil, center 1". Length 6 13/16", width 6 3/8", depth 2 13/16". Time mainspring No. 131, 9' x 11/16" x .015. Strike mainspring No. 131. — $75.00

**No. 44.**—8 Day, large, pendulum, strike, spring, Square Frame. Pendulum, from center of movement to end of rating screw, 9 1/2", with No. 12 ball; diameter of ball 2 7/32". Crown 46, recoil, center 1". Length, 7 1/4", width 6 1/4", depth 2 7/8". Time mainspring No. 131, 9' x 11/16" x .015. Strike mainspring No. 131. — $75.00

**No. 44 A.**—8 Day, large, pendulum, strike, spring, Square Frame, Celestial. Pendulum, from center of movement to end of rating screw, 9 1/2", with No. 12 Ball; diameter of ball 2 7/32". Crown 46, recoil, center 1". Length 7 1/4", width 6 1/4", depth 2 7/8". Time mainspring No. 131, 9' x 11/16" x .015. Strike mainspring No. 131. — $75.00

**No. 44 B.**—8 Day, large, pendulum, strike, spring, Square Frame, V. P., G. B. Pendulum, from center of movement to end of rating screw, 9 3/4", with No. 36 ball; diameter of ball 2 7/16". Crown 46, recoil, center 1". Length 6 7/8", width 6 1/4", depth 2 7/8". Time mainspring No. 41, 10' x 11/16" x .018. Strike mainspring No. 41. — $75.00

**No. 44 C.**—8 Day, large, pendulum, strike, spring, Square Frame, V. P., W. B. Pendulum, from center of movement to end of rating screw, 9 3/4", with No. 36 ball; diameter of ball 2 7/16". Crown 46, recoil, center 1". Length 6 7/8", width 6 1/4", depth 2 7/8". Time mainspring No. 131, 9' x 11/16" x .015. Strike mainspring No. 131. — $75.00

# 1907 SETH THOMAS CLOCK MOVEMENT DESCRIPTIONS

**No. 44 D.**—8 Day, large, pendulum, strike, spring, Square Frame, V. P. W. B. Pendulum, from center of movement to end of rating screw, 8 1/2", with No. 36 ball; diameter of ball 2 7/16". Crown 50, recoil, center 1". Length 6 7/8", width 6 1/4", depth 2 7/8". Time mainspring No. 131, 9' x 11/16" x .015. Strike mainspring No. 131.

$75.00

**No. 44 E.**—8 Day, large, pendulum, strike, spring, Square Frame, Panama, V. P., G. B. Pendulum, from center of movement to end of rating screw, 13 1/4", with No. 33 ball; diameter of ball 2 11/16". Rod in two sections, lower part wood. Crown 38, recoil, center 1". Length 6 7/8", width 6 1/4", depth 2 7/8". Time mainspring No. 41, 10' x 11/16" x .018. Strike mainspring No. 41

$85.00

**No. 44 F.**—8 Day, large, pendulum, strike, spring, Square Frame, Globe Strike. Pendulum, from center of movement to end of rating screw, 17 3/4", with No. 22 ball; diameter of ball 3 11/16". Rod in two sections, lower part wood. Crown 32, dead-beat, center 2". Length 7 1/4", width 6 1/4", depth 3 7/8". Time mainspring No. 131, 9' x 11/16" x .015. Strike mainspring No. 131.

$85.00

**No. 44 G.**—8 Day, large, pendulum, strike, spring, Square Frame, 12" Drop Strike. Pendulum, from center of movement to end of rating screw, 13", with No. 12 ball; diameter of ball 2 7/32". Crown 38, dead-beat, center 2". Length 7 1/4", width 6 1/4", depth 3 7/8". Time mainspring No. 131, 9' x 11/16" x .015. Strike mainspring No. 131.

$85.00

**No. 44 H.**—8 Day, large, pendulum, strike, spring, Square Frame, 10" Drop Strike. Pendulum, from center of movement to end of rating screw, 12", with No. 12 ball; diameter of ball 2 7/32". Crown 40, dead-beat, center 2". Length 7 1/4", width 6 1/4", depth 3 7/8". Time mainspring No. 131, 9' x 11/16" x .015. Strike mainspring No. 131.

$85.00

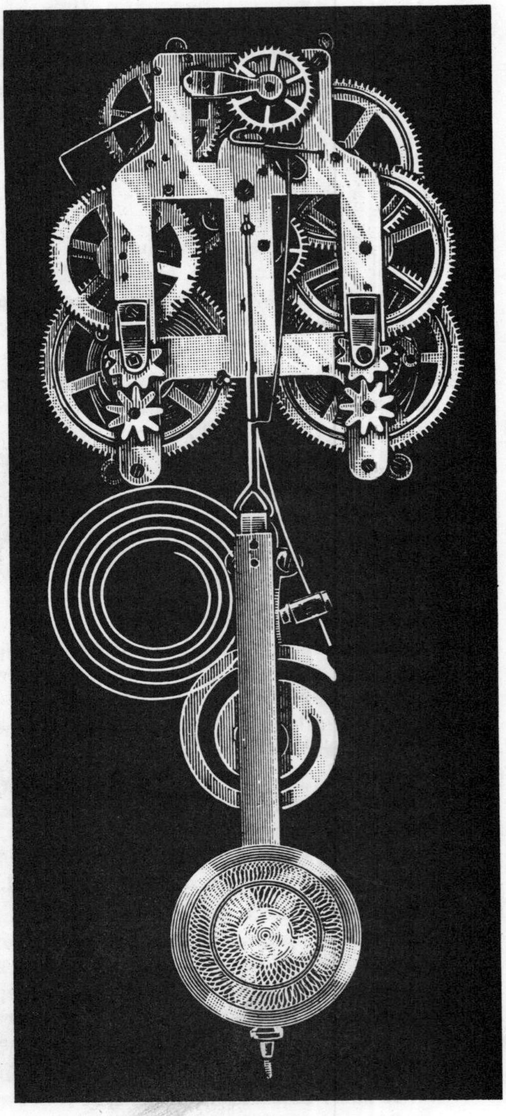

No. 44 E.

**No. 44 I.**—8 Day, large, pendulum, strike, spring, Square Frame, English. Pendulum, from center of movement to end of rating screw, 13", with No. 12 ball; diameter of ball 2 7/32". Crown 38, recoil, center 2 1/4". Length 7 1/4", width 6 1/4", depth 4 3/8". Time mainspring No. 131, 9' x 11/16" x .015. Strike mainspring No. 131. $85.00

**No. 44 J.**—8 Day, large, pendulum, strike, spring, Square Frame, English. Pendulum, from center of movement to end of rating screw, 7 3/4", with No. 12 ball; diameter of ball 2 7/32". Crown 53, recoil, center 2 1/4". Length 7 1/4", width 6 1/4", depth 4 3/8". Time mainspring No. 131, 9' x 11/16" x .015. Strike mainspring No. 131. $85.00

**No. 44 K.**—8 Day, large, pendulum, strike, spring, Square Frame, English. Pendulum, from center of movement to end of rating screw, 6 1/2", with No. 12 ball; diameter of ball 2 7/32". Crown 55, recoil, center 2 1/4". Length 7 1/4", width 6 1/4", depth 4 3/8". Time mainspring No. 131, 9' x 11/16" x .015. Strike mainspring No. 131. $85.00

*To receive top prices, movements must be complete and original, and they must be in good running or repairable condition. Damaged, worn-out or missing parts will reduce prices accordingly.*

**No. 44 L.** No information at this time.

# 1907 SETH THOMAS CLOCK MOVEMENT DESCRIPTIONS

**No. 44 M.**—8 Day, large, pendulum, strike, spring, Square Frame, old Angelus. Pendulum, from center of movement to end of rating screw, 9 1/2", with No. 12 ball; diameter of ball 2 7/32". Crown 46, recoil, center 1". Length 7 1/4", width 6 1/2", depth 2 7/8". Time mainspring No. 131, 9' x 11/16" x .015. Strike mainspring No. 131. $85.00 9 1/2

**No. 44 N.**—8 day, large, pendulum, strike, spring, Square Frame, Jackson Calendar. Pendulum, from center of movement to end of rating screw, 13", with No. 12 ball; diameter of ball 2 7/32". Crown 38, recoil, center 1 1/2". Length 7 1/4", width 6 1/4", depth 4 3/8". Time mainspring No. 131, 9' x 11/16" x .015. Strike mainspring No. 131. $100.00 Unusual

**No. 44 O.**—8 day, large, pendulum, strike, spring, Square Frame, Drop Octagon Calendar. Pendulum, from center of movement to end of rating screw, 13", with No. 12 ball; diameter of ball 2 7/32". Crown 38, deadbeat, center 2". Length 7 1/4", width 6 1/4", depth 3 7/8". Time mainspring No. 131, 9' x 11/16" x .015. Strike mainspring No. 131. $85.00

**No. 44 P.**—8 Day, large, pendulum, strike, spring, Square Frame, with calendar attachments, Nos. 6 and 7 Office Calendar. Pendulum, from center of movement to end of rating screw, 9 1/2", with No. 12 ball; diameter of ball 2 7/32". Crown 46, recoil, center 1 1/4". Length 7 1/4", width 6 1/4", depth 3 3/8". Time mainspring No. 41, 10' x 11/16" x .018. Strike mainspring No. 41. $100.00

**No. 44 Q.**—8 Day, large, pendulum, strike, spring, Square Frame, with calendar attachments, Nos. 3, 4, 6 and 7 Parlor Calendars. Pendulum, from center of movement to end of rating screw, 8 5/16", with No. 12 ball; diameter of ball 2 7/32". Crown 50, recoil, center 1 1/4". Length 7 1/4", width 6 1/4", depth 3 3/8". Time mainspring No. 41, 10' x 11/16" x .018. Strike mainspring No. 41. $100.00

**No. 44 R.**—8 Day, large, pendulum, strike, spring, Square Frame, with calendar attachments, No. 5 Parlor Calendar. Pendulum, from center of movement to end of rating screw, 8", with No. 12 ball; diameter of ball 2 7/32". Crown 52, recoil, center 1 1/4". Length 7 1/4", width 6 1/4", depth 3 3/8". Time mainspring No. 41, 10' x 11/16" x .018. Strike mainspring No. 41. $100.00

**No. 44 S.**—8 Day, large, pendulum, strike, spring, Square Frame, with calendar attachments, Slow Strike, G. B. Nos. 8 and 9 Parlor Calendars. Pendulum, from center of movement to end of rating screw, 8 5/16", with No. 12 ball; diameter of ball 2 7/32". Crown 50, recoil, center 1 1/4". Length 7 1/4", width 6 1/4", depth 3 3/8". Time mainspring No. 41, 10' x 11/16", x .018. Strike mainspring No. 41. $100.00

*To receive top prices, movements must be complete and original, and they must be in good running or repairable condition. Damaged, worn-out or missing parts will reduce prices accordingly.*

**No. 44 T.** No information at this time.

**No. 44 U.**—8 Day, large, pendulum, strike, spring, Square Frame, Baird Strike, no stops. Discontinued. Pendulum, from center of movement to end of rating screw, 17 1/4", with No. 12 ball · diameter of ball 2 7/32". Crown 32, recoil, center 1 3/4". Length 7 1/4", width 6 1/4", depth 3 7/8". Time mainspring No. 131, 9' x 11/16" x .015. Strike mainspring No. 131. $100.00

**No. 44 V.**—8 Day, large, pendulum, strike, spring, Square Frame, short hammer, Chatham. Pendulum, from center of movement to end of rating screw, 6 1/2", with No. 12 ball; diameter of ball 2 7/32". Crown 55, recoil, center 2 1/4". Length 7 1/4", width 6 1/4", depth 4 3/8". Time mainspring No. 131, 9' x 11/16" x .015. Strike mainspring No. 131. $85.00

**No. 44 W.**—8 Day, large, pendulum, half-hour strike, spring, Slow Strike, Square Frame, with drop calendar attachments, new Dixie. Pendulum, from center of movement to end of rating screw, 9 1/2", with No. 12 ball; diameter of ball 2 7/32". Crown 46, recoil, center 2". Length 7 1/4", width 6 1/4", depth 3 7/8". Time mainspring No. 41, 10' x 11/16" x .018. Strike mainspring No. 41. $100.00

# 1907 SETH THOMAS CLOCK MOVEMENT DESCRIPTIONS

**No. 44 X.**—8 Day, large, pendulum, strike, spring, Square Frame, with drop calendar attachments, Globe. Pendulum, from center of movement to end of rating screw, 17 3/4", with No. 22 ball; diameter of ball 3 11/16". Rod in two sections, lower part wood. Crown 32, dead-beat, center, 2". Length 7 1/4", width 6 1/4", depth 3 7/8". Time mainspring No. 131, 9′ x 11/16" x .015. Strike mainspring No. 131. $100.00

**No. 45.**—8 Day, small, pendulum, half-hour strike, spring, Carson. Discontinued. Crown 28, center 7/8". Length 5 3/4", width 4 7/16", depth 2 7/8". Time mainspring No. 33, 6′ x 3/4" x .015. Strike mainspring No. 39, 4 1/2′ x 3/4" x .012. $55.00

**No. 45 A.**—8 day, small, pendulum, half-hour strike, spring, plain escapement. Discontinued. Length 5 3/4", width 4 7/16", depth 2 7/8". Time mainspring No. 33, 6′ x 3/4" x .015. Strike mainspring No. 39, 4 1/2′ x 3/4" x .012. $55.00

**No. 45 B.**—8 Day, small, pendulum, half-hour strike, spring, V. E. Discontinued. Length 4 13/16", width 4 5/16", depth 3 1/8". Time mainspring No. 33, 6′ x 3/4" x .015. Strike mainspring No. 39, 4 1/2′, x 3/4" x .012. $55.00

**No. 46.**

*To receive top prices, movements must be complete and original, and they must be in good running or repairable condition. Damaged, worn-out or missing parts will reduce prices accordingly.*

**No. 47.**—8 Day, small, pendulum, strike, spring, Mitchel, Vance & Co. Discontinued. Pendulum, from center of movement to bottom of ball, 3 5/8", with follower in center of regulator; diameter of ball 1 9/16". Crown 24, center 5/8". Length 4", width 3 7/16", depth 3 1/8". Time mainspring No. 38, 4 1/2′ x 3/4" x .013 (loop). Strike mainspring No. 38. $55.00

**No. 48.**—8 Day, small, pendulum, half-hour strike, spring, steel cut pinions, V. E. Pendulum, from center of movement to bottom of ball, 4 3/4", with follower in center of regulator. Crown 28, center 15/16". Length 4 1/2", width 4", depth 2 7/8". Time mainspring No. 36, 7′ x 3/4" x .013 (loop). Strike mainspring No. 32, 5′ x 3/4" x .016 (loop). $90.00

**No. 48 A.**—8 Day, small, pendulum, half-hour strike, spring (No. 48 1/2), P.E., N. Y. Store, Low Regulator. Discontinued. Pendulum, from center of movement to bottom of ball, 4", with follower in center of regulator. Crown 28, center 15/16". Length 4", width 4", depth 2 3/4". Time mainspring No. 36, 7′ x 3/4" x .013 (loop). Strike mainspring, No. 32, 5′ x 3/4" x .016 (loop). $85.00

**No. 48 B.**—8 Day, small, pendulum, half-hour strike, spring (No. 48 1/2, long rod), P. E., Low Regulator. Discontinued. Pendulum, from center of movement to bottom of ball, 5 1/16", with follower in center of regulator. Crown 28, center 15/16". Length 4 1/2", width 4", depth 2 3/4". Time mainspring No. 36, 7′x 3/4"x .013 (loop). Strike mainspring No. 32, 5′ x 3/4" x .016 (loop). $85.00

**No. 48 C.**—8 Day, small, pendulum, half-hour strike, spring (No. 49, short rod), P.E., thin plates, iron case. Discontinued. Pendulum, from center of movement to bottom of ball, 4", with follower in center of regulator. Crown 28, center 15/16". Length 4 1/2", width 4 3/4", depth 2 3/4". Time mainspring No. 36, 7′ x 3/4" x .013 (loop). Strike mainspring No. 32, 5′ x 3/4" x .016 (loop). $85.00

**No. 48 D.**—8 Day, small, pendulum, half-hour strike, spring (No. 49, long rod), P. E., thin plates, Fenton. Pillar brackets, high regulator. Discontinued. Pendulum, from center of movement to bottom of ball, 5 1/16", with follower in center of regulator. Crown 28, center 15/16". Length 4 1/2", width 4 3/4", depth 2 3/4". Time mainspring No. 36, 7′ x 3/4" x .013 (loop). Strike mainspring No. 32, 5′x 3/4" x .016 (loop). $85.00

# 1907 SETH THOMAS CLOCK MOVEMENT DESCRIPTIONS

**No. 48 E.**—8 Day, small, pendulum, half-hour strike, spring (No. 48 1/2, change in train), Calais. Discontinued. Pendulum, from center of movement to bottom of ball, 8 5/8", with follower in center of regulator. Crown 28, center 15/16". Length 4 1/2", width 4 3/4", depth 2 3/4". Time mainspring No. 36, 7' x 3/4" x .013 (loop). Strike mainspring No. 32, 5' x 3/4" x .016 (loop). $85.00

**No. 48 F.**—8 Day, small, pendulum, half-hour strike, spring (No. 48 1/2, long rod), P. E., high regulator, pillar brackets. Discontinued. Pendulum, from center of movement to bottom of ball, 5 1/16", with follower in center of regulator. Crown 28, center 15/16". Length 4 1/2", width 4 3/4", depth 2 3/4". Time mainspring No. 36, 7' x 3/4" x .013 (loop). Strike mainspring No. 32, 5' x 3/4" x .016 (loop). $85.00

**No. 48 G.**—8 Day, small, pendulum, half-hour strike, spring, P. E., thin plates, buffed back plate, position of dial stud and hammer changed, Kroeber. Discontinued. Pendulum, from center of movement to bottom of ball, 4", with follower in center of regulator. Crown 28, center 15/16". Length 8", width 4 3/4", depth 2 3/4". Time mainspring No. 36, 7' x 3/4" x .013 (loop). Strike mainspring No. 32, 5' x 3/4" x .016 (loop). $85.00

**No. 48 H.**—8 Day, small, pendulum, half-hour strike, spring, P. E., thin plates, buffed back plate. Discontinued. Pendulum, from center of movement to bottom of ball, 4", with follower in center of regulator. Crown 28, center 15/16". Length 4 1/2", width 4 3/4", depth 2 3/4". Time mainspring No. 36, 7' x 3/4" x .013 (loop). Strike mainspring No. 32, 5' x 3/4" x .016 (loop). $85.00

**No. 48 I.**—8 Day, small, pendulum, half-hour strike, spring, P. E., thin plates, buffed back plate, Kroeber. Discontinued. Pendulum, from center of movement to bottom of ball, 5 1/16", with follower in center of regulator. Crown 28, center 15/16". Length 4 1/2", width 4 3/4", depth 2 3/4". Time mainspring No. 36, 7' x 3/4" x .013 (loop). Strike mainspring No. 32, 5' x 3/4" x .016 (loop). $85.00

**No. 48 J.**—8 Day, small, pendulum, half-hour strike, spring, solid back plate buffed, brass pillars, spring boxes, long pillar brackets. Used in place of No. 48 C, low regulator. Pendulum, from center of movement to bottom of ball, 4", with follower in center of regulator; diameter of ball 1 11/16". Crown 28, center 15/16". Length 4 1/2", width 4 3/4", depth 2 3/4". Time mainspring No. 150, 7' x 3/4" x .013. Strike mainspring No. 151, 5' x 3/4" x .016. $85.00

**No. 48 K.**—8 Day, small, pendulum, half-hour strike, spring, solid back plate buffed, brass pillars, spring boxes. Used in place of No. 48 H, low regulator. Pendulum, from center of movement to bottom of ball, 4", with follower in center of regulator; diameter of ball 1 11/16". Crown 28, center 15/16". Length 4 1/2", width 4 3/4", depth 2 3/4". Time mainspring No. 150, 7' x 3/4" x .013. Strike mainspring No. 151, 5' x 3/4" x .016. $85.00

**No. 48 L.**—8 Day, small, pendulum, half-hour strike, spring, solid back plate buffed, brass pillars, spring boxes, low regulator, P. E. Used in place of No. 48 A, N. Y. Store. Pendulum, from center of movement to bottom of ball, 4", with follower in center of regulator; diameter of ball 1 11/16". Crown 28, center 15/16". Length 4 1/2", width 4 3/4", depth 2 3/4". Time mainspring No. 150, 7' x 3/4" x .013. Strike mainspring No. 151, 5' x 3/4" x .016. $85.00

**No. 48 M.**—8 Day, small, pendulum, half-hour strike, spring, solid back plate buffed, brass pillars, spring boxes, train changed low regulator, P. E. Made for Empire. Pendulum, from center of movement to bottom of ball, 6 5/8" with follower in center of regulator; with Wilkinson's mercurial pendulum. Crown 28, center 15/16". Length 4 1/2", width 4 3/4", depth 2 3/4". Time mainspring No. 150, 7' x 3/4" x .013. Strike mainspring No. 151, 5' x 3/4" x .016. $85.00

**No. 48 N.**—8 Day, small, pendulum, half-hour strike, spring, solid back plate buffed, brass pillars, spring boxes, low regulator, P. E. Used in Empire. Pendulum, from center of movement to bottom of ball, 5 1/16" with follower in center of regulator; with Empire mercury pendulum, 5 7/8" to end of rod; diameter of ball 1 11/16". Crown 28, center 15/16". Length 4 1/2", width 4 3/4", depth 2 3/4". Time mainspring No. 150, 7' x 3/4" x .013. Strike mainspring No. 151, 5' x 3/4" .016. $85.00

# 1907 SETH THOMAS CLOCK MOVEMENT DESCRIPTIONS

**No. 48 O.**—8 Day, small, pendulum, half-hour strike, spring, solid back plate buffed, brass pillars, spring boxes, train changed, pillar brackets, high regulator, P. E. Used in place of No. 48 E. Pendulum, from center of movement to bottom of ball, 8 5/8", with follower in center of regulator; diameter of ball 2 1/4". Crown 28, center 15/16". Length 4 1/2", width 4 3/4", depth 2 3/4". Time mainspring No. 150, 7′ x 3/4" x .013. Strike mainspring No. 151, 5′ x 3/4" x .016. — $85.00

**No. 48 P.**—8 Day, small, pendulum, half-hour strike, spring, solid back plate buffed, brass pillars, high regulator, pillar brackets. To be used in place of No. 48 D and No. 48 F. Pendulum, from center of movement to bottom of ball, 5 1/16", with follower in center of regulator; diameter of ball 1 11/16". Crown 28, center 15/16". Length 4 1/2", width 4 3/4", depth 2 3/4". Time mainspring No. 150, 7′ x 3/4" x .013. Strike mainspring No. 151, 5′ x 3/4" x .016. — $85.00

**No. 48 Q.**—8 Day, small, pendulum, half-hour strike, spring, solid back plate buffed, brass pillars, spring boxes, low regulator, P. E. Used in Tory. Pendulum, from center of movement to bottom of ball, 5 1/16", with follower in center of regulator. Crown 28, half dead-beat, center 15/16". Length 4 1/2", width 4 3/4", depth 2 3/4". Time mainspring No. 150, 7′ x 3/4" x .013. Strike mainspring No. 151, 5′ x 3/4" x .016. — $85.00

**No. 48 R.**—8 Day, small, pendulum, half-hour strike, spring, solid back plate buffed, brass pillars, spring boxes, high regulator, P. E. Used in Wales, Parma, Modena and Touraine. Pendulum from center of movement to bottom of ball, 5 1/16", with follower in center of regulator. Crown 28, half dead-beat, center 15/16". Length 4 1/2", width 4 3/4", depth 2 3/4". Time mainspring No. 150, 7′ x 3/4" x .013. Strike mainspring No. 151, 5′ x 3/4", x .016. — $85.00

*To receive top prices, movements must be complete and original, and they must be in good running or repairable condition. Damaged, worn-out or missing parts will reduce prices accordingly.*

**No. 49.** No information at this time.

**No. 50.**—15 Day, fine, pendulum, time, spring. Discontinued. — $55.00

**No. 51.**—15 Day, fine, pendulum, half-hour strike, spring, steel cut pinions, G. B. Used with cylindrical ball for Nymph. Pendulum, from center of movement to bottom of ball, 3 5/8" with follower in center of regulator, with flat ball; size of ball 1 9/16". Crown 24, center 5/8". Length 4", width 3 3/8", depth 3". Time mainspring No. 37, 4 1/2′ x 3/4" x .013. Strike mainspring No. 37. — $90.00

**No. 51 A.**—15 Day, fine, pendulum, half-hour strike, steel cut pinions, spring, cup bell. Pendulum, from center of movement to bottom of ball, 3 5/8" with follower in center of regulator, flat ball; size of ball 1 9/16". Crown 24, center 5/8". Length 3 13/16", width 3 3/8", depth 3". Time mainspring No. 37, 4 1/2′ x 3/4" x .013. Strike mainspring No. 37. — $80.00

**No. 51 B.**—15 Day, fine, pendulum, half-hour strike, spring, steel cut pinions, low escapement, G. B. Pendulum, from center of movement to bottom of ball, 3 5/8" with follower in center of regulator, flat ball; size of ball 1 9/16". Crown 24, center 5/8". Length 3 13/16", width 3 3/8", depth 3". Time mainspring No. 37, 4 1/2′ x 3/4" x .013. Strike mainspring No. 37. — $65.00

**No. 51 C.**—15 Day, fine, pendulum, half-hour strike, spring, steel cut pinions, high regulator, G. B. Pendulum, from center of movement to bottom of ball, 3 5/8" with follower in center of regulator, flat ball; size of ball 1 9/16". Crown 24, center 5/8". Length 4", width 3 3/8" depth 3". Time mainspring No. 37, 4 1/2′ x 3/4" x .013. Strike mainspring No. 37. — $75.00

No. 51.

**No. 51 D.**—15 Day, fine, pendulum, half-hour strike, spring, steel cut pinions, G. B. Pendulum, from center of movement to bottom of ball, 3 5/8" with follower in center of regulator, flat ball; size of ball 1 9/16". Crown 24, center 11/16". Length 4", width 3 5/8", depth 3 1/8". Time mainspring No. 37, 4 1/2′ x 3/4" x .013. Strike mainspring No. 37. — $75.00

# 1907 SETH THOMAS CLOCK MOVEMENT DESCRIPTIONS

**No. 51 E.**—15 Day, fine, pendulum, half-hour strike, spring, steel cut pinions. Orchid, change in train, center wheel 46 teeth and third pinion 8 teeth large, open crutch. Used in Empire. Pendulum from center of movement to bottom of ball, 5", with follower in center of regulator, with flat ball. When using mercury pendulum it is 5 13/16". Crown 24, center 5/8". Length 4", width 4 1/8", depth 3". Time mainspring No. 37, 4 1/2' x 3/4" x .013. Strike mainspring No. 37. $75.00

**No. 51 F.**—15 Day, fine, pendulum, half-hour strike, spring, steel cut pinions, change in train, center wheel 42 teeth, third pinion 10. G. B., Empire. Pendulum, from center of movement to bottom of ball, 9 5/8", with follower in center of regulator; cylindrical ball used. Crown 24, center 5/8". Length 4", width 3 3/8", depth 3". Time mainspring No. 37, 4 1/2' x 3/4" x .013. Strike mainspring No. 37. $75.00

**No. 51 G.**—15 Day, fine, pendulum, half-hour strike, spring, steel cut pinions, change in train, fifth wheel 57 teeth, G. B., open crutch, Empire, No. 40 case. Pendulum, from center of movement to bottom of rating screw, 4 3/4", with follower in center of regulator. Mercury pendulum. Crown 24, center 5/8". Length 4", width 4 1/8", depth 3". Time mainspring, No. 37, 4 1/2' x 3/4" x .013. Strike mainspring No. 37. $75.00

**No. 51 H.**—15 Day, fine, pendulum, strike, spring, steel cut pinions, plates same as regular No. 51, train same as Orchid, G. B., open crutch, Empire. Pendulum (Empire mercurial), from center of movement to end of rod, 5 13/16", with follower in center of regulator. Crown 24, half dead-beat, center 5/8". Length 4", width 4 1/8", depth 3". Time mainspring No. 37, 4 1/2' x 3/4" x .013. Strike mainspring No. 37. $75.00

**No. 51 I.**—15 Day, fine, pendulum, half-hour strike, spring, steel cut pinions, G. B., change in train, center wheel 41 teeth, third pinion same as in No. 51 F, open crutch. Made for Empire. Pendulum, from center of movement to bottom of ball, 11 3/8", with follower in center of regulator, fancy ball. Crown 24, half dead-beat, center 5/8". Length 4", width 4 1/8", depth 3". Time mainspring No. 37, 4 1/2' x 3/4" x .013. Strike mainspring No. 37. $75.00

**No. 51 J.**—15 Day, fine, pendulum, half-hour strike, spring, steel cut pinions, same plates as Orchid with brackets, train same as regular No. 51, G. B., for Empire. Pendulum, from center of movement to bottom of ball, 3 5/8", with follower in center of regulator, flat ball; size of ball 1 9/16". Crown 24, half dead-beat, center 5/8". Length 4", width 3 3/8", depth 3". Time mainspring No. 37, 4 1/2' x 3/4" x .013. Strike mainspring No. 37. $75.00

**No. 51 K.**—15 Day, fine, pendulum, half-hour strike, spring, steel cut pinions, high regulator, brackets for holding movement in case screwed to plate, Rivoli. Pendulum, from center of movement to bottom of ball, 3 5/8", with follower in center of regulator, with flat ball; size of ball 1 9/16". Crown 24, center 5/8". Length 4", width 3 3/8", depth 3". Time mainspring No. 37, 4 1/2' x 3/4" x .013. Strike mainspring No. 37. $75.00

**No. 51 L.**—15 Day, fine, pendulum, half-hour strike, spring, steel cut pinions, low regulator bridge, Tambour. Pendulum, from center of movement to bottom of ball, 3 5/8", with follower in center of regulator, with flat ball; size of ball 1 9/16". Crown 24, center 5/8". Length 4", width 3 3/8", depth 3", Time mainspring No. 37, 4 1/2' x 3/4" x .013. Strike mainspring No. 37. $75.00

**No. 51 M.**—15 Day, fine, pendulum, half-hour strike, spring, steel cut pinions. Plates for low escapement with No. 51 E brackets. Used in Empire. Train same as regular No. 51. Pendulum, from center of movement to bottom of ball, 3 5/8", with follower in center of regulator, with flat ball; size of ball 1 9/16". Crown 24, center 5/8". Length 4", width 3 3/8", depth 3". Time mainspring No. 37, 4 1/2' x 3/4" x .013. Strike mainspring No. 37. $75.00

**No. 51 N.**—15 Day, fine, pendulum, half-hour strike, spring, steel cut pinions, cup bell, low regulator. Change in train, center wheel 41 teeth, third pinion 8 leaves, open crutch, for No. 50 Empire. Pendulum, from center of movement to bottom of ball, 7 3/8" with follower in center of regulator. Fancy ball. Crown 24, half dead-beat, center 5/8". Length 4", width 4 1/8", depth 3 1/2". Time mainspring No. 37, 4 1/2' x 3/4" x .013. Strike mainspring No. 37. $75.00

# 1907 SETH THOMAS CLOCK MOVEMENT DESCRIPTIONS

**No. 52.**—15 Day, fine, pendulum, strike, spring, cup bell, Andrews. Discontinued. Pendulum, from center of movement to bottom of ball, 3 1/2", with follower in center of regulator; diameter of ball 1 9/16". Crown 24, center 3/4". Length 4 1/4", width 3 5/8", depth 3 3/4". Time mainspring No. 37, 4 1/2' x 3/4" x .013. Strike mainspring No. 37. $75.00

**No. 53.**—8 Day, fine, pendulum, strike, spring, V. E., G. B. Discontinued. Pendulum, from center of movement to bottom of ball, 3 5/8", with follower in center of regulator; diameter of ball 1 9/16". Crown 28, center 7/8". Length 4 7/16", width 4", depth 3 1/2". Time mainspring No. 37, 4 1/2' x 3/4" x .013. Strike mainspring No. 37. $60.00

**No. 53 1/2.**—15 Day, fine, pendulum, half-hour strike, spring, V. E., steel cut pinions, jeweled pallets, G. B. Pendulum, from center of movement to bottom of ball, 4 5/8", with follower in center of regulator; diameter of ball 1 9/16", flat ball. Crown 28, center 1". Length 4 7/16", width 4", depth 3 1/2". Time mainspring No. 37, 4 1/2' x 3/4" x .013. Strike mainspring No. 37. $100.00

**No. 54.** No information at this time.

**No. 55.**—8 Day, time, weight, No. 10 Office Calendar, lantern pinions. Discontinued. Crown 44, dead-beat, center 1 1/2". Length 6 3/8", width 4 9/16", depth 4". $225.00

**No. 55 A.**—8 Day, time, weight, Nos. 12 and 13 Office Calendars, lantern pinions. Pendulum, from center of movement to end of rating screw, 22 1/8", with No. 27 ball; diameter of ball 4 9/16". Wood rod. Crown 40, dead-beat, center 1". Length 7 3/16", width 5 3/8", depth 4 7/16". $225.00

**No. 56.**
**No. 57.**
**No. 58.** No information at this time.
**No. 59.**
**No. 60.**

*To receive top prices, movements must be complete and original, and they must be in good running or repairable condition. Damaged, worn-out or missing parts will reduce prices accordingly.*

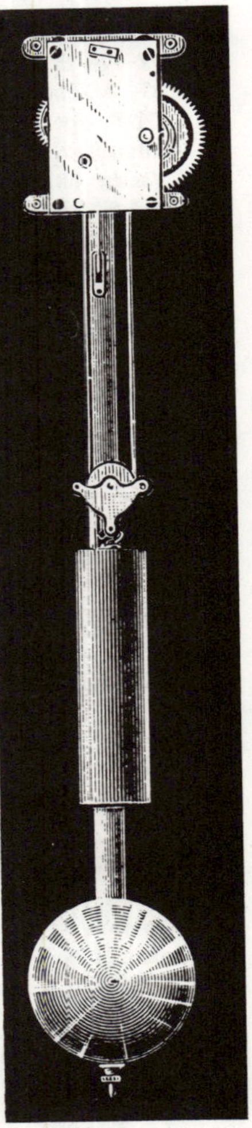

No. 61.

**No. 61.**—8 Day, time, weight, Nos. 2, 6, 9 and 30 regulators, lantern pinions, maintaining power. Pendulum, from center of movement to end of rating screw, 22 1/2", with No. 27 ball; diameter of ball 4 9/16". Wood rod. Crown 40, dead-beat, center 1". Length 4 1/2", width 4 7/8", depth 4 3/16". $200.00 w/Bracket

**No. 61 A.**—8 Day, time, weight, No. 18 regulator, lantern pinions, large barrel, maintaining power. Pendulum, from center of movement to end of rating screw, 41", with No. 32 ball; diameter of ball 8". Wood rod. Crown 30, dead-beat, center 1". Length 4 1/2", width 4 7/8", depth 4 3/16". $275.00

**No. 61 B.**—8 Day, time, weight, No. 6 Double Time, lantern pinions. Pendulum, from center of movement to end of rating screw, 22 1/2", with No. 27 ball; diameter of ball 4 9/16". Wood rod. Crown 40, dead-beat, center 1". Length 4 1/2", width 4 7/8", depth 4 3/16". Unusual $300.00 w/Bracket

**No. 62.**—8 Day, time, weight, Nos. 3 and 7 regulators, keystone plate, steel cut pinions, maintaining power. Pendulum, from center of movement to end of rating screw, 26 3/4", with No. 27 ball; diameter of ball 4 9/16". Wood rod. Crown 36, dead-beat, center 1 1/2". Length 5 3/4", width 5 9/16", depth 4 1/2". $300.00

# 1907 SETH THOMAS CLOCK MOVEMENT DESCRIPTIONS

| | |
|---|---|
| **No. 62 A.**—8 Day, time, weight, No. 1 Regulator, keystone plate, steel cut pinions, maintaining power. Pendulum, from center of movement to end of rating screw, 21 1/2", with No. 27 ball; diameter of ball 4 9/16". Wood rod. Crown 40, dead-beat, center 1 1/2". Length 5 3/4", width 5 9/16", depth 4 1/2". | $200.00 w/Bracket |
| **No. 62 B.**—8 Day, time, weight, No. 17 Regulator, keystone plate, steel cut pinions, large barrel, adjustable bracket, maintaining power. Pendulum, from center of movement to end of rating screw, 40 3/8", with No. 42 ball; diameter of ball 8". Wood rod. Crown 30, dead-beat, center 1 1/2". Length 7 11/16", width 7 1/4", depth 5 9/16" and 6 1/16". | Unusual $300.00 w/Bracket |
| **No. 62 C.**—8 Day, time, weight, No. 8 Regulator, keystone plate, steel cut pinions, maintaining power. Pendulum, from center of movement to end of rating screw, 31 3/8", with No. 32 ball; diameter of ball 8". Wood rod. Crown 34, dead-beat, center 1 1/2". Length 5 3/4", width 5 9/16", depth 4 1/2". | $325.00 w/Bracket |
| **No. 62 D.**—8 Day, time, weight, No. 5 Office Calendar, keystone plate, steel cut pinions, calendar attachments. Pendulum, from center of movement to end of rating screw, 26 3/4", with No. 27 ball; diameter of ball 4 9/16". Wood rod. Crown 36, dead-beat, center 1 1/2". Length 7 3/16", width 5 5/8", depth 5 13/16". | $350.00 w/Bracket |
| **No. 62 E.**—8 Day, time, weight, Nos. 9 and 11 Office Calendars, keystone plate, steel cut pinions, calendar attachments. Pendulum, from center of movement to end of rating screw, 40 3/8", with No. 42 ball; diameter of ball 8". Wood rod. Crown 30, center 1 1/2". Length 7 3/16", width 5 3/4", depth 5 3/4". | $350.00 w/Bracket |
| **No. 62 F.**—8 Day, time, weight, No. 17 Regulator, double time, keystone plate, steel cut pinions, maintaining power. Pendulum, from center of movement to end of rating screw, 40 3/8", with No. 42 Ball; diameter of ball 8". Wood rod. Crown 30, dead-beat, center 1 1/2". Length 5 3/4", width 5 1/2", depth 5 13/16". | Rare $750.00 |
| **No. 62 G.**—8 Day, time weight, keystone plate, steel cut pinions, adjustable bracket; studs for holding dial to movement are in the dial. Pendulum, from center of movement to end of rating screw, 40 3/8", with No. 42 ball; diameter of ball 8". Wood rod. Crown 30, dead-beat, center 1 1/2". Length 5 3/4", width 5 1/2", depth 5 13/16". | $275.00 w/Bracket |
| **No. 63.**—8 Day, strike, weight, No. 3 Strike Regulator, lantern pinions, maintaining power. Pendulum, from center of movement to end of rating screw, 26 3/4", with No. 27 ball; diameter of ball 4 9/16". Wood rod. Crown 36, dead-beat, center 1 5/16". Length 6", width 5 5/8", depth 4 7/16". | $375.00 w/Bracket |
| **No. 63 A.**—8 Day, strike, weight, No. 3 Strike Regulator, lantern pinions, maintaining power. Pendulum, from center of movement to end of rating screw, 39 5/8", with No. 27 ball; diameter of ball 4 9/16". Wood rod. Crown 30, dead-beat, center 1 5/16". Length 6", width 5 5/8", depth 4 7/16". | $275.00 w/Bracket |
| **No. 64.**—8 Day, time, weight, Nos. 4 and 5 Regulators, small keystone plate, steel cut pinions, hardened and polished, adjustable bracket, maintaining power. Pendulum, from center of movement to end of rating screw, 22 5/16", with No. 22 ball; diameter of ball 3 11/16". Wood rod. Crown 40 dead-beat, center 15/16". Length 5 11/16", width 4 3/4", depth 3 3/4" and 3 7/8". | $250.00 w/Bracket |
| **No. 65.**—8 Day, strike, weight, old No. 7 Regulator. Discontinued. | $275.00 w/Bracket |
| **No. 66.**—8 Day, strike, weight, Lincoln. Pendulum, from center of movement to end of rating screw, 14 1/2", with No. 22 ball; diameter of ball 3 11/16". Wood rod. Crown 34, half dead-beat, center 1". Length 6 1/2", width 4 3/4", depth 2 3/4". | $95.00 |
| **No. 66 A.**—8 Day, strike, weight, Flora. Pendulum, from center of movement to end of rating screw, 17 1/4", with No. 22 ball; diameter of ball 3 11/16". Wood rod. Crown 30, half dead-beat, center 1". Length 6 1/2", width 4 3/4", depth 2 3/4". | $95.00 |

# 1907 SETH THOMAS CLOCK MOVEMENT DESCRIPTIONS

**No. 67.**—8 Day, time, weight, sweep seconds, No. 16 Regulator, keyston plate, steel cut pinions, adjustable bracket, maintaining power. Pendulum, from center of movement to end of rating screw, 42 1/2", with No. 42 ball; diameter of ball 8". Wood rod. Crown 30, dead-beat, center 1". Length 7 11/16", width 7 5/16", depth 5 5/16" and 5 3/4". $600.00

**No. 67 A.**—8 Day, time, weight, sweep seconds, No. 19 Regulator, keystone plate, steel cut pinions, crutch for mercurial pendulum, adjustable bracket, maintaining power. Pendulum, from center of movement to end of rating screw, 42 1/4", with double jar mercurial pendulum. Crown 30, dead-beat, center 1". Length 8 5/16", width 7 5/16", depth 5 5/16" and 5 9/16". $750.00 w/Bracket & Mercury Pendulum

**No. 67 B.**—8 Day, time, weight, sweep seconds, keystone plate, steel cut pinions, adjustable bracket; studs for holding the dial to movement are riveted or screwed in the front plate; the dial is screwed to the studs after the movement is in the case. Pendulum, from center of movement to end of rating screw, 42 1/2", with No. 42 ball; diameter of ball 8". Wood rod. Crown 30, dead-beat, center 1". Length 7 11/16", width 7 5/16", depth 5 5/16" and 5 3/4".

**No. 68.**—8 Day, time, weight, No. 31 Regulator, large keystone plate, steel cut pinions, adjustable bracket, seconds pendulum, maintaining power. Pendulum, from center of movement to end of rating screw, 40 1/8", with No. 42 ball; diameter of ball 8". Wood rod. Crown 30, dead-beat, center 1 1/2". Length 10 1/2", width 8 3/16", depth 5 11/16" and 6 1/8". $375.00

**No. 68 A.**—8 Day, time, weight, large keystone plate, steel cut pinions, small barrel, adjustable bracket; studs for holding the dial to movement are riveted or screwed in front plate; the dial is screwed to the studs after the movement is in the case. Pendulum, from center of movement to end of rating screw, 40 1/8", with No. 42 ball; diameter of ball 8". Wood rod. Crown 30, dead-beat, center 1 1/2". Length 10 1/2", width 8 3/16", depth 5 11/16" and 6 1/8". $275.00 w/Bracket

**No. 69.**—8 Day, time, weight, No. 32 Regulator, large keystone plate, lantern pinions, seconds pendulum, adjustable bracket, maintaining power. Pendulum, from center of movement to end of rating screw, 40 1/8", with No. 32 ball; diameter of ball 8". Wood rod. Crown 30, dead-beat, center 1 1/2". Length 10 1/2", width 8 3/16", depth 5 3/4" and 6 1/8". $300.00 w/Bracket

**No. 69 A.**—8 Day, time, weight, large keystone plate, lantern pinions, seconds pendulum, adjustable bracket; studs for holding the dial to movement are riveted or screwed in the front plate; the dial is screwed to the studs after the movement is in the case. Pendulum, from center of movement to end of rating screw, 40 1/8", with No. 32 ball; diameter of ball 8". Wood rod. Crown 30, dead-beat, center 1 1/2". Length 10 1/2", width 8 3/16", depth 5 3/4" and 6 1/8". $275.00 w/Bracket

**No. 70.**—8 Day, time, weight, sweep seconds, Nos. 10 and 14 Regulators, steel cut pinions, hardened and polished, jeweled pallets, mercurial pendulum, adjustable bracket, maintaining power. Pendulum, from center of movement to end of rating screw, 44 3/8", with single jar mercurial pendulum. Crown 30, dead-beat, center 1 1/8". Length 10", width 9", depth 6 5/8" and 7 3/8". $800.00 w/Bracket & Mercury Pendulum

**No. 71.**—8 Day, time, weight, independent seconds, Nos. 12 and 15 Regulators, steel cut pinions, hardened and polished, jeweled pallets, adjustable bracket, maintaining power. Pendulum, from center of movement to end of rating screw, 41 1/2", with single jar mercurial pendulum. Crown 30, dead-beat, center 1 1/8". Length 12 1/2", width 9 5/16", depth 7" and 7 7/16". $400.00 w/Bracket

**No. 72.**—8 Day, weight, hour strike, cathedral bell, front plate polished, seconds pendulum, lantern pinions, adjustable bracket, maintaining power. Pendulum, from center of movement to end of rating screw, 40 1/4", with No. 50 ball; diameter of ball 6". Wood rod. Crown 30, dead-beat, center 1 7/8". Length 10", width 9 3/4", depth 8 7/16" and 8 7/8. $200.00

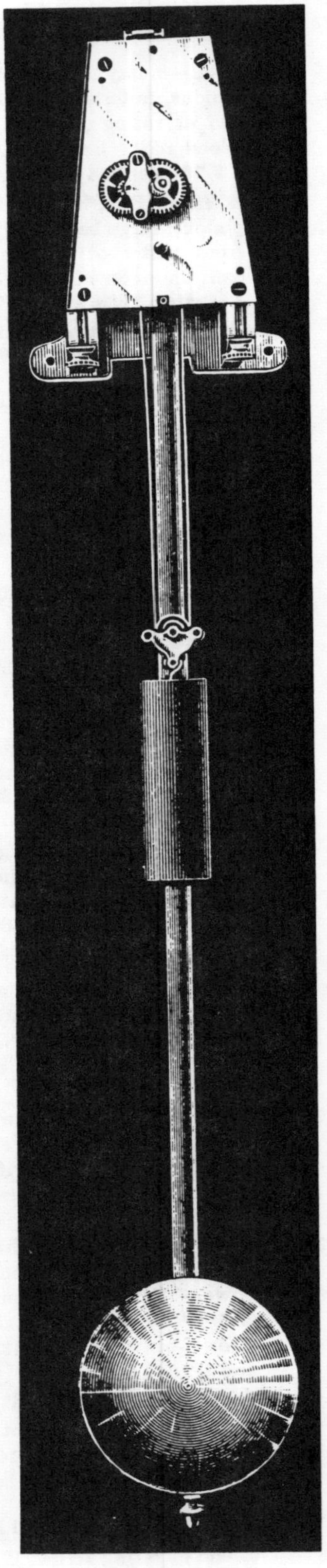

No. 68.

# 1907 SETH THOMAS CLOCK MOVEMENT DESCRIPTIONS

**No. 72 A.**—8 Day, weight, half-hour strike, cathedral bell, plates rayed, seconds pendulum, lantern pintons, adjustable bracket, maintaining power. Pendulum, from center of movement to end of rating screw, 40 1/4", with No. 50 ball; diameter of ball 6". Wood rod. Crown 30, dead-beat, center 1 7/8". Length 10", width 9 3/4", depth 8 7/16" and 8 7/8". $250.00 w/Bracket

**No. 72 B.**—8 Day, weight, half-hour strike, cathedral bell, moon phases, plates rayed, seconds pendulum, adjustable bracket. Movement arranged so that a seconds hand can be used. Moon disc on dial. Maintaining power. Pendulum, from center of movement to end of rating screw, 40 1/4", with No. 50 ball; diameter of ball 6". Wood rod. Crown 30, dead-beat, center 1 7/8". Length 10", width 9 3/4", depth 8 7/16" and 8 7/8". $250.00 w/Bracket

**No. 72 C.**—8 Day, weight, half-hour strike, cathedral bell, moon phases, plates rayed, seconds pendulum, adjustable bracket. No seconds hand can be used with this movement. Moon disc on movement. Maintaining power. Pendulum, from center of movement to end of rating screw, 40 1/4", with No. 50 ball; diameter of ball 6". Wood rod. Crown 30, dead-beat, center 1 7/8". Length 14 7/8", width 9 3/4", depth 8 7/16" and 8 7/8". $275.00 w/Bracket

**No. 72 D.**—8 Day, weight, half-hour strike, cathedral bell, moon phases, plates rayed, seconds pendulum. Movement arranged so that a seconds hand can be used. Movement rests on seat board; the pendulum is suspended from a stud secured to the back plate. Moon on dial. Maintaining power. Pendulum, from center of movement to end of rating screw, 40 1/4", with No. 50 ball; diameter of ball 6". Wood rod. Crown 30, dead-beat, center 1 7/8". Length 10", width 9 3/4", depth 5 1/2". $200.00

**No. 73.**—8 Day, weight, quarter strike, 4 bells, No. 21 Regulator, seconds pendulum, lantern pinions, maintaining power. Discontinued. Crown 30, dead-beat. $350.00

**No. 74.**—8 Day, weight, quarter strike, Westminster chimes, 4 bells, moon phases, seconds pendulum, lantern pinions, adjustable bracket. No seconds hand used with this movement. Maintaining power. Discontinued. Pendulum, from center of movement to end of rating screw, 40", with No. 50 ball; diameter of ball 6". Wood rod. Crown 30, dead-beat, center 1 7/8". Length 13 3/4", width 10 3/4", depth 8 7/16" and 8 11/16". $350.00

**No. 74 A.**—8 Day, weight quarter strike, Westminster chimes, 4 bells, seconds pendulum, lantern pinions. Can use seconds hand with this movement. Adjustable bracket. Maintaining power. Discontinued. Pendulum, from center of movement to end of rating screw, 40", with No. 50 ball; diameter of ball 6". Wood rod. Crown 30, dead-beat, center 1 7/8". Length 13 3/4", width 10 3/4", depth 8 7/16" and 8 11/16". $350.00

**No. 74 B.**—8 Day, weight, quarter strike, Westminster chimes, 4 bells, moon phases, seconds pendulum, lantern pinions, rayed plates, adjustable bracket. No seconds hand can be used with this movement. Moon disc on movement. Maintaining power. Pendulum, from center of movement to end of rating screw, 40", with No. 50 ball; diameter of ball 6". Wood rod. Crown 30, dead-beat, center 1 7/8". Length 15 3/8", width 10 3/4", depth 8 7/16" and 8 11/16". $375.00 w/Bracket

**No. 74 C.**—8 Day, weight, quarter strike, Westminster chimes, 4 bells, seconds pendulum, lantern pinions, rayed plates, adjustable bracket, maintaining power. Can use seconds hand with this movement. Pendulum, from center of movement to end of rating screw, 40", with No. 50 ball; diameter of ball 6". Wood rod. Crown 30, dead-beat, center 1 7/8". Length 13 3/4", width 10 3/4", depth 8 7/16" and 8 11/16". $350.00 w/Bracket & Bell

**No. 74 D.**—8 Day, weight, quarter strike, Westminster chimes, 4 bells, moon phases, seconds pendulum, lantern pinions, rayed plates, adjustable bracket. Movement arranged so that a seconds hand can be used. Moon on dial. Maintaining power. Pendulum, from center of movement to end of rating screw, 40", with No. 50 ball; diameter of ball 6". Wood rod. Crown 30, dead-beat, center 1 7/8". Length 13 3/4", width 10 3/4", depth 8 7/16" and 8 11/16". $350.00

No. 72 A.

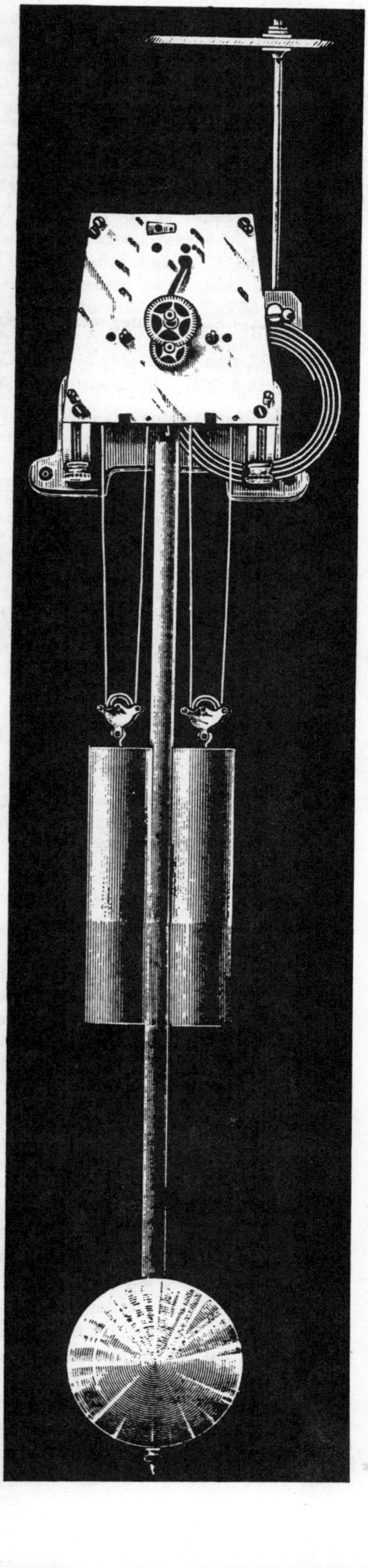

# 1907 SETH THOMAS CLOCK MOVEMENT DESCRIPTIONS

No. 74 B.

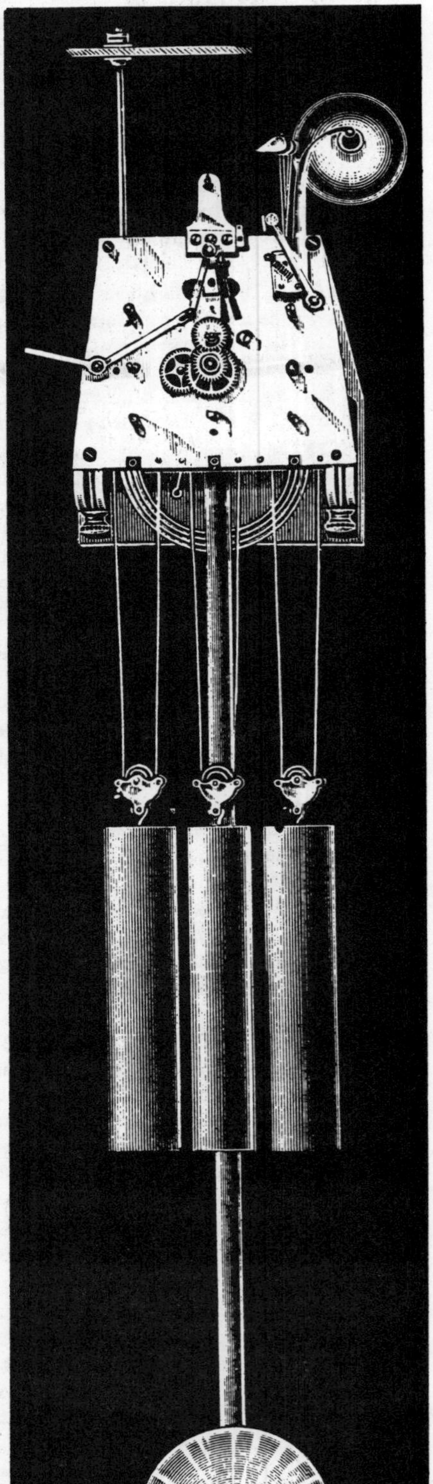

**No. 74 E.**—Same No. 74 B, with the following exceptions: The movement rests on a seat board; the pendulum is suspended from a stud secured to the back of the movement. Moon disc on movement. Maintaining power. Pendulum, from center of movement to end of rating screw, 40", with No. 50 ball; diameter of ball 6". Wood rod. Crown 30, dead-beat, center 1 7/8". Length 15 3/4", width 10 3/4", depth 5 1/2". $350.00

**No. 74 F.**—Same as No. 74 E, without moon phases. Can use seconds hand. Maintaining power. Pendulum, from center of movement to end of rating screw, 40", with No. 50 ball; diameter of ball 6". Wood rod. Crown 30, dead-beat, center 1 7/8". Length 13 3/4", width 10 3/4", depth 5 1/2". $350.00

**No. 74 G.**—Same as No. 74 D, with the following exceptions: The movement rests on a seat board; the pendulum is suspended from a stud secured to the back of the movement. Moon disc on dial. Maintaining power. Pendulum, from center of movement to end of rating screw, 40", with No. 50 ball; diameter of ball 6". Wood rod. Crown 30, dead-beat, center 1 7/8". Length 13 3/4", width 10 3/4", depth 5 1/2." $350.00

**No. 75.**—Large Turntable Motor. Center 1 1/4". Length 9 5/8", width 10 3/4", depth 5 1/2". Mainspring No. 139, 18' x 1 5/16" x .025. $165.00

(See illustration on following page)

**No. 76.**—Precision Clock, Graham escapement, old model, adjustable bracket, maintaining power. Pendulum, from center of movement to point at bottom of ball, 39". Crown 30, Length 12 1/2", width 9 5/16", depth 7" and 7 7/16". Rare $750.00

**No. 76 A.**—Precision Clock, Gravity escapement, old model, adjustable bracket, maintaining power. Pendulum, from center of movement to point at bottom of ball, 39". Crown 30. Length 12 1/2", width 9 5/16", depth 7" and 7 7/16". Rare $750.00

**No. 77.**
**No. 78.** No information at this time.
**No. 79.**

*To receive top prices, movements must be complete and original, and they must be in good running or repairable condition. Damaged, worn-out or missing parts will reduce prices accordingly.*

**No. 80.**—1 Day, strike, weight, common. Pendulum 14 5/16", with No. 14 ball. Crown 42, recoil, center 1". Length 6 1/8", width 6 1/2", depth 3 1/8". (See illustration on following page) $50.00

**No. 81.**—8 Day, strike, weight, small. Discontinued. Crown 44, recoil, center 1". Length 6 1/4", width 5 9/16", depth 2 7/8". $60.00

**No. 82.**—8 Day, strike, weight, common, large column. Pendulum, from center of movement to end of rating screw, 21 3/8", with No. 12 ball; diameter of ball 2 7/32". Crown 36, recoil, center 15/16". Length 7 3/8", width 7 1/8", depth 3 5/16". $70.00

**No. 82 A.**—8 Day, strike weight, common, large O. G. Pendulum, from center of movement to end of rating screw, 19 3/8", with No. 12 ball; diameter of ball 2 7/32". Crown 38, recoil, center 15/16". Length 7 3/8", width 7 1/8", depth 3 5/16". $110.00

**No. 82 B.**—8 Day, strike, weight, common large Nos. 1 and 10 Parlor Calendars. Pendulum, from center of movement to end of rating screw, 11 7/8", with No. 12 ball; diameter of ball 2 7/32". Crown 50, recoil, center 1 3/16". Length 7 3/8", width 7 1/8", depth 3 7/16". $200.00

**No. 82 C.**—8 Day, strike, weight, common large Lunar. Pendulum, from center of movement to end of rating screw, 16 3/8", with No. 43 ball; diameter of ball 3 3/8". Wood rod in two sections. Crown 40, recoil, center 1 3/16". Length 9 1/2", width 8 5/8", depth 3 5/8". $200.00

# 1907 SETH THOMAS CLOCK MOVEMENT DESCRIPTIONS

No. 75.

*To receive top prices, movements must be complete and original, and they must be in good running or repairable condition. Damaged, worn-out or missing parts will reduce prices accordingly.*

**No. 82 D.**—8 Day, strike, weight, common large Colonial, slow strike, Putnam and Franklin. Discontinued. Pendulum, from center of movement to end of rating screw, 21 3/8", with No. 12 ball; diameter of ball 2 7/32". Crown 36, recoil, center 15/16". Length 7 3/8", width 7 1/8", depth 3 5/16". — $75.00

**No. 83.** No information at this time.

**No. 84.**—8 Day, weight, quarter strike, Westminster chimes on 4 cathedral bells, cathedral hour bell, seconds pendulum, steel-cut pinions, rack and snail strike; can use seconds hand. Pendulum, from center of movement to end of rating screw, 41 1/2", with No. 54 ball; diameter of ball 5 11/16". Steel rod. Crown 30, center 2 5/8". Length 8 1/2", width 9", depth 10 1/2" including bells, without bells 9". — $700.00 Complete with Dial & Gongs

**No. 85.**—8 Day, extra large, pendulum, strike, spring, Queen Anne. Pendulum, from center of movement to end of rating screw, 17 1/2", with No. 22 ball; diameter of ball 3 11/16". Wood rod. Crown 44, dead-beat, center 1 7/16". Length 7 7/8", width 7 1/4", depth 4 1/4". Time mainspring No. 136, 11 1/2' x 3/4" x .018. Strike mainspring No. 136. — $150.00

**No. 85 A.**—8 Day, extra large, pendulum, strike, spring. Pendulum, from center of movement to end of rating screw, 38 7/8", with No. 27 ball; diameter of ball 4 9/16". Wood rod. Crown 30, dead-beat, center 1 7/16". Length 7 7/8", width 7 1/4", depth 4 1/4". Time mainspring No. 136, 11 1/2' x 3/4" x .018. Strike mainspring No. 136. — $200.00 w/Bracket

**No. 85 B.**—8 Day, extra large, pendulum, strike, spring, Suez. Pendulum, from center of movement to end of rating screw, 21 5/16", with No. 22 ball; diameter of ball 3 11/16". Wood rod. Crown 40, dead-beat, center 1 7/16". Length 7 7/8", width 7 1/4", depth 4 1/4". Time mainspring No. 136, 11 1/2' x 3/4" x .018. Strike mainspring No. 136. — $150.00

**No. 85 C.**—8 Day, extra large, pendulum, strike, spring, Hotel. Pendulum, from center of movement to end of rating screw, 8 5/8", with No. 22 ball; diameter of ball 3 11/16". Wood rod. Crown 60, dead-beat, center 1 7/16". Length 7 7/8", width 7 1/4", depth 4 1/4". Time mainspring No. 136. 11 1/2' x 3/4" x .018. Strike mainspring No. 136. — $75.00

# 1907 SETH THOMAS CLOCK MOVEMENT DESCRIPTIONS

No. 80.

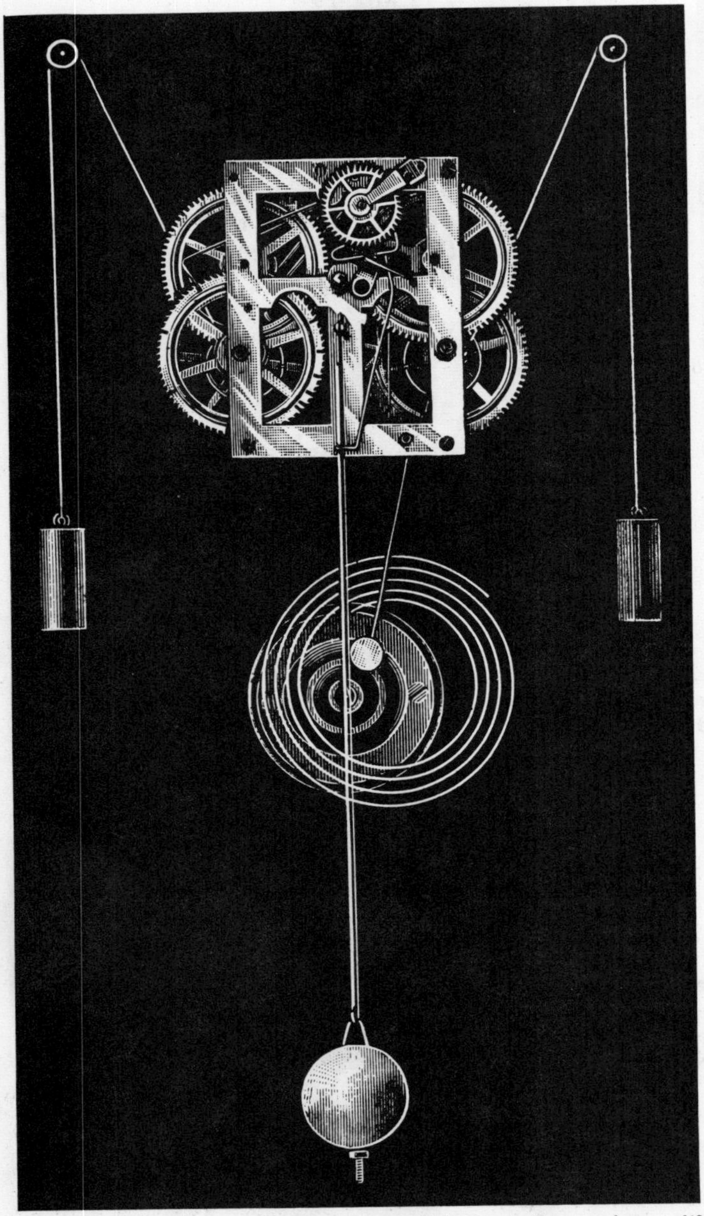

**No. 85 D.**—8 Day, extra large, pendulum, strike, spring, with Calendar attachments, new Fashion. Pendulum, from center of movement to end of rating screw, 17 13/16", with No. 22 ball; diameter of ball 3 11/16". Wood rod. Crown 44, dead-beat, center 1 7/16". Length 7 7/8", width 7 1/4", depth 4 5/8". Time mainspring No. 136, 11 1/2' x 3/4" x .018. Strike mainspring No. 136. $400.00

**No. 85 E.**—8 Day, extra large, pendulum, strike, spring, World. Pendulum, from center of movement to end of rating screw, 21 5/16", with No. 22 ball; diameter of ball 3 11/16". Wood rod. Crown 40, dead-beat, center 1 1/8". Length 7 7/8", width 7 1/4", depth 4". Time mainspring No. 136, 11 1/2' x 3/4" x .018. Strike mainspring No. 136. $150.00

**No. 85 F.**—8 Day, extra large, pendulum, strike, spring, with calendar attachments, old Fashion. Pendulum, from center of movement to end of rating screw, 9 3/16", with No. 33 ball; diameter of ball 2 11/16". Wood rod. Crown 60, dead-beat, center 1 5/8". Length 7 7/8", width 7 1/4", depth 3 5/8". Time mainspring No. 136, 11 1/2' x 3/4" x .018. Strike mainspring No. 136. $400.00

# 1907 SETH THOMAS CLOCK MOVEMENT DESCRIPTIONS

**No. 85 G.**—8 day, large, pendulum, strike, spring, with calendar attachments, Southern Calendar. Pendulum, from center of movement to end of rating screw, 17 13/16", with No. 22 ball; diameter of ball 3 11/16". Wood rod. Crown 44, dead-beat, center 1 7/16". Length 7 7/8", width 7 1/4", depth 4 1/4". Time mainspring No. 136, 11 1/2' x 3/4" x .018. Strike mainspring No. 136. $400.00

**No. 85 H.**—8 Day, large, pendulum, strike, spring, pendulum in front of movement, pillar brackets, Yorktown. Pendulum, from center of movement to end of rating screw, 17 13/16", with No. 22 ball; diameter of ball 3 11/16". Wood rod. Crown 44, dead-beat, center 1 7/16". Length, 7 7/8", width 7 1/4", depth 3". Time mainspring No. 136, 11 1/2' x 3/4" x .018. Strike mainspring No. 136. $75.00

**No. 85 I.**—8 Day, large, pendulum, strike, spring, wind and set at back. Pendulum, from center of movement to end of rating screw, 12 1/2", with No. 22 ball; diameter of ball 3 11/16". Wood rod. Crown 50, dead-beat, center 1 7/16" on front. Length 7 7/8", width 7 1/4", depth 4 1/4". Time mainspring No. 136, 11 1/2' x 3/4" x .018. Strike mainspring No. 136. $75.00

**No. 85 J.**—8 Day, extra large, pendulum, strike, spring, World. Pendulum, from center of movement to end of rating screw, 26 3/4", No. 22 ball; diameter of ball 3 11/16". Wood rod. Crown 36, dead-beat, center 1 1/8". Length 7 7/8", width 7 1/4", depth 4". Time mainspring No. 136, 11 1/2' x 3/4" x .018. Strike mainspring No. 136. $100.00

**No. 86.**—15 Day, extra large, pendulum, time, spring, double spring, World. Pendulum, from center of movement to end of rating screw, 21", with No. 22 ball; diameter of ball 3 11/16". Wood rod. Crown 40, dead-beat, center 1 1/8". Length 7 13/16", width 7 1/16", depth 4 1/8". Mainspring No. 137, 10' x 3/4" x .015. $150.00

**No. 86 A.**—15 Day, extra large, pendulum, time, spring, double spring, 14" and 18" Lobby. Pendulum, from center of movement to end of rating screw, 12 1/2", with No. 22 ball; diameter of ball 3 11/16". Wood rod. Crown 50, dead-beat, center 1 1/8". Length 7 13/16", width 7 1/16", depth 4 1/8". Mainspring No. 137, 10' x 3/4" x .015. $85.00

**No. 86 B.**—15 Day, extra large, pendulum, time, spring, double spring. Pendulum, from center of movement to end of rating screw, 8 1/4", with No. 22 ball; diameter of ball 3 11/16". Wood rod. Crown 60, dead-beat, center 1 1/8". Length 7 13/16", width 7 1/16", depth 4 1/8". Mainspring No. 137, 10' x 3/4" x .015. $85.00

**No. 86 C.**—15 Day, extra large, pendulum, time, spring, double spring. Pendulum, from center of movement to end of rating screw, 38 9/16", with No. 27 ball; diameter of ball 4 9/16". Wood rod. Crown 30, dead-beat, center 1 1/8". Length 7 13/16", width 7 1/16", depth 4 1/8". Mainspring No. 137, 10' x 3/4" x .015. $125.00

**No. 86 D.** No information at this time.

**No. 86 E.**—15 Day, extra large, pendulum, time, spring, double spring. Pendulum, from center of movement to end of rating screw, 26 5/8", with No. 27 ball; diameter of ball 4 9/16". Wood rod. Crown 36, dead-beat, center 1 1/8". Length 7 13/16", width 7 1/16", depth 4 1/8". Mainspring No. 137, 10' x 3/4" x .015. $85.00

**No. 86 F.**—15 Day, extra large, pendulum, time, spring, double spring, 18" Gallery. Pendulum, from center of movement to end of rating screw, 8 1/4", with No. 22 ball; diameter of ball 3 11/16". Wood rod. Crown 60, dead-beat, center 1 1/2". Length 7 13/16", width 7 1/16", depth 4 1/2". Mainspring No. 137, 10' x 3/4" x .015. $75.00

**No. 86 G.**—15 Day, extra large, pendulum, time, spring, double spring, 14" and 18" Lobby, and 24" Gallery. Pendulum, from center of movement to end of rating screw, 12 1/2", with No. 22 ball; diameter of ball 3 11/16". Wood rod. Crown 50, dead-beat, center 1 1/2". Length 7 13/16", width 7 1/16", depth 4 1/2". Mainspring No. 137, 10' x 3/4" x .015. $75.00

No. 86 F.

# 1907 SETH THOMAS CLOCK MOVEMENT DESCRIPTIONS

*To receive top prices, movements must be complete and original, and they must be in good running or repairable condition. Damaged, worn-out or missing parts will reduce prices accordingly.*

**No. 86 H.**
**No. 86 I.** No information at this time.
**No. 86 J.**

**No. 86 K.**—15 Day, extra large, pendulum, time, spring, double spring, wind and set at back. Pendulum, from center of movement to end of rating screw, 12 1/2", with No. 22 ball; diameter of ball 3 11/16". Wood rod. Crown 50, dead-beat, center 1 1/2". Length 7 13/16", width 7 1/16", depth 4 1/2". Mainspring No. 136, 11 1/2' x 3/4" x .018. — $85.00

**No. 86 L.**—15 Day, extra large, pendulum, time, spring, double spring, large winding arbors for marble dial. Pendulum, from center of movement to end of rating screw, 8 1/4", with No. 22 ball; diameter of ball 3 11/16". Crown 60, dead-beat, center 1 7/8". Length 7 13/16", width 7 1/16", depth 4 7/8". Mainspring No. 136, 11 1/2' x 3/4" x .018. — $85.00

**No. 86 M.**—15 Day, extra large, pendulum, time, spring, double spring, large winding arbors, for marble dial. Pendulum, from center of movement to end of rating screw, 12 1/2", with No. 22 ball; diameter of ball 3 11/16". Wood rod. Crown 50, dead-beat, center 1 7/8". Length 7 13/16", width 7 1/16", depth 4 7/8". Mainspring No. 136, 11 1/2' x 3/4" x .018. — $85.00

**No. 86 N.**—15 Day, extra large, pendulum, time, spring, double spring, large winding arbors, for marble dial. Pendulum, from center of movement to end of rating screw, 16 7/8", with No. 22 ball; diameter of ball 3 11/16". Wood rod. Crown 44, dead-beat, center 1 7/8". Length 7 13/16", width 7 1/16", depth 4 7/8". Mainspring No. 136, 11 1/2' x 3/4" x .018. — $85.00

**No. 86 O.**—15 Day, extra large, pendulum, time, spring, double spring, for double dial. Pendulum, from center of movement to end of rating screw, 10 1/8", with No. 22 ball; diameter of ball 3 11/16". Yoke rod in two sections; lower part wood. Crown 56, dead-beat, center-front, 1", center-back, 2". Length 7 13/16", width 10", depth 5 1/4". Mainspring No. 136, 11 1/2' x 3/4" x .018. — $85.00

**No. 86 P**—30 Day, extra large, pendulum, time, spring, double spring, solid front plate. Pendulum, from center of movement to end of rating screw, 16 7/8", with No. 22 ball; diameter of ball 3 11/16". Wood rod. Crown 44, dead-beat; center 1 1/8". Length 7 13/16", width 7 1/16", depth 4 1/8". Mainspring No. 137, 10' x 3/4" x .015. — $115.00

**No. 87.**—8 Day, extra large, pendulum, half-hour strike, weight, chain wind. Pendulum, from center of movement to end of rating screw, 39", with No. 56 ball; diameter of ball 5 1/8". Wood rod. Crown 30, dead-beat, center 1 11/16". Length 12 1/4", width 7 1/16", depth 4 11/16". — $250.00 Complete with Pendulum & Weights

**No. 87 A.**—8 Day, extra large, pendulum, half-hour strike, weight, chain wind, moon phases. Pendulum, from center of movement to end of rating screw, 17 1/2", with No. 56 ball; diameter of ball 5 1/8". Wood rod in two sections. Crown 44, dead-beat, center 1 11/16". Length 14 1/4", width 7 1/16", depth 4 11/16". — $250.00

**No. 88.**—8 Day, extra large, pendulum, quarter strike, spring, F. E., Hecla. Pendulum, from center of movement to end of rating screw, 10 13/16", with No. 36 ball; diameter of ball 2 7/16". Crown 40, dead-beat, center 1". Length 7", width 7", depth 2 7/8". Time mainspring No. 24, 9' x 3/4" x .020. H. Strike mainspring No. 21, 9' x 3/4" x .022. Q. Strike mainspring No. 21. — $225.00

**No. 88 A.**—8 Day, extra large, pendulum, quarter strike, spring, B. E., Marcy Pendulum, from center of movement to end of rating screw, 21 1/4", with No. 27 ball; diameter of ball 4 9/16". Wood rod. Crown 34, dead-beat, center 1". Length 7", width 7", depth 3 3/8". Time mainspring No. 24, 9' x 3/4" x .020". H. Strike mainspring No. 21, 9' x 3/4" x .022. Q. Strike mainspring No. 21. — $250.00

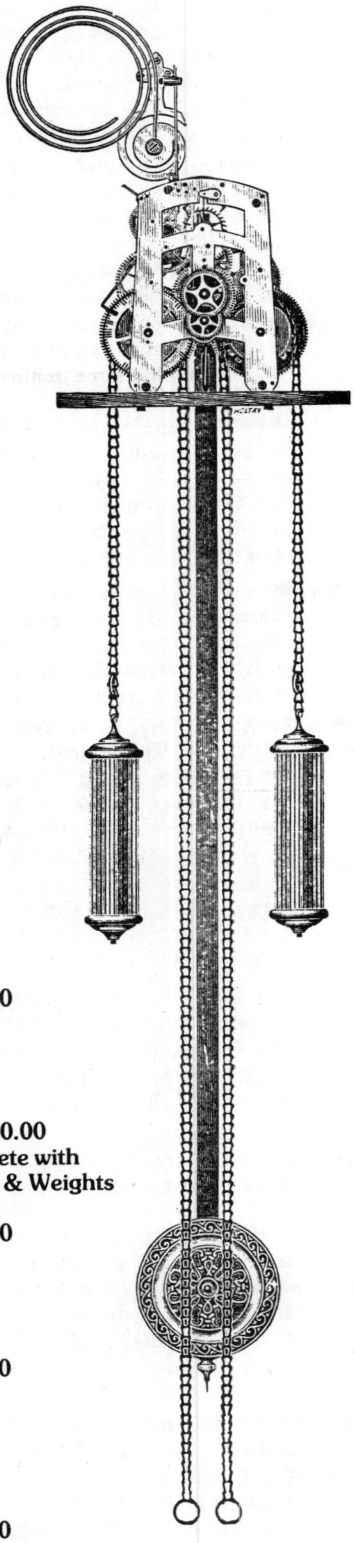

No. 87.

# 1907 SETH THOMAS CLOCK MOVEMENT DESCRIPTIONS

**No. 88 B.**—8 Day, extra large, pendulum, quarter strike, spring, chime, 4 bells, F. E. Pendulum, from center of movement to end of rating screw, 10 13/16", with No. 36 ball; diameter of ball 2 7/16". Crown 40, dead-beat, center 1". Length 7", width 7", depth 2 7/8". Time mainspring No. 24, 9' x 3/4" x .020. H. Strike mainspring No. 21, 9' x 3/4" x .022. Q. Strike mainspring No. 21. $225.00

**No. 89.**—8 Day, large, pendulum, half-hour strike spring, B. E., G. B., Kroeber. Pendulum, from center of movement to bottom of ring, 4 9/16"; special ball. Crown 39, dead-beat, center 13/16". Length 7 1/4", width 6", depth 3". Time mainspring No. 29, 9' x 3/4" x .017. Strike mainspring No. 29. $80.00

**No. 89 A.**—8 Day, large, pendulum, half-hour strike, spring, B. E., G. B., same as Kroeber except marked Seth Thomas. Pendulum, from center of movement to end of rating screw, 3 5/8", with No. 46 ball; diameter of ball 1 11/16". Crown 39, dead-beat, center 13/16". Length 7 1/4", width 6", depth 3". Time mainspring No. 29, 9' x 3/4" x .017. Strike mainspring No. 29. $60.00

**No. 89 B.**—8 Day, large, pendulum, half-hour strike, spring, B. E., G. B., with attachments for music box, Eichler. Pendulum, from center of movement to end of rating screw, 4", with No. 47 ball; diameter of ball 1 1/2". Crown 39, dead-beat, center 1". Length 7 1/4", width 7 1/2", depth 3 1/4". Time mainspring No. 29, 9' x 3/4" x .017. Strike mainspring No. 29. $80.00

**No. 89 C.**—8 Day, large, pendulum, half-hour strike, spring, B. E., G. B., change in train, third wheel 42 teeth. Pendulum, from center of movement to end of rating screw, 5", with No. 57 ball; diameter of ball 1 11/16". Crown 39, half dead-beat, center 1". Length 7 1/4", width 6", depth 3 1/2". Time mainspring No. 29, 9' x 3/4" x .017. Strike mainspring No. 29. $70.00

No. 89 D.

**No. 89 D.**—8 Day, large, pendulum, half-hour strike, spring, F. E., W. B., change in train, third and fourth wheels 40 teeth, third pinion 8 leaves. Pendulum, from center of movement to end of rating screw, 9 5/8", with No. 36 ball; diameter of ball 2 7/16". Crown 39, half dead-beat, center 1". Length 8 3/8", width 6", depth 2 3/4". Time mainspring No. 29, 9' x 3/4" x .017. Strike mainspring No. 29. $80.00

**No. 89 E.**—8 Day, large, pendulum, half-hour strike, spring, F. E., G. B., change in train third and fourth wheels 40 teeth; third pinion 8 leaves. Pendulum, from center of movement to end of rating screw, 9 5/8", with No. 36 ball; diameter of ball 2 7/16". Crown 39, half dead-beat, center 1". Length 8 3/8", width 6", depth 2 3/4". Time mainspring No. 29, 9' x 3/4" x .017. Strike mainspring No. 29. $70.00

**No. 89 F.**—8 Day, large, pendulum, time, spring, B. E., Reminder Alarm, with regulator same as No. 89 C. Pendulum, from center of movement to end of rating screw, 9 5/8", with No. 36 ball; diameter of ball 2 7/16". Crown 39, half dead-beat, center 1". Length 8 3/8", width 7 1/4", depth 3". Time mainspring No. 29, 9' x 3/4" x .017. Alarm mainspring No. 29. $70.00

**No. 89 G.**—8 Day, large, pendulum, half-hour strike, spring, Arts and Crafts. This has the same time train as No. 89 D and the same center as No. 89 C; B. E. Pendulum, from center of movement to end of rating screw, 9 5/8", with No. 36 ball; diameter of ball 2 7/16". Crown 39, half dead-beat, center 1". Length 8 3/8", width 6", depth 2 3/4". Time mainspring No. 29, 9' x 3/4" x .017. Strike mainspring No. 29. $70.00

**No. 90.** No information at this time.

**No. 91.**—1 Day, lever, Nutmeg, time. Crown 15, center 19/32". Length 2 7/8", width 3 1/8", depth 1 7/8". Mainspring No. 15, 5' x 3/8" x .016. $20.00

**No. 92.**—1 Day, lever, Nutmeg, time, alarm. Crown 15, center 19/32". Length 3 1/4", width 3 1/8", depth 2". Mainspring No. 17, 5' x 3/8" x .015. Alarm mainspring No. 83, 2 1/2' x 3/16" x .015. $25.00

**No. 92 A.**—1 Day, lever, Nutmeg, time, alarm, Elk, long hammer. Crown 15, center 19/32". Length 3 1/4", width 3 1/8", depth 2". Mainspring No. 17, 5' x 3/8" x .015. Alarm mainspring No. 83, 2 1/2' x 3/16" x .015. $30.00

**No. 92 B.**—1 Day, lever, Anvil, time. Crown 15, center 19/32". Length 3 1/4", width 3 1/8", depth 2". Mainspring No. 17, 5' x 3/8" x .015. $25.00

# 1907 SETH THOMAS CLOCK MOVEMENT DESCRIPTIONS

**No. 92 C.**—1 Day, lever, Anvil, time, alarm Crown 15, center 19/32". Length 3 1/4", width 3 1/8", depth 2". Mainspring No. 17, 5' x 3/8" x .015. Alarm mainspring No. 83, 2 1/2' x 3/16" x .015. $30.00

**No. 93.**—1 Day, lever, Joker, time, Crown 15, center 19/32". Length 4 1/4", width 2 1/2", depth 1 7/8". Mainspring No. 66, 5' x 3/8" x .016. $40.00

**No. 94.**—1 Day, lever, Joker, time, alarm. Crown 15, center 19/32". Length 4 1/4", width 3 1/8", depth 1 7/8". Mainspring No. 68, 5' x 3/8" x .015. Alarm mainspring No. 80, 2 1/2' x 1/4", x .016. $45.00

**No. 94 A.**—1 Day, lever, Joker, time, alarm, Artist. Crown 15, center 19/32". Length 4 1/4", width 3 1/8", depth 1 7/8". Time mainspring No. 66, 5' x 3/8" x .016. Alarm mainspring No. 80, 2 1/2' x 1/4" x 016. $45.00

**No. 95.**—1 Day, lever, Joker, strike. Crown 15, center 19/32". Length 4 1/4", width 3 1/2", depth 2 1/16". Time mainspring No. 66, 5' x 3/8" x .016. Strike mainspring No. 67, 3' x 3/8" x .015. $45.00

**No. 95 A.**—1 Day, lever, Joker, strike, Artist. Crown 15, center 19/32". Length 4 1/4", width 3 1/2", depth 2 1/16". Time mainspring No. 66, 5' x 3/8" x .016. Strike mainspring No. 67, 3' x 3/8" x .015. $45.00

**No. 96.**—1 Day, lever, time, Pansy, pillars drilled, round plates. Discontinued. Crown 15, center 19/32". Length 2 3/16", width 2 1/4", depth 1 5/16". Mainspring No. 99, 3' x 5/16" x .012 (loop). $25.00

**No. 96 A.**—1 Day, lever, time, Paper Weight. Discontinued. Crown 15, center 9/32". Length 2 3/16", width 2 1/4", depth 1 5/16". Mainspring No. 99, 3' x 5/16" x .012 (loop). $25.00

**No. 96 B.**—1 Day, lever, time, Shetland. Discontinued. Crown 15, center 9/32". Length 2 3/16", width 2 1/4", depth 1 5/16". Mainspring No. 99 3' x 5/16" x .012 (loop). $25.00

**No. 96 C.**—1 Day, lever, time, Vulcan. Discontinued. Crown 15, center 9/32". Length 2 1/4", width 2 3/16", depth 3/4". Mainspring No. 134, 27" x 3/16" x .015. $25.00

**No. 96 D.**—1 Day lever, time, Pony Porcelain, with long threaded stud for different depth of cases. Discontinued. Crown 15, center 9/32". Length 2 1/4", width 2 3/16", depth 1 1/2" to 2 7/8". Mainspring No. 99, 3' x 5/16" x .012 (loop). $25.00

**No. 96 F.**—1 Day, lever, time, Colt model, pillars drilled for feet and ring; also for case back studs. Used in place of No. 96 movement. This movement has a spring box, light balance and light center friction. Change made March 1905. Crown 15, center 5/16". Length 2 3/16", width 2 1/4", depth 1 5/16". Mainspring No. 154, 32" x 5/16" x .012. $25.00

**No. 96 G.**—1 Day, lever, time, Colt model, Paper Weight. This movement requires different pillars. Crown 15, center 5/16". Length 2 3/16", width 2 1/4", depth 1 5/16". Mainspring No. 154, 32" x 5/16" x .012. $25.00

**No. 96 H.**—1 Day, lever, time, Colt model. This movement is the same as No. 96 F, with the exception of regulator and addition of case back stud, used in fancy cases and in place of No. 96 D. Crown 15, center 5/16". Length 2 3/16", width 2 1/4", depth 1 5/16". Mainspring No. 154, 32" x 5/16" x .012. $25.00

**No. 96 I.**—1 Day, lever, time, Colt model. Same as No. 96 H, with the exception of regulator and long case back studs, Soldier Boy. Crown 15, center 5/16". Length 2 3/16", width 2 1/4", depth 1 5/16". Mainspring No. 154, 32" x 5/16", x .012. $25.00

**No. 96 J.**—1 Day, lever, time, Colt model. Same as No. 96 F, except the center, which is longer in front, Night Clock. Crown 15, center 5/8". Length 2 3/16", width 2 1/4", depth 1 5/16". Mainspring No. 154, 32" x 5/16" x .012. $25.00

**No. 97.**—8 Day, time, pendulum, spring, Baird. Pendulum, from center of movement to end of rating screw, 17 1/4", with No. 12 ball; diameter of ball 2 7/32". Crown 36, recoil, center 1 3/4". Length 4 11/16", width 4 1/4", depth 3 5/8". Mainspring No. 131, 9' x 11/16" x .015. $80.00

**No. 97 A.**—8 day, time, pendulum, spring, Baird, Drop Calendar. Pendulum, from center of movement to end of rating screw, 17 1/4", with No. 12 ball; diameter of ball 2 7/32". Crown 36, recoil, center 1 3/4". Length 4 11/16", width 4 1/4", depth 3 5/8". Mainspring No. 131, 9' x 11/16" x .015. $125.00

# 1907 SETH THOMAS CLOCK MOVEMENT DESCRIPTIONS

No. 98.

**No. 97 B.**—8 Day, time, pendulum, spring, Baird. Pendulum, from center of movement to end of rating screw, 17 1/4", with No. 12 ball; diameter of ball 2 7/32". Crown 36, recoil, center 1 1/4." Length 4 11/16", width 4 1/4", depth 3 1/8". Mainspring No. 131, 9' x 11/16" x .015.

$80.00

**No. 97 C.**—8 Day, time, pendulum, spring, Baird. Pendulum, from center of movement to end of rating screw, 15 3/8", with No. 12 ball; diameter of ball 2 7/32". Crown, recoil, center 1 1/4". Length 4 11/16", width 4 1/4" depth 3 1/8". Mainspring No. 131, 9' x 11/16" x .015.

$80.00

**No. 97 D.**—8 Day, time, pendulum, spring, Baird. Pendulum, from center of movement to end of rating screw, 15 3/8", with No. 12 ball; diameter of ball 2 7/32". Crown 38, recoil, center 1 3/4". Length 4 11/16", width 4 1/4", depth 3 5/8". Mainspring No. 131, 9' x 11/16" x .015.

$80.00

**No. 97 E.**—8 Day, time, pendulum, spring, Baird. Discontinued. Nos. 4 and 5 Office Time, train changed. Pendulum, from center of movement to end of rating screw, 8 7/8", with No. 12 ball; diameter of ball 2 7/32". Crown 40, recoil, center 1". Length 4 11/16", width 4 1/4", depth 2 7/8". Mainspring No. 131, 9' x 11/16" x .015.

$80.00

**No. 97 F.**—8 Day, time, pendulum, spring, Baird. Pendulum, from center of movement to end of rating screw, 21 7/8", with No. 12 ball; diameter of ball 2 7/32". Crown 31, recoil, center 1 1/4". Length 4 11/16", width 4 1/4", depth 3 1/8". Mainspring No. 131, 9' x 11/16" x .015.

$80.00

**No. 97 G.**—8 Day, time, pendulum, spring Baird. Pendulum, from center of movement to end of rating screw, 14 3/8", with No. 12 ball; diameter of ball 2 7/32". Crown 40, recoil, center, 1 1/4". Length 4 11/16", width 4 1/4", depth 3 1/8". Mainspring No. 131, 9' x 11/16" x .015.

$80.00

**No. 97 H.**—8 Day, time, pendulum, spring, with calendar attachments, Saturn Calendar. Pendulum, from center of movement to end of rating screw, 13", with No. 12 ball; diameter of ball 2 7/32". Crown 42, recoil, center 1 3/4". Length 4 11/16", width 4 1/4", depth 3 5/8". Mainspring No. 131, 9' x 11/16" x .015.

$100.00

**No. 98.**—Turntable. Center 7/8". Length 6 13/16", width 5 1/2", depth 2 11/16". Mainspring No. 8, 7' x 1" x .012.

$60.00

**No. 99.**—Paper Winder, 1" wood roll. Center 2". Length 7 1/2", width 5", depth 4 7/8". Mainspring No. 29, 9' x 3/4" x .018.

$40.00

**No. 99 A.**—Paper winder, 1/2" wood roll. Center 2". Length 7 1/2", width 5", depth 4 7/8". Mainspring No. 29, 9' x 3/4" x .018.

$40.00

**No. 100.**—Precision Clock, Graham escapement, new model maintaining power. Pendulum, from center of movement to point at bottom of ball, 39". Crown 30. Length 15", width 17 1/2", depth 8".

Rare $750.00

**No. 100 A.**—Precision Clock, Gravity escapement, new model maintaining power. Pendulum, from center of movement to point at bottom of ball, 39". Crown 30. Length 15", width 17 1/2", depth 8".

Rare $750.00

**No. 101.**—8 Day, time, large pendulum, spring, steel cut pinions, wheels and barrel gilded, plates damaskeened, Bank movement. Pendulum, from center of movement to end of rating screw, 9", cylindrical ball. Crown 30, center 1 7/8". Length 9 3/4", width 7 1/8", depth 6". Mainspring No. 148, 10' x 1 1/4" x .020.

$85.00

All Movements made by the Seth Thomas Clock Co. are stamped with their Trade Mark, thus:

ST

# 1915 SETH THOMAS CLOCK MOVEMENT DESCRIPTIONS

**No. 2.—1-Day, Lever, Time. 240 Beats.** Space required, 4⅜ inches high, 5 inches wide, and 2½ inches between back of case and glass. Suitable for use with dials from 6 to 12 inches in diameter. Price includes key and hands. List........................**$4.00**

$30.00

No. 2.

**No. 10.—8-Day, Lever, Time. 240 Beats.** Space required, 5 inches high, 6 inches wide, and 2½ inches between back of case and glass. Has two springs, in going barrels; is a substantial Movement; excellent for electrical purposes or for registering devices, having a surplus of power. Price includes key and hands. List ..................................................**$7.00**

$45.00

**No. 10A.**—Same as No. 10 with addition of four jewels (i. e., 2 hole jewels and 2 end jewels) for the special style balance pivots. Price includes key and hands. List.........................**$9.00**

$50.00

**No. 22.—8-Day, Lever, Time. 300 Beats.** Space required, 4½ inches high, 4⅛ inches wide, and 2½ inches between back of case and glass. Will fit in circular case having inside diameter of 5 inches. Has two springs in barrels; steel cut pinions—tempered, and chronometer balance. The escape wheel, pallets and balance run in jewels and the pallets are jeweled—which, with the impulse pin, makes eleven jewels in all. Has an excellent regulator, which will admit of close adjustment. All parts polished, and the plates damaskeened. We use this movement for dials up to 12 inches in diameter. An excellent time-piece. Price includes key and hands. List..............................................**$40.00**

This movement can be furnished with special length center post for gallery clocks. Write for prices.

$60.00

No. 10.

**No. 35C.—Detached Alarm Part, with Fixed Key.** Runs about 20 seconds. Space required, 2½ inches high, 1½ inches wide and 2½ inches from back to end of key. Can be applied to any one of the pendulum movements from No. 44E inclusive; also No. 89D. Price includes cup bell, brass alarm disk, latch and connecting wire. List..........................................**70c**

$15.00

**No. 41A.—8-Day, Pendulum, Time, Large, Spring.** Space required, 6 inches high, 5½ inches wide, and 4 inches between back of case and glass. Distance between center and lowest point of pendulum, 13 inches. Price includes key, hands (not larger than 12 inches), rod and ball. List................................................**$4.00**

$55.00

No. 22.

**No. 41H.**—Same as No. 41A, but with escapement on back and with longer rod of wood and 1-lb. ball. Distance from center to lowest point of pendulum, 15 inches. Space required for swing of pendulum, 8 inches. Price includes key, hands (not larger than 12 inches), rod and ball. List.................................**$5.00**

$95.00

**No. 41M.**—Same as No. 41A, but with shorter rod. Distance from center to lowest point of pendulum, 7 inches. Price includes key, hands (not larger than 12 inches), rod and ball. List.......**$4.00**

$75.00

**No. 44.—8-Day, Pendulum, Spring Strike.** Strikes the hours and is suitable for a wire or cup bell only as it strikes too fast for cathedral bell. Space required, 7¼ inches high, 6¼ inches wide, and 3 inches between back of case and glass. Distance between center of movement and lowest point of pendulum, 9½ inches. Price includes key, hands (not larger than 12 inches), rod and ball. List ..................................................**$6.00**

$75.00

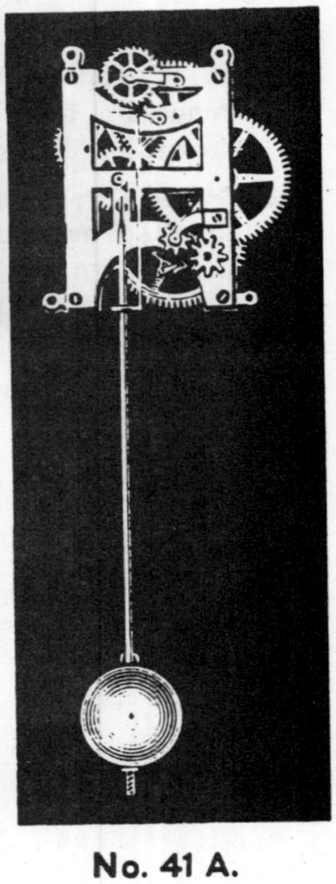

No. 41 A.

# 1915 SETH THOMAS CLOCK MOVEMENT DESCRIPTIONS

**No. 44E.—8-Day, Pendulum, Spring Strike.** Strikes the hours and half hours. Suitable for cathedral bell. Space required, 10 inches high, 6½ inches wide, and 3 inches between back of case and glass. Distance from center of movement to lowest point of pendulum, 13⅜ inches. Space required for swing of pendulum, 7 inches. The lower section of the pendulum rod is of wood, with a 10-ounce ball. Price includes key, hands (not larger than 12 inches), rod and ball. List..............................**$7.00**

**$85.00**

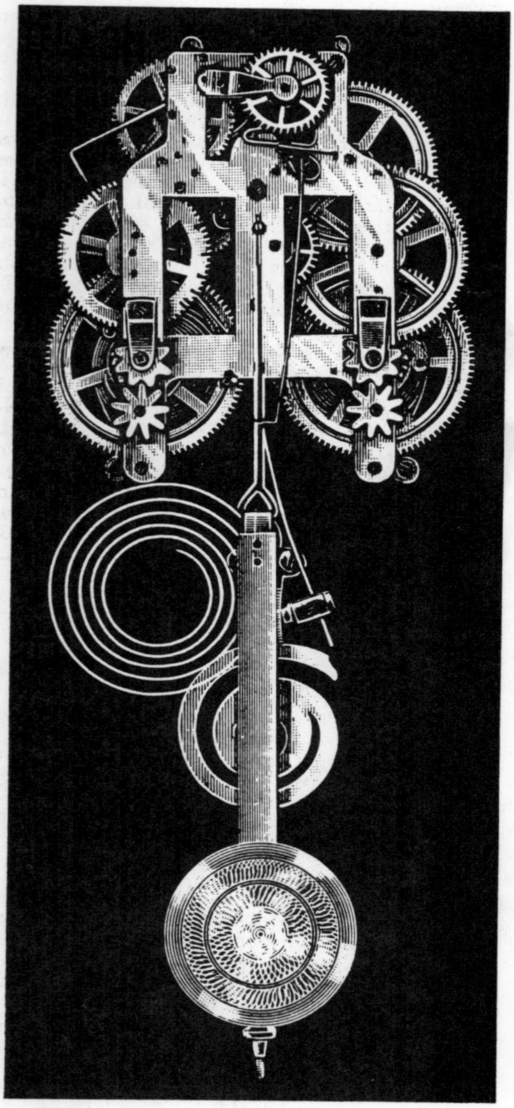

**No. 44 E.**

**No. 48L.—8-Day, Pendulum, Spring Strike.** Lantern pinions, springs in boxes, and rack striking. Strikes hours and half hours on cathedral bell. Fitted to sash for 4½-inch opening only. Space required from front of case outside to back of case inside, 4¼ inches. Distance from center to lowest point of pendulum, 4 inches. Price includes cathedral bell and stand, ornamented front sash, with beveled glass, porcelain dial, with face-plate and hands, back sash with straps, key and pendulum. List...........**$12.00**

**$85.00**

**No. 48Q.—8-Day, Pendulum, Spring Strike.** Same as No. 48L, but with longer pendulum. Distance from center to bottom of ball, 5½ inches. Fitted to sash for 6¼-inch opening only. Price includes cathedral bell and stand, front sash with convex beveled glass, convex porcelain dial with Arabic figures only; face-plate, straps with angle pieces to secure the movement in case; hands, key and pendulum. No back sash. List.................**$14.00**

**$85.00**

**No. 48R.—8-Day, Pendulum, Spring Strike.** Same as No. 48Q, but fitted to sash for 5¼-inch opening only. Price includes cathedral bell and stand, front sash with convex beveled glass, convex porcelain dial with Arabic figures only; face-plate and hands, back sash with straps, key and pendulum. List................**$14.00**

**$85.00**

**No. 51.—15-Day, Pendulum Strike.** A fine polished Movement, with springs in going barrels and steel cut pinions, rack striking. Strikes the hours and half hours on cathedral bell. Fitted to sashes for 4 and 4½ and 5-inch openings. Space required from front of case outside to back of case inside, 4 inches. Distance from center to lowest point of pendulum, 4 inches. Price includes 3-inch cathedral bell and stand, ornamented front sash with beveled glass porcelain dial, back sash, straps, face-plate, key and pendulum. List.........................................**$25.00**

NOTE.—Give size of opening for sash when ordering.

**$90.00**

**No. 68.—8-Day, Weight, Time. 60 Beats (seconds).** Space required, 10¼ inches high, 8¼ inches wide and 5⅞ inches between back of case and glass. Has polished plates and cut-steel pinions, sapphire jeweled Graham pallets, maintaining power and wood rod with 10-pound ball 8 inches in diameter. The Movement rests on a heavy iron bracket from which the pendulum is also suspended. Distance from center to lowest point of pendulum, 40½ inches. Space for drop of weight, measured from center of dial to bottom of case, 42½ inches. Space for swing of pendulum, 12 inches. Price includes key, metal cord, weight, rod, ball, and iron bracket. List.........................................**$40.00**

**$375.00**

*To receive top prices, movements must be complete and original, and they must be in good running or repairable condition. Damaged, worn-out or missing parts will reduce prices accordingly.*

# 1915 SETH THOMAS CLOCK MOVEMENT DESCRIPTIONS

No. 72A.

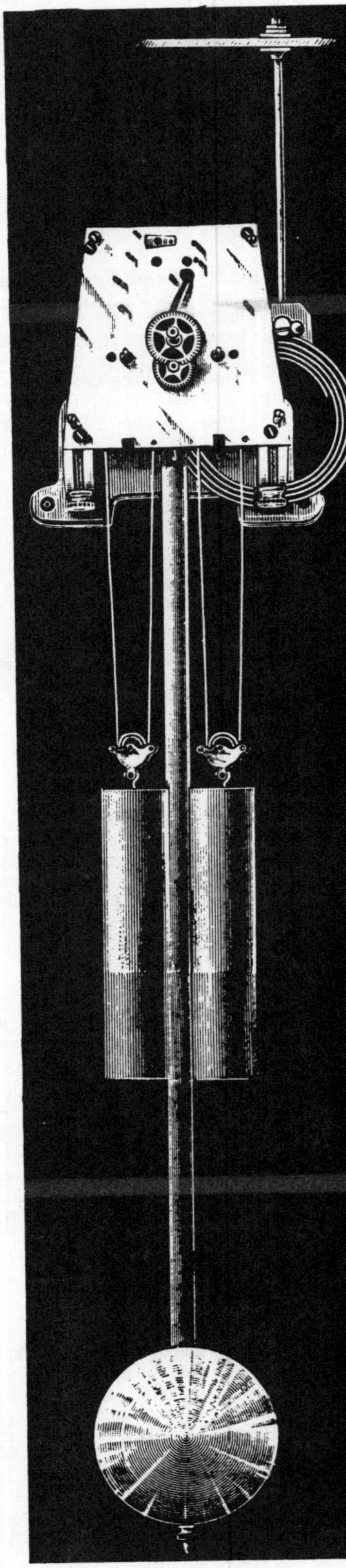

**No. 72A.—8-Day, Weight, Striking Hours and Half Hours.** A good Movement for hall clocks. Has Graham escapement and lantern pinions, also maintaining power. Plates are damaskeened. Movement rests on an iron bracket from which the pendulum is also suspended. Supplied with long brackets or short brackets. Space required, 10 inches high, 10 inches wide, and distance between back of case and glass with long bracket, 8¾ inches; with short bracket, 6¾ inches. Distance from center to lowest point of pendulum, 41 inches. Distance required for drop of weights from center of dial to bottom of weights, 56 inches. Four posts project from front plate for support of dial. The rod is of wood with brass-covered ball weighing 4 pounds and measuring 6 inches in diameter. Weights are brass shells, 10 inches long and 2¼ inches in diameter, filled with lead, and weigh, respectively, strike side, 11½ pounds; time side, 5½ pounds. Space for swing of pendulum, 11½ inches; or, if a 4½-inch ball be used, 9½ inches. Price includes rod, ball, metal cords, weights, key, 12-inch hands, 6-inch cathedral bell and stand. When ordering, state if wanted with long bracket or with short bracket, also size of pendulum ball. Dials suitable, No. 366, No. 373 and No. 435. Can be fitted with seat-board, 13$\frac{3}{16}$ or 15 inches long, in place of bracket, if desired. For clock case with removable head, the seat-board is more suitable than the bracket. List........**$50.00**

NOTE.—Dials are extra.

**$250.00 w/Bracket**

**No. 72B.**—Same as No. 72A, but fitted with attachments for moon. Dials suitable, No. 437 and No. 441. Can be fitted with seat-board, 13$\frac{3}{16}$ or 15 inches long, in place of bracket, if desired. List ........................................................**$50.00**

NOTE.—Dials are extra.

**$250.00 w/Bracket**

**No. 72C.**—Same as No. 72A, but fitted with moon disk 8⅝ inches in diameter, showing phases of the moon, and without seconds. Dial suitable, No. 405. Can be fitted with seat-board 13$\frac{3}{16}$ inches long, in place of bracket, if desired. List..........**$55.00**

NOTE.—Dials are extra.

**$275.00 w/Bracket**

**No. 75.—Motor Movement.** An extra heavy train of six wheels, driven by two springs, each 1$\frac{5}{16}$ inches wide, 18 feet long, 23 gauge, ending with a fly. Intended for turning a table in show window, or for any other purpose for which a large spring motor is required. Table and contents should not weigh more than 50 pounds, well balanced. Will run 7 hours with one turn of table per minute. Size, 10½ inches high, 10 inches wide, 5½ inches deep. List..............................................**$24.00**

**$75.00**

**Disk.**—For use with No. 75 when required. A circular iron frame, 4 inches in diameter, for mounting on center shaft, to which a table may be secured. List.................................**60c**

**$25.00**

# 1915 SETH THOMAS CLOCK MOVEMENT DESCRIPTIONS

**No. 84.—8-Day, Weight, Quarter Strike, Westminster Chimes, with Disk Showing Moon Phases.** Space required, 8 inches high, 8 inches wide, and 11¾ inches between back of case and glass. Space required for swing of pendulum, 11½ inches. Distance from center for drop of weights, 71 inches. Distance from center to lowest point of pendulum, 42 inches. The Movement rests on a seat-board 15 inches long. The pendulum is suspended from a stud screwed to the back plate of movement. The chimes are struck on four cathedral bells, and the hours on a large cathedral bell. Can be made silent or chime at will. Has Graham escapement, steel-cut pinions tempered and maintaining power. The lower arms of the hour and quarter racks are provided with springs, which prevent any straining or bending of those parts in case the hands should be turned backward. The crutch wire is also provided with beat screws, by which the beat can be adjusted with precision. All of the parts are highly finished, and the plates are damaskeened. The rod is of soft steel. The ball is of zinc covered with brass, 5¾ inches in diameter, and weighs 3½ pounds. The weights are brass shells filled with lead, and weigh, respectively, counting them from left to right: 1st, Hour, 20 pounds; 2d, Time, 14½ pounds; 3d, Chime, 32½ pounds. Price includes the movement with all bells, with their stands, seat-boards, metal cords, weights, key, hands, rod and ball, also No. 443 dial and moon disk. This Movement is equal in every detail to the finest of English makes. List ................................**$350.00**

$700.00
Complete with
Dial & Gongs

No. 84.

**No. 86U.—30-Day, Pendulum, Time, Double Springs. 120 Beats (half seconds).** Space required, 8 inches high, 7 inches wide, and 4½ inches between back of case and glass. Distance from center to lowest point of pendulum, 9 inches. Space required for swing of pendulum, 6 inches. Price includes key, rod and ball, and hands up to 12 inches, inclusive. List....................**$15.00**

$125.00
w/Bracket

**No. 86V.—30-Day, Pendulum, Time, Double Springs. 100 Beats.** Having longer rod, otherwise same as No. 86U. Distance from center to lowest point of pendulum, 13½ inches. Space required for swing of pendulum, 7 inches. Price includes key, rod and ball, and hands up to 12 inches, inclusive. List................**$15.00**

$125.00
w/Bracket

**No. 86W.—30-Day, Pendulum, Time, Double Springs. 80 Beats.** Space required, 8 inches high, 7 inches wide and 4 inches between back of case and glass. Distance from center to lowest point of pendulum, 21½ inches. Space required for swing of pendulum, 8 inches. Movement and rod are hung on an iron bracket. Price includes key, rod and ball, and hands up to 12 inches, inclusive. List ..........................................**$15.00**

$125.00
w/Bracket

**No. 86X.—30-Day, Pendulum, Time, Double Springs. 60 Beats (seconds).** Same as No. 86W, but with longer rod and heavier ball. Distance from center to lowest point of pendulum, 39 inches. Space required for swing of pendulum, 10 inches. Price includes key, rod and ball, and hands up to 12 inches, inclusive. List, **$20.00**

$150.00
w/Bracket

**No. 86Y.—30-Day, Pendulum, Time, Double Springs for Double Dials.** Suitable for any size not larger than 30 inches. Space required, 8 inches high, 10 inches wide, and 5¼ inches between front and back ends of center posts, making about 4¼ inches between the outside face of dials. Distance from center to lowest point of pendulum, 10⅛ inches. Price includes key, rod and ball, and hands up to 12 inches, inclusive. List................**$25.00**

$125.00

**Dial No. 443.**

Suitable for No. 84 movement only.

# 1915 SETH THOMAS CLOCK MOVEMENT DESCRIPTIONS

**No. 89A.D.—8-Day, Pendulum, Spring Strike.** Strikes the hours and half hours on cathedral bell. The brackets for securing Movement to case are attached to the front plate. The escapement is between the plates and pendulum is on the back of Movement. Space required, 8 inches high, 6 inches wide, and 4¾ inches between back of case and glass. Distance from center of Movement to lowest point of pendulum, 5 inches. Price includes key, hands, rod, ball and small cathedral bell, with short iron stand. List ........................................................ **$5.00**

$50.00

*To receive top prices, movements must be complete and original, and they must be in good running or repairable condition. Damaged, worn-out or missing parts will reduce prices accordingly.*

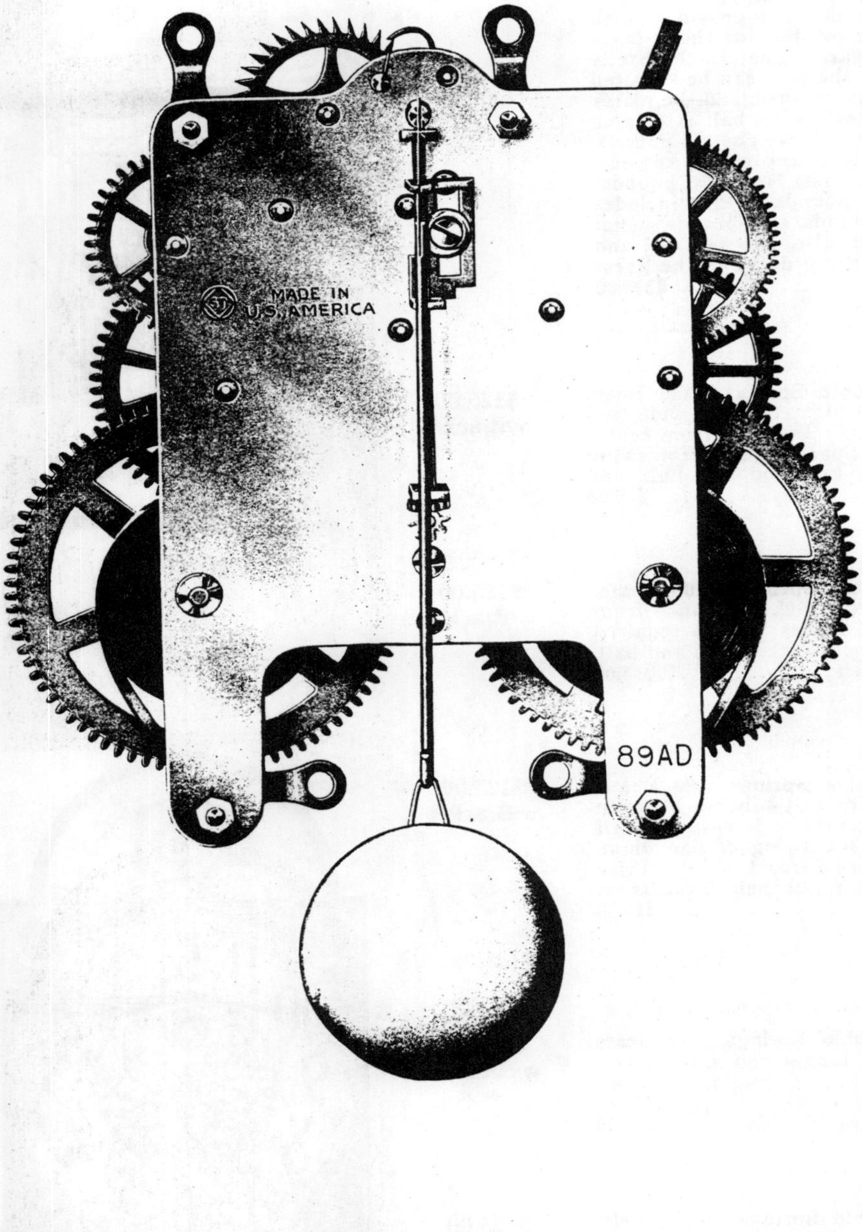

No. 89 D.

**No. 89D.**—Same as No. 98A.D., but with longer rod. The escapement is between the plates and pendulum is on the front of Movement. The brackets are on the back of plate. Space required, 8 inches high, 6 inches wide and 2¾ inches between back of case and glass. Distance from center to lowest point of pendulum, 9¾ inches. Price includes key, hands, rod and ball, wire bell and stand. List ........................................................ **$4.50**

$50.00

# 1915 SETH THOMAS CLOCK MOVEMENT DESCRIPTIONS

**No. 98.—Turn-table Movement.** A heavy train of four wheels driven by a spring 1 inch wide, 10 feet long, 25 gauge, ending with a fly. Intended for turning a small table in show windows. Table and contents should not weigh more than 8 pounds, well balanced. Will run 10 hours with one turn of table per minute. Size, 6½ inches high, 4¾ inches wide and 3 inches deep. List........**$5.00** **$60.00**

NOTE.—Note carefully what above prices include, and which dials can be used with hall-clock movements. Movements can be supplied only as specified. Alterations cannot be undertaken.

No. 98.

**All Movements made by the Seth Thomas Clock Co. are stamped with their Trade Mark, thus:**

No. 373

No. 366

No. 405

**Dial No. 366.**—Suitable for Movement No. 72A only. Made of brass, silvered. The center section is polished, and the outer section dead finish. Black Arabic numerals, ornamental lines being in gilt. Total diameter, 12½ inches. Minute circle, 11¼ inches. List ..................................................**$14.00** **$40.00**

**Dial No. 373.**—Suitable for Movement No. 72A only. Made of brass, silvered. The center section is polished, and the outer section dead finish. Black Arabic numerals, trimmed with gilt. Ornamental lines are gilt. 15⅝ inches high, 11⅜ inches wide. Minute circle, 9⅝ inches. List..................................**$17.00** **$75.00**

NOTE.—Similar dial can be supplied at same price to fit any other movements, provided no seconds circle required.

**Dial No. 405.**—Suitable for Movement No. 72C only. Made of brass, silvered. The center section being polished, and the remainder of Dial dead finish. Black Arabic numerals trimmed with gilt. Ornamental lines are gilt. 15⅝ inches high, 11⅜ inches wide. Minute circle, 9⅝ inches. List........................**$17.00** **$75.00**

Nos. 435, 437

**Dial No. 433.**—12 inches, Arabic "Continental" Dial, made of metal, painted white, 12 inches square, 9½-inch minute circle. Stenciled gilt corners and Arabic numerals. Not suitable for hall-clock movements. List.........................................**$2.00** **$40.00**

**Dial No. 435.**—Suitable for Movement No. 72A only. The arched section is solid—showing a mock moon. List.............**$16.00** **$125.00**

NOTE.—Similar Dials can be supplied at same price to fit any other movement, provided no seconds circles required.

**Dial No. 437.**—Suitable for Movement No. 72B only. Is made of heavy iron, painted cream white, with floral corners in gilt and color. Lithographed moon disk. 18⅞ inches high, 13⅜ inches wide, 12⅜-inch minute circle. List........................**$17.00** **$150.00**

**Dial No. 441.**—Suitable for Movement No. 72B only. Is made of heavy brass, with gilt raised numerals on raised silvered circle; gilt ornamental pierced center and corners. Hand-painted moon disk. Well finished throughout. 19 inches high, 13¾ inches wide, 12-inch minute circle. List.................................**$60.00** **$200.00**

**Dial No. 443.**—Same pattern as No. 441 with addition of hour and chime shut-off indications in corner over numerals 2 and 10. For No. 84 movement only. List..............................**$70.00** **$200.00**

NOTE.—The cost of repainting dials of old hall clocks varies from $13.00 to $35.00 each, list.

Nos. 441, 443

# 1922 SETH THOMAS CLOCK MOVEMENT DESCRIPTION

$90.00

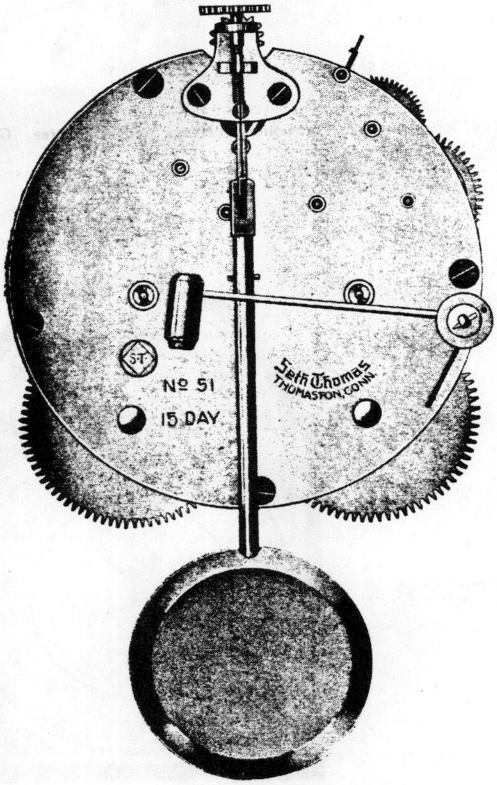


**No. 51 MOVEMENT**

Highest grade 15-day pendulum strike movement for fine mantel clocks, with heavy hard brass plates and cut wheels, all brass parts gold plated. Mainsprings in going barrels and steel cut polished pinions, rack striking. Strikes the hour and half hour on cathedral bell.

# 1924 SETH THOMAS CLOCK MOVEMENT DESCRIPTION

$375.00

## No. 68 MOVEMENT

EIGHT-DAY, Weight, Time Sixty beats (second). Heavy polished plates. Cut steel pinions. Graham dead beat escapement with sapphire jeweled pallets. Maintaining power. Movement fastened to heavy iron bracket from which the pendulum is also suspended. Movement throughout is made of the highest grade materials and is of the very best workmanship.

Fitted in Regulator Nos. 70 and 80

$200.00

# 1921 SETH THOMAS CLOCK MOVEMENT DESCRIPTION

## NO. 77B MOVEMENT

New Keystone shape movement. Heavy polished plates. Graham dead-beat escapement. Maintaining power. Grooved barrel with a new special click, which makes it easy to unwind cord from barrel without injury to click spring.

The dial wheels are outside the plates, making a more perfect bearing for the center post than in the old pattern. There is also more space between center and seconds staff which allows the use of a larger seconds circle.

Fitted in Regulators Nos. 2 and 4

Iron bracket with suspension spring fastened to same. Movement is secured to bracket with four long screws between the plates so it can be removed easily and the screws cannot be lost or mislaid. A very strong and durable movement.

*Complete as Shown*

**$115.00**

# 1922 SETH THOMAS CLOCK MOVEMENT DESCRIPTION

## No. 86 MOVEMENT

THIS movement is of exceptional value, and is considered the strongest and best spring movement ever made for Office Clocks.

It is of large size, made of heavy stock with six large pillars, making a very strong and inflexible frame.

The motive power is furnished by two springs 10 feet long, 3/4 inch wide and .017 inch thick.

The winding arbors are fitted with stop work which permits more even tension of springs and prevents overwinding. Is fitted with Graham deadbeat escapement (between the plates), wood rod and brass-covered zinc ball. Made with pendulums of different lengths. Pallets are hardened, ground and polished. Polished pivots and pivot holes all countersunk to retain the oil. All wheels are cut from hard, rolled brass plate and attached to staffs with brass collets. Cut is exact size and accurately represents heavy plates, wheels, staffs, etc.

# 1922 SETH THOMAS CLOCK MOVEMENT DESCRIPTION

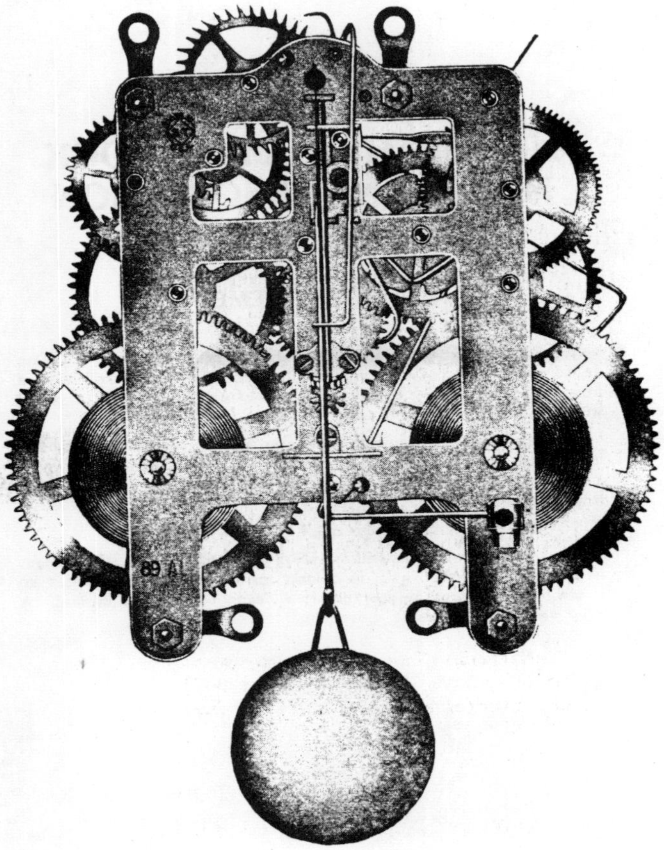


**$50.00**

**No. 89 MOVEMENT**

8-day pendulum strike movement for mantel clocks, with heavy hard brass, rectangular plates and cut wheels. Lacquered. Strikes the hour and half hour on cathedral bell.

# 1928 SETH THOMAS CLOCK MOVEMENT DESCRIPTION

**$50.00**

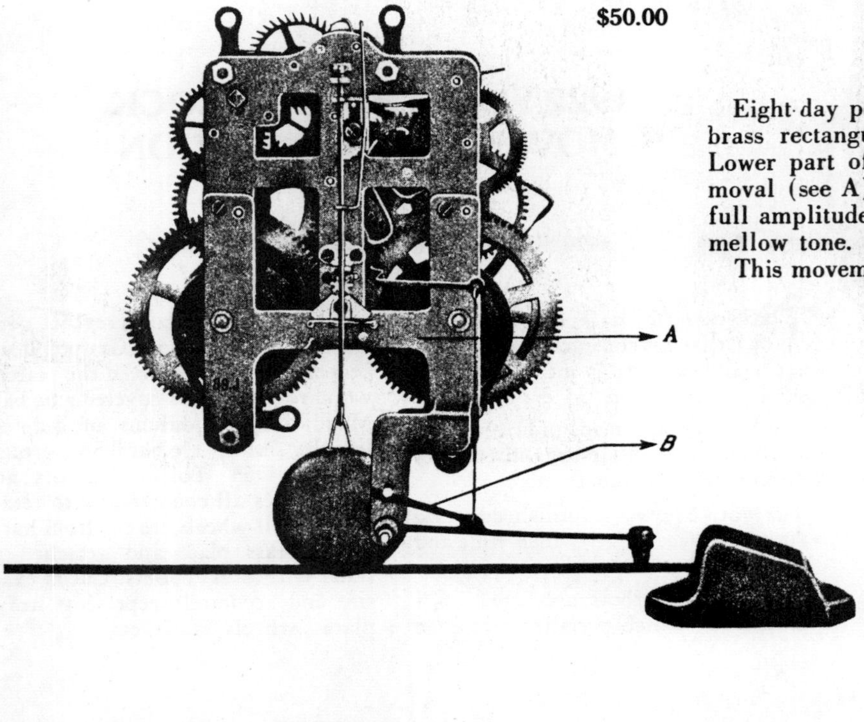


*No. 89 Movement*

Eight-day pendulum strike movement. Heavy hard brass rectangular plates and cut gears throughout. Lower part of back plate is detachable for easy removal (see A). (B) rebound lever which allows for full amplitude of rod vibrations thus assuring a very mellow tone.

This movement is front wind.

# Ca. 1915 SETH THOMAS CLOCK MOVEMENT DESCRIPTIONS

**$155.00**

Seth Thomas Sonora Chime movement No.104, Front Plate for Eight Bell Whittington and Westminster Chimes (H. Kirk Fallin)

Seth Thomas Sonora Chime movement No.104, Back Plate for Eight Bell Whittington and Westminster Chimes (H. Kirk Fallin)

**$85.00**

Movement No.90, Front Plate 4 or 5 Bell Chime Clocks. (H. Kirk Fallin)

Movement No.90, Back Plate 4 or 5 Bell Chime Clocks. (H. Kirk Fallin)

# 1928 SETH THOMAS CLOCK MOVEMENT DESCRIPTION

$40.00

No. 103-A Movement

Eight-day lever time movement, pin escapement. Balance wheel screws are fitted with bearing and cap jewels. Cut gears and pinions in hard rolled brass plates. Has detachable mainspring box. Back wind and set. Diameter 2 9/16 inches. Depth 2¼ inches. Very good timekeeper.

Used in all eight-day Boudoir and Desk Clocks.

# 1886 SETH THOMAS CLOCK MOVEMENT DESCRIPTION

No. 94.

$45.00
Movement only

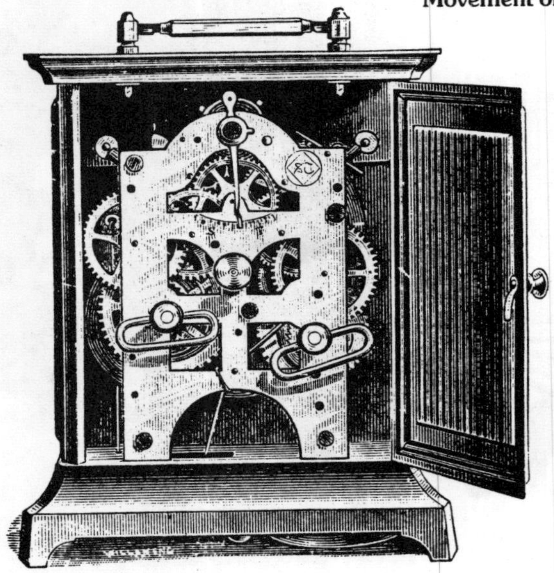

Back view of
JOKER, CRYSTAL
MIKADO, ARTIST

# 1926 SETH THOMAS CLOCK MOVEMENT DESCRIPTIONS

$50.00

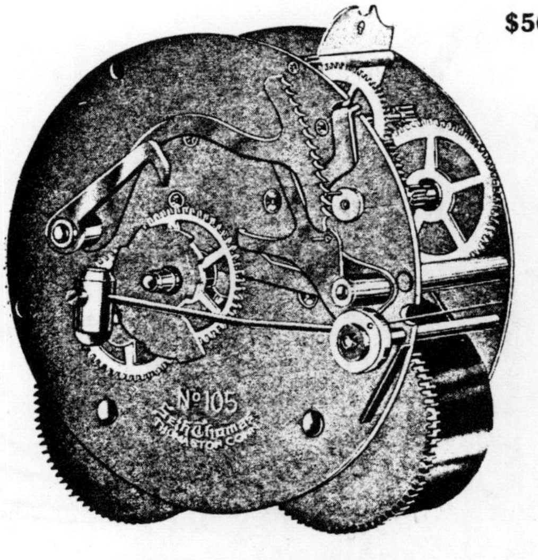


**No. 105 MOVEMENT**

Highest grade 8-day lever strike movement for fine mantel clocks, with heavy hard brass plates and cut wheels. All brass parts gold plated. Mainsprings in going barrels and steel cut polished pinions, rack striking. Striking the hours and half hours on cathedral bell. Movement fitted with detachable 7-jeweled escapement, Breguet hairspring and cut compensated balance.

$75.00

No. 112 MOVEMENT

Eight day fine lever time movement with cut steel pinions throughout and highly burnished pivots. Is fitted with seven jewels, Breguet hairspring and compensated bi-metal balance wheel cut for expansion.

Movement is 2 11/16 inches in diameter; 1 15/16 inches deep. Used in Eden, Enfield, Ensign clocks and Fine Lever Clock No. 3 and No. 4 as illustrated in Catalog No. 776.

# 1928 SETH THOMAS CLOCK MOVEMENT DESCRIPTION

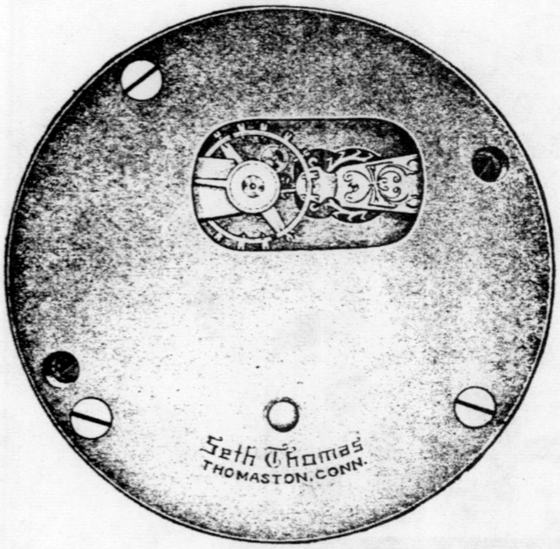


**No. 112, 7-JEWEL MOVEMENT**
Actual Size.

A very accurate 8-day lever time movement. It has a cut compensation balance wheel with timing screws and cut steel pinions, hardened and polished, and is fitted with dust cap. Cut shows jeweled watch escapement and solid construction. Highest grade material.

**$75.00**

# 1926 SETH THOMAS CLOCK MOVEMENT DESCRIPTION

## WESTMINSTER CHIME MOVEMENT No. 113

$165.00

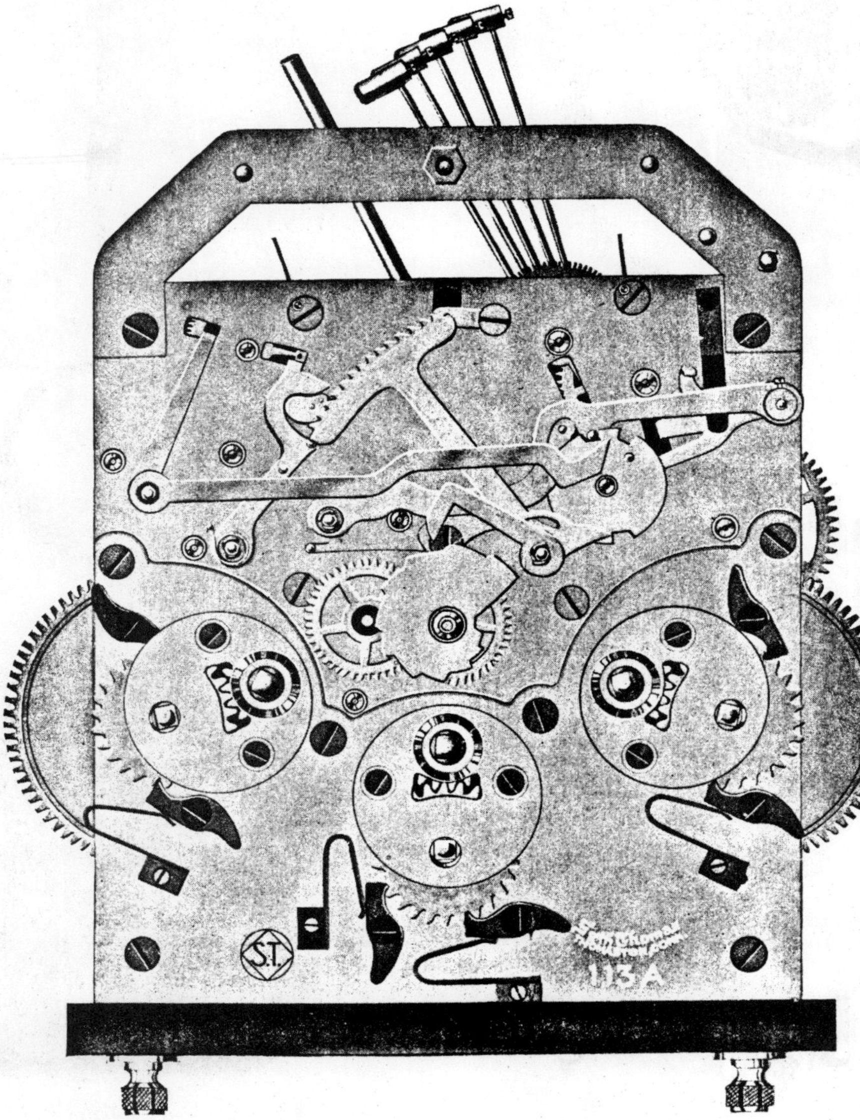


A HIGH-GRADE Westminster chime movement consisting of three separate trains to operate the hour strike, the time and the chime. All wheels are accurately cut of heavy brass, solid cut steel pinions, burnished and polished; pivots hardened and burnished. Heavy cut steel pallets perfectly ground, hardened and polished to reduce friction and insure long life.

The plates made of heavy hard rolled brass. All pivot holes provided with deep cut oil cups for retaining oil. Fly staff pivots and pallet staff are adjustable by moving adjusting buttons set in front plate. Large mainspring barrels which can be easily removed by detaching lower section of front plate without taking movement apart.

Each winding ratchet fitted with double clicks.

A reliable chime adjustment or self-correcting lever which regulates or corrects the chime and hour strike automatically within the hour. The chimes may be silenced by moving the indicator hand, allowing the hours only to strike.

The movement is rigidly mounted on a seatboard securely fastened to clock case, and may be removed by detaching hands and loosening two thumb screws.

A movement of the finest design and workmanship acquired by our experience for over 100 years in clock manufacturing.

In Chime Nos. 72, 73, 74, 75 and 78.

# 1922 SETH THOMAS CLOCK MOVEMENT DESCRIPTION

**$115.00**

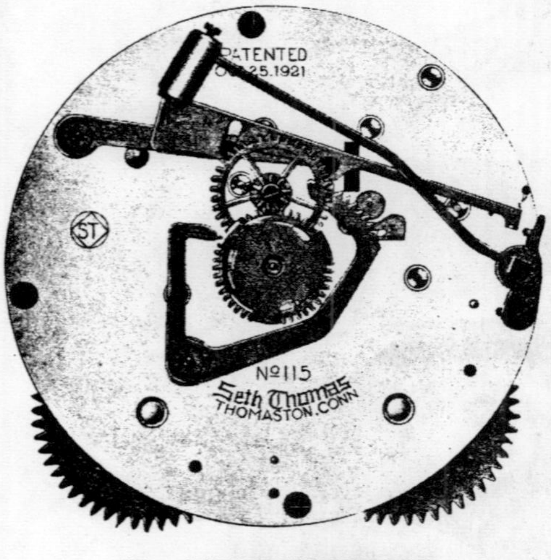


### No. 115 MOVEMENT

Highest grade 8-day lever "Ship's Bell" strike movement for fine mantel clocks, with heavy hard brass plates and cut wheels, all brass parts gold plated. Mainsprings in going barrels with detachable barrel plate, steel cut polished pinions, rack striking, cathedral bell. Movement fitted with detachable 7-jeweled escapement, Breguet hairspring and cut compensated balance. This movement has our patented Ship's Bell striking device, which is the most reliable and the simplest on the market.

# 1926 SETH THOMAS CLOCK MOVEMENT DESCRIPTION

**$55.00**

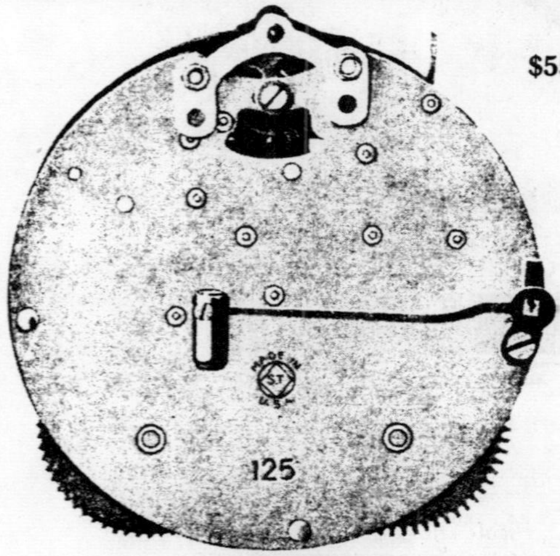


### No. 125 MOVEMENT

Eight day lever strike movement of the pin escapement type with detachable mainspring barrels. Plates of hard brass, lacquered. Cut wheels throughout and solid cut steel pinions in the time train. Movement fitted with compensated bi-metal balance wheel cut for expansion. Jeweled balance bearings. Rack striking. Most reliable and particularly recommended for clocks which are not to be permanently stationary.

# 1928 SETH THOMAS CLOCK MOVEMENT DESCRIPTION

**$65.00**
*Complete as Shown*

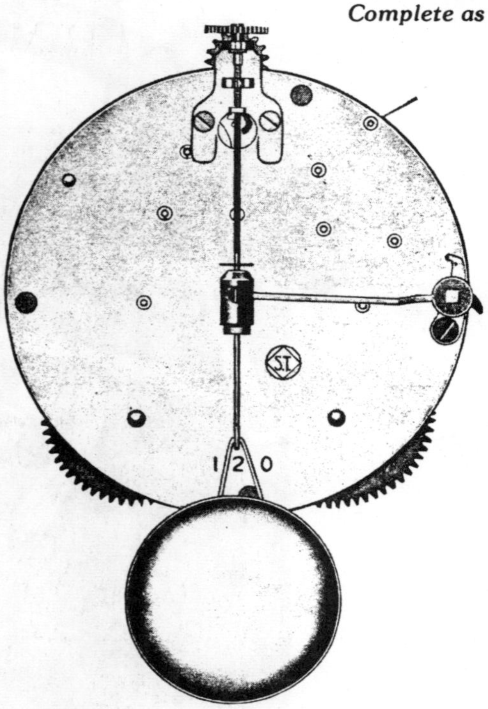


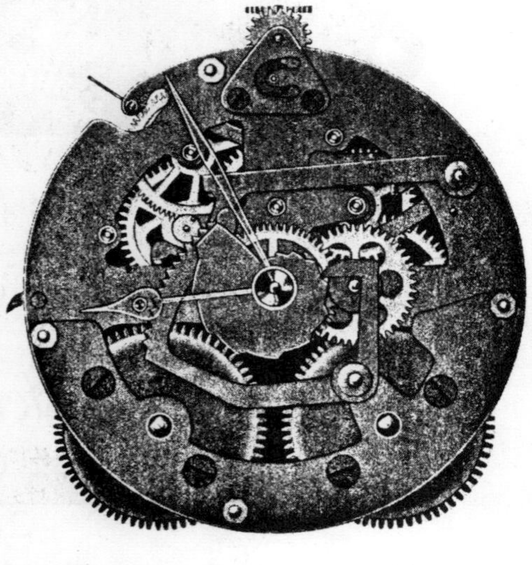

### No. 120 Movement

Eight-day, pendulum, hour and half hour rack strike. Lantern and cut steel pinions throughout. Solid back plate dipped and lacquered. Mainspring in spring boxes attached to front plate made in two sections. The lower section can be detached to remove mainspring boxes leaving the rest of the movement intact.

# 1926 SETH THOMAS CLOCK MOVEMENT DESCRIPTION

**$85.00**

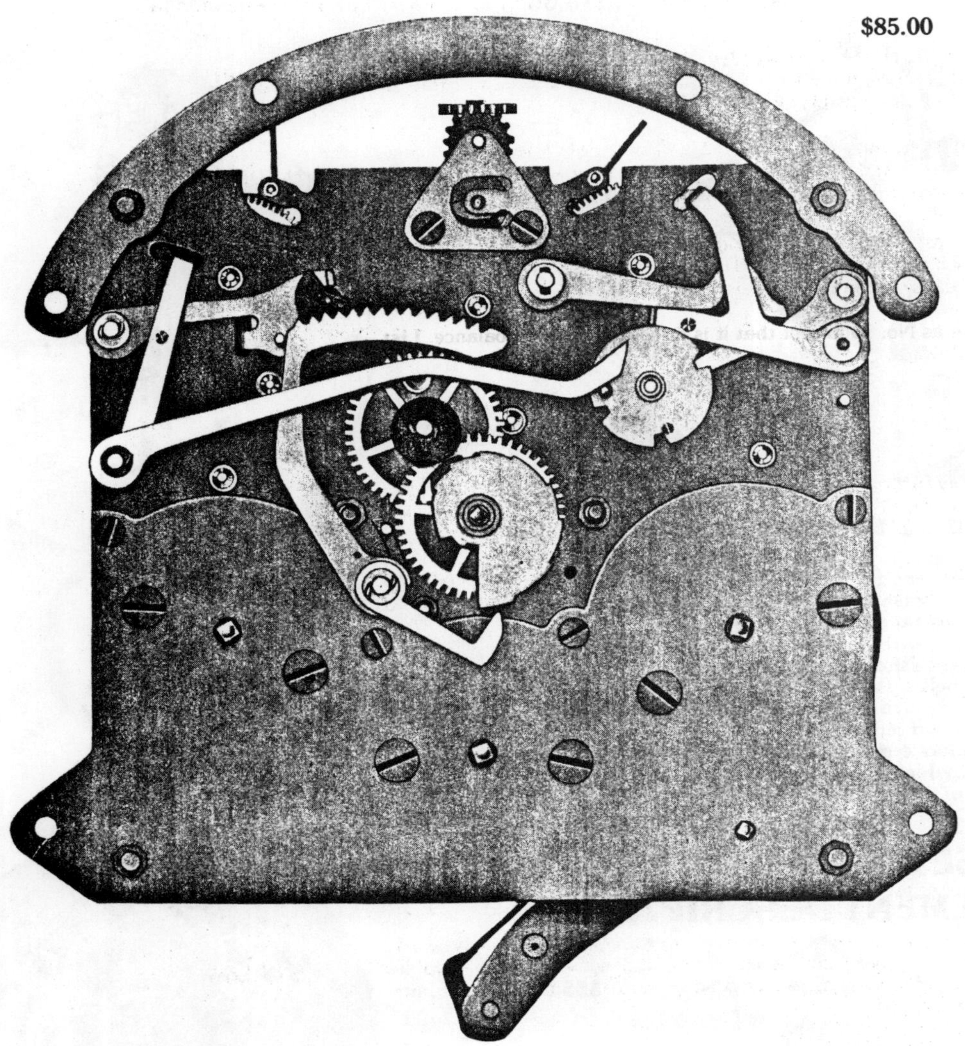

## Westminster Chime Movement No. 124

RELIABLE 8-day Westminster Chime Movement chiming each quarter hour and striking the hours on rods.

This movement has three separate trains to operate the hour strike, time and chime trains.

All wheels are made of hard brass accurately machine cut. All fine lantern pinions except the first pinions and fly pinions which are of solid cut steel.

Solid heavy brass plates dipped and lacquered. Deep countersunk oil cups with retaining rings.

Mainsprings are in boxes attached to the front plate made in two sections. The lower section can be detached to remove main spring boxes, leaving the rest of the movement intact.

Dead beat escapement, the pallets of solid steel, impulse faces ground and tempered and highly polished.

Should the chimes be disarranged when setting the clock, a self-correcting lever will automatically correct the chimes and hour strike *within the next hour.*

The chimes may be silenced by using the smaller end of the winding key, inserting it in the shut-off arbor. A half turn toward "silent" will shut off the chimes and allow the hours only to strike. The shut-off arbor will be found at the bottom of the dial between numerals "4" and "5."

This movement is fitted up with a front bracket and screwed in position.

The chime rods are of a special metal alloy of superior quality and perfectly tuned. The chimes are played on four rods and the fifth rod is used for the hour strike.

In chimes Nos. 91, 92, 93, 94, 95 and 96.

### SIZE

Length, 8 inches; width, 6¾ inches; depth, 5½ inches. Pendulum from center to end of ball, 4¼ inches.

# 1933 SETH THOMAS CLOCK MOVEMENT DESCRIPTIONS

*To receive top prices, movements must be complete and original, and they must be in good running or repairable condition. Damaged, worn-out or missing parts will reduce prices accordingly.*

**No. 7** - 1-day lever, ship bell strike; length, 7¾ inches, width 5¾ inches; depth, 2⅜ inches; fitted with 6 inch silvered dial and hands; runs 50 hours on one winding. **List Price $13.50** — **$80.00**

**No. 7-D** - 1-day lever, ship bell strike used in "Boat Clock"; length, 5 inches, width, 5¾ inches; depth over bell 3½ inches; fitted with 6 inch silvered dial and hands; runs 50 hours on one winding. **List Price $13.50** — **$80.00**

**No. 10** - 8-day lever time, double spring movement; length 4-15/16 inches; width, 5¾ inches; depth, 2-7/16 inches; used in 8-day wood lever clocks; will run 9 days on one winding. **List Price $11.00** — **$45.00**

**No. 10A** - Same as No. 10 except that it is fitted with jeweled balance. **List Price $14.00** — **$50.00**

**No. 10J** - 8-day spring lever time, double spring movement; length 4-15/16 inches; width 5 inches; depth 2⅜ inches; fitted with 6 inch silvered dial and hands; used in 8-day engine lever clocks. **List Price $17.00** — **$50.00**

**No. 41AC** - 8-day spring time, pendulum movement; length 5⅝ inches; width, 5⅞ inches; depth, 2¾ inches. Pendulum from center to end of rating screw 13 inches. Used in No. 1 Office Time. **List Price $5.00** — **$55.00**

**No. 41AE** - Same as in No. 41AC except depth is 3¾ inches and pendulum from center to end of rating screw is 11⅞ inches. Used in 10 inch Drop Octagon Time **List Price $5.25** — **$55.00**

**No. 41AF** - Same as No. 41AC except depth is 3¾ inches. Used in 12 inch Drop Octagon Time. **List Price $5.25** — **$55.00**

**No. 41AN** - Same as No. 41AC except depth is 3¾ inches and pendulum from center to end of rating screw is 17⅞ inches. Used in Globe. **List Price $4.75** — **$55.00**

**No. 41S** - Same as No. 41AC except depth is 3¾ inches and pendulum from center to end of rating screw is 10⅛ inches. Used in 8 inch Drop Octagon **List Price $5.25** — **$55.00**

**No. 68** - High grade 8-day weight time movement; length 10½ inches; width, 8¼ inches; large keystone plates; steel cut pinions; polished sapphire pallet stones; maintaining power; used in No. 70 and No. 80 Regulators- **List Price $75.00** — **$375.00**

**No. 77A** - High grade 8-day weight time movement; keystone plates, steel cut pinions; polished; maintaining power; pendulum, 22½ inches; length 5¼ inches; width 5 inches; depth, 4 5/16 inches; used in No. 21 Regulator. **List Price $35.00** — **$200.00**

**No. 86T** - Thirty-day double spring, time pendulum movement; length 8⅛ inches; width 7⅛ inches; depth 4 inches; pendulum 8¾ inches to end of rating screw; used in Hudson and No. 1 Arcade. **List Price $20,00** — **$115.00**

**No. 86W** - Same as No. 86T, except that pendulum is 21⅜ inches to end of rating screw; used in World and No. 16 Office. **List Price $20.00** — **$110.00**

**No. 89** - 8-day spring, hour and half hour pendulum strike movement; back escapement; length 6 inches; width 6 inches; depth 3½ inches; open plates; pendulum to end of ball 4⅝ inches; used in Adamantines and cheap cabinets. **List Price $8.00** — **$50.00**

**No. 89A** - 8-day spring; hour and half hour pendulum strike on gong; open plates; length, 6 inches; width 6 inches; depth 3 3/16 inches; pendulum to end of rating screw 6⅝ inches; used in Cottage, Plymouth, Arctic Strike, Chatham Strike, No. 1 and No. 4 Office Strike. **List Price $8.00** — **$55.00**

**No. 89B** - Same as No. 89A except pendulum from center to end of rating screw, 9 13/16 inches; used in 8 inch Drop Octagon Strike and 8 inch Bradford Strike. **List Price $8.00** — **$65.00**

# 1933 SETH THOMAS CLOCK MOVEMENT DESCRIPTIONS

*To receive top prices, movements must be complete and original, and they must be in good running or repairable condition. Damaged, worn-out or missing parts will reduce prices accordingly.*

**No. 89C** - Same as No. 89A except depth is 3 11/16 inches and pendulum from center to end of rating screw is 11 5/16 inches; used in 10 inch Drop Octagon Strike and 10 inch Bradford Strike. **List Price $8.00** — **$65.00**

**No. 89D** - Same as 89C except pendulum is 13¼ inches from center to end of rating screw; used in 12 inch Drop Octagon Strike and 12 inch Bradford Strike. **List Price $8.00** — **$65.00**

**No. 89E** - Same as 89A except length 6 3/16 inches; width 6 inches; depth 3 15/16 inches; pendulum from center to end of rating screw 22 inches; used in No. 6 Office Strike and Globe Strike. **List Price $8.00** — **$65.00**

**No. 89F** - Same as 89D, but furnished with calendar attachment. **List Price $12.00** — **$80.00**

**No. 103A** - 8-day lever time, 4 jewel escapement, round plates; diameter, 2 9/16 inches; depth 2¼ inches; back wind and set; complete with 2½ inch silver dial, sash, glass, hands and back. **List Price $8.00**
Above with 3 inch dial fit up. **List Price $10.00.** Used in cheap 8-day Novelties. — **$40.00**

**No. 110** - 8-day automatic adjustable alarm, cut pinions; length 4 inches; width 4 inches; depth 2½ inches; complete with dial and hands. Used in 8-day alarm. **List Price (plain dial) $9.00**
**List Price (radium dial) $10.00** — **$50.00**

**No. 112** - 8-day fine lever time, 7 jewel movement, cut steel pinions, Bregnet hairspring, compensated balance; diameter 2 11/16 inches; depth 1 15/16 inches; used in fine 3 inch Novelties; complete with 3 inch silvered dial, sash, glass and hands. **List Price $30.00** — **$75.00**

**No. 113** - Highest grade 8-day, three train, Westminster chime movement, cut steel pinions, automatic chime control, removable barrels; length 8 inches; width 6¾ inches; depth 5½ inches; pendulum from center to end of ball 4¼ inches; used in 74-75 and 77 Chimes. **List Price $50.00** — **$165.00**

**No. 113A** - Same as No. 113 except fitted with extension bridge for "Silent" and Regulating arbors; used in No. 71-72-73 chimes **List Price $50.00** — **$165.00**

**No. 113B** - Same as No. 113 except pendulum from center to end of rating screw is 15 inches; used in No. 101-102-103 chimes. **List Price $50.00** — **$165.00**

**No. 117** - Highest grade 8-day lever time, cut steel pinions, 7 jewel escapement, Bregnet hairspring, compensated balance; length 4 5/16 inches; width 2 25/32 inches; depth 1⅝ inches; used in No. 6-60-80 Fine Marine Clocks. **List Price No. 6 Dial $35.00 No. 8 Dial $37.50**

**No. 120** - 8-day hour and half hour strike, pendulum movement, round plates removable barrels, steel cut pinions, silent fly and rack; length 4 inches; width 3⅝ inches; depth 3 3/16 inches; pendulum from center of movement to end of ball 4 5/16 inches. **List Price $12.00** — **$65.00**

**No. 120F** - Same as No. 120 except pendulum from center to end of ball, 3 1/16 inches; used in No. 1 and No. 2 Tambour clocks. **List Price $12.00** — **$65.00**

**No. 124** - 8-day medium grade, three train, Westminster chime movement, automatic chime control, silent rack, cut steel pinions, removable spring barrels, pendulum from movement center to bottom of ball, 4 inches; used in 5 inch dial chime clocks. **List Price $20.00** — **$85.00**

**No. 124A** - Same as No. 124 with exception of arrangement of strike hammers; used in upright 5 inch Dial Chime Clocks. **List Price $20.00** — **$85.00**

# 1936 SETH THOMAS CLOCK MOVEMENT DESCRIPTIONS

**No. 7**–1-day lever, ship bell strike; fitted with 6-inch silvered dial and hands; runs 50 hours on one winding; length, 8 inches; width, 5¼ inches; depth, 2 inches; used in Monitor. Dealer Price $8.00 **$80.00**

**No. 7D**–1-day lever, ship bell strike; fitted with 6-inch silvered dial and hands; runs 50 hours on one winding; length, 5 inches; width, 5¼ inches; depth over bell, 3¼ inches; used in Merrimac and Boat Clock. Dealer Price $8.50 **$80.00**

**No. 10**–8-day lever time, double spring movement; will run 9 days on one winding; length, 4 15/16 inches; width, 5¾ inches; depth, 2 7/16 inches; used in 8-day wood lever clocks. Dealer Price $5.50 **$45.00**

**No. 10A**–Same as No. 10 except that it is fitted with jeweled balance. Dealer Price $7.00 **$50.00**

**No. 10AC**–Same as No. 10 except with compensated jeweled balance. Dealer Price $10.00 **$65.00**

**No. 10J**–8-day lever time, double spring movement; fitted with 6 inch silvered dial and hands; length, 4 15/16 inches; width, 5¾ inches; depth, 2¼ inches; used in 8-day engine lever clocks. Dealer Price $8.00 **$50.00**

**No. 10AJ**–Same as No. 10J except with jeweled balance. Dealer Price $7.50 **$50.00**

**No. 41AC**–8-day spring time, back mounting, pendulum movement; escapement in front; length, 5⅝ inches; width, 5⅞ inches; depth, 2¾ inches. Pendulum from center to end of rating screw 7¼ inches; used in Principal and Chairman Time. Dealer Price $2.50 **$55.00**

**No. 41AE**–Same as No. 41AC except depth is 3½ inches and pendulum from center to end of rating screw is 12⅜ inches. Used in 10" Drop Octagon Time. Dealer Price $2.75 **$55.00**

**No. 41AF**–Same as No. 41AC except depth is 3½ inches and pendulum from center to end of rating screw is 13⅜ inches. Used in 12" Drop Octagon Time. Dealer Price $2.75 **$55.00**

**No. 41S**–Same as No. 41AC except depth is 3½ inches and pendulum from center to end of rating screw is 9¾ inches. Used in 8" Drop octagon. Dealer Price $2.75 **$55.00**

**No. 68**–High grade 8-day weight time movement; heavy keystone plates; polished steel cut pinions; sapphire pallet stones; maintaining power; length, 10½ inches; width, 8¼ inches; depth, 5½ inches; pendulum from center to end of rating screw 40⅛ inches; used in No. 70 and No. 80 Regulators. Dealer Price $40.00 **$375.00**

**No. 77**–High grade 8-day weight time movement; keystone plates; polished steel cut pinions; maintaining power; length, 5¼ inches; width, 5 inches; depth, 4 5/16 inches; pendulum, 22½ inches; used in No. 2 Regulator. Dealer Price $20.00 **$200.00**

**No. 86T**–Thirty-day double spring, time pendulum movement; length, 8⅛ inches; width 7⅛ inches; depth, 3½ inches; pendulum 8½ inches to end of rating screw; used in Hudson.Arcade. Dealer Price $9.25 **$115.00**

# 1936 SETH THOMAS CLOCK MOVEMENT DESCRIPTIONS

*To receive top prices, movements must be complete and original, and they must be in good running or repairable condition. Damaged, worn-out or missing parts will reduce prices accordingly.*

**No. 86W**–Same as No. 86T except that pendulum is 21⅜ inches to end of rating screw. Dealer Price $10.00 **$110.00**

**No. 89**–8-day spring, hour and half hour pendulum strike movement; open plates; length, 6 inches; width, 6 inches; depth, 4½ inches; pendulum to end of ball 4¾ inches; used in No. 17 Sentinel. Dealer Price $3.75 **$50.00**

**No. 89A**–8-day spring, hour and half hour pendulum strike on gong; open plates; length, 6 inches; depth, 3 3/16 inches; width 6 inches; pendulum to end of rating screw 5⅝ inches; used in Sharon and Kingsbury, also in Principal and Chairman Strike. Dealer Price $3.75 **$55.00**

**No. 89B**–Same as No. 89A except depth, 3⅜ inches and pendulum from center to end of rating screw 9 13/16 inches; used in 8" Drop Octagon Strike. Dealer Price $4.50 **$65.00**

**No. 89C**–Same as No. 89A except depth is 3⅜ inches; and pendulum from center to end of rating screw is 11 inches; used in 10" Drop Octagon Strike. Dealer Price $4.50 **$65.00**

**No. 89D**–Same as No. 89C except pendulum is 12¼ inches from center to end of rating screw; used in 12" Drop Octagon Strike. Dealer Price $4.50 **$65.00**

**No. 89E**–Same as No. 89A except length is 6 3/16 inches; width is 6 inches; depth is 3¾ inches and pendulum from center to end of rating screw is 22 inches; used in No. 6 Office Strike and Globe Strike. Dealer Price $5.25 **$65.00**

**No. 89F**–Same as No. 89D, but furnished with calendar attachment. Dealer Price $5.00 **$70.00**

**No. 89L**–Similar to No. 89 except with rack and snail and ding-dong strike every quarter hour on two-rod bell; length, 7½ inches; width, 7¼ inches; depth, 5 inches; pendulum length, 4¾ inches. Dealer Price $4.50 **$55.00**

**No. 89LA**–Same as No. 89M except each quarter hour is announced on three-rod gong. A single note followed by a chord marks the quarter hour; two notes and chords the half hour, and three the three quarter hour. The hour is then announced in a chord. Dealer Price $4.50 **$55.00**

**No. 89M**–Same as No. 89 except chord strike at hour and half hour on two-rod bell; length, 7½ inches; width, 7¼ inches; depth, 5 inches; pendulum length, 4¾ inches. Dealer Price $4.00 **$50.00**

**No. 112**–8-day fine lever time, 7-jewel movement, cut steel pinions; compensated balance; diameter 2 11/16 inches; depth, 1 15/16 inches. Dealer Price $15.00 **$75.00**

# 1936 SETH THOMAS CLOCK MOVEMENT DESCRIPTIONS

*To receive top prices, movements must be complete and original, and they must be in good running or repairable condition. Damaged, worn-out or missing parts will reduce prices accordingly.*

**112A**–Same as No. 112 except front wind and front regulator; complete with 4-inch silvered dial and hands; used in No. 4 Fine Lever clocks. Dealer Price $17.00 **$75.00**

**112B**–Same as No. 112A; complete with 3-inch silvered dial and hands; used in No. 3 Fine Lever clocks. Dealer Price $16.50 **$75.00**

**113B**–Same as No. 113A except no top bridge or regulating arbors; pendulum from center to end of rating screw is 15 inches; used in No. 3W Hall clocks. Dealer Price $24.95 **$165.00**

**113E**–Same as No. 113A except pendulum from center to end of ball 4⅜ inches; used in No. 81 Chime and No. 4 Hall clock. Dealer Price $24.95 **$165.00**

**115A**–High grade 8-day fine lever ship bell movement, cut steel pinions, removable barrel, 7-jewel detachable escapement, compensated balance; diameter 3⅝ inches; overall 4⅛ inches; depth, 2⅜ inches; used in No. 2 Mayflower,, No. 44 and No. 66 Ship Bell clocks. Dealer Price $22.50 **$115.00**

**No. 117**–High grade 8-day lever time movement, cut steel pinions; 7-jewel escapement, compensated balance; length 4 5/16 inches; width 2 25/32 inches; depth, 1⅝ inches; used in No. 6, 60 and 80 Fine Lever clocks. Dealer Price $15.00 **$90.00**

with 6-inch Dial $17.50

with 8-inch Dial $20.00

**No. 120**–8-day hour and half hour strike, pendulum movement, round plates, removable barrels, steel cut pinions, rack and snail; length, 4 inches; width, 3⅝ inches; depth 3⅝ inches; pendulum from center of movement to end of ball 4 5/16 inches. Dealer Price $5.00 **$55.00**

**No. 120F**–Same as No. 120 except pendulum from center to end of ball, 3 1/16 inches; used in Dunbar and Thames clocks. Dealer Price $5.00 **$55.00**

**No. 120K**–Same as No. 120 except pendulum from center to end of rating screw, 17 inches; used in Crandall. Dealer Price $5.00 **$55.00**

**No. 120L**–Same as No. 120F except with bell pillar in back plate; used in No. 1 and No. 2 Console. Dealer Price $5.00 **$55.00**

**No. 120M**–Same as No. 120 except with bell pillar in back plate; used in No. 2 and No. 3 Century. Dealer Price $5.00 **$55.00**

**No. 124**–8-day medium grade, horizontal, three train, Westminister chime movement, automatic chime control, cut steel pinions, removable spring barrels; length, 6⅜ inches; width, 6 inches; depth, 4½ inches; pendulum from movement center to bottom of ball, 4 inches; used in 5-inch dial chime clocks. Dealer Price $10.00 **$85.00**

# 1936 SETH THOMAS CLOCK MOVEMENT DESCRIPTIONS

*To receive top prices, movements must be complete and original, and they must be in good running or repairable condition. Damaged, worn-out or missing parts will reduce prices accordingly.*

**No. 124A**–Same as No. 124 except vertical arrangement of strike hammers; top bracket; length, 8 inches; width, 5½ inches; depth, 4⅞ inches; pendulum length, 4 inches; used in upright 5-inch dial chime clocks. Dealer Price $10.00 **$85.00**

**No. 124B**–Same as No. 124 except with top and bottom mounting brackets; used in No. 52 Chime. Dealer Price $10.00 **$85.00**

**No. 124C**–Same as No. 124A except mounting brackets; used in No. 53 Chime. Dealer Price $10.00 **$85.00**

**No. 124D**–Same as No. 124 except top mounting brackets; length, 7¼ inches; used in No. 56 Chime. Dealer Price $10.00 **$85.00**

**No. 891**–8-day spring, hour and half hour pendulum strike on gong, recoil escapement; open plates; length, 6 inches; width, 6 inches; depth, 4½ inches; center to end of pendulum, 4¾ inches; used in No. 5585 and No. 5595 Plymouth. Dealer Price $3.50 **$40.00**

**No. 891 L**–Similar to No. 891 except rack and snail and ding-dong strike every quarter hour on two-rod bell; used in Plymouth No. 5584 and No. 5592. Dealer Price $4.25 **$45.00**

**No. 891 M**–Same as No. 891 except chord strike at hour and half hour on two rod bell; used in No. 5591 and No. 5593 Plymouth. Dealer Price $3.75 **$40.00**

**No. 1120J**–8-day hour and half hour strike, pendulum movement, round plates, removable barrels, steel cut pinions, rack and snail; length, 4 inches; width, 3⅝ inches; depth, 3⅝ inches; pendulum from center to end of rating screw, 17 inches; used in No. 5596 Plymouth. Dealer Price $5.00 **$45.00**

**No. 5100**–8-day non-jeweled time movement, double spring, back wind, metal balance with Elinvar hairspring for temperature compensation; round plates; 2⅝ inches in diameter; overall depth, 1½ inches. Dealer Price $2.25 **$30.00**

With 3¼ inch sash, dial, hands, etc. as in Delmore, Delrod and Delkirk 3.50 **$40.00**

**No. 5101**–Same as No. 5100 except front wind; used in Precise and No. 5573 Plymouth. Dealer Price $2.25 **$30.00**

**No. 5102**–Same as No. 5100 except long center; used in Delray. Dealer Price $2.25 **$30.00**

**No. 5103**–Same as No. 5100 except short center. Dealer Price $2.25 **$30.00**

With 2½ inch sash, dial, hands, etc., as in No. 1 and No. 3 Delta. $3.10 **$35.00**

# 1936 SETH THOMAS CLOCK MOVEMENT DESCRIPTIONS

## SYNCHRONOUS SELF-STARTING TIME MOVEMENTS

Diameter 3⅜ inches, Depth 2⅜ inches
Motor 240 R.P.H.—Back Set-Use 5001 oil

**No. 1500**-60 cycle 110-120 volt, as in Judson. Dealer Price $3.50

**No. 1510**-60 cycle 110-120 volt, Bottom set, as in President. Dealer Price $3.50

**No. 1520**-60 cycle 110-120 volt, Front set, as in Senate. Dealer Price $3.50

## SYNCHRONOUS HOUR AND HALF HOUR STRIKE MOVEMENTS

4 5/16 inches Wide, 4 7/16 inches Long, and 3⅜ inches Deep
Motor 240 R.P.M.

**No. 1600**-60 cycle 110-120 volt. Dealer Price $6.50

**No. 1601**-60 cycle 110-120 volt with brackets, as in Middleton. Dealer Price $6.50

**No. 1602**-60 cycle 110-120 volt hammer shaft for Banjos (as in Brookfield). Dealer Price $6.50

**No. 1604** -160 cycle 110-120 volt with bell pillar in back plate (as in No. 2 and No. 3 Century and No. 1 and No. 2 Console). Dealer Price $6.50

## WESTMINSTER SYNCHRONOUS ELECTRIC CHIME MOVEMENTS

5¼ inches Wide, 7 1/16 inches Long, and 4⅝ inches Deep

**No. 1700**-60 cycle 110-120 volt Horizontal as in Roxbury, etc. Dealer Price $10.00

**No. 1702**-60 cycle 110-120 volt as in Medbury. Dealer Price $10.00

**No. 1750**-60 cycle 110-120 volt vertical as in Westbury. Dealer Price $10.00

**Listed below are the original Seth Thomas catalogues which were used in compiling this guide. In a continuing program to make the guide complete, I would appreciate receiving any information from readers about additional catalogues that are available.**

| YEAR | CATALOGUE NUMBER | PAGES | NOTE |
|---|---|---|---|
| 1863 | | 68 | |
| Ca. 1870 | | 71 | Seth Thomas "and Sons" |
| 1875 | 127 | 40 | |
| 1881 | 342 | 32 | |
| 1884-85 | 387 | 128 | |
| 1886 | 411 | 128 | |
| 1890 | 495 | 63 | Tower |
| 1892-93 | 531 | 136 | |
| 1894-95 | 553 | 120 | |
| 1895 | 577 | 16 | Supplement |
| 1896 | 583 | 71 | Tower |
| 1900-01 | 625 | 119 | |
| 1901 | 629 | 15 | Movements and Parts |
| 1902-03 | 633 | 154 | |
| 1904-05 | 646 | 144 | |
| Ca. 1904 | 653 | 16 | Movements and Parts |
| 1906-07 | 660 | 80 | |
| 1907 | | 25 | Movements and Parts |
| 1909-10 | 675 | 96 | |
| 1911 | 686 | 16 | Supplement |
| Ca. 1912 | 6 | 76 | Tower |
| 1912 | 7 | 72 | Tower |
| 1913 | 700 | 80 | Centennial |
| Ca. 1913 | 701 | 24 | Electric |
| 1915 | 707 | 60 | |
| 1917-18 | 735 | 48 | |
| 1921-22 | 760 | 48 | |
| | | 12 | Electric |
| 1922 | 765 | 48 | |
| 1922 | | 24 | Electric |
| 1923 | 765 | 4 | Supplement "B" |
| 1924 | 766 | 24 | |
| 1924 | 775 | 36 | |
| 1925 | 775 | 6 | Supplement "A" |
| 1924 | 776 | 35 | |
| 1924 | 776 | 22 | Hanging and Wall Clocks |
| 1926 | 777 | 39 | |
| 1926 | 777 | 6 | Supplement "A" |
| 1927 | 777 | 4 | Supplement "B" |
| 1928 | 777 | 4 | Supplement "C" |
| Ca. 1928 | | 4 | Special Supplement of Seth Thomas Clocks |
| 1928 | 780 | 55 | |
| 1929 | 780 | 4 | Supplement to Catalog 780 |
| Ca. 1930 | 8 | 56 | Tower |
| 1931 | 785 | 15 | |
| 1931 | 790 | 32 | Electric |
| 1932 | 795 | 20 | |
| 1933 | | 20 | Movements and Parts |
| 1934 | 797 | 28 | |
| 1935 | 798 | 44 | |
| 1936 | | 26 | Movements and Parts |
| 1936 | 799 | 24 | |
| 1938 | 800 | 24 | |
| 1938 | 800-A | 24 | |
| 1939 | 802 | 24 | Keywound and Electric |
| 1941 | 804 | 24 | Electric and Keywound |
| 1947 | 805 | 16 | |
| 1948 | 806 | 19 | |
| 1949 | 807 | 19 | Electric and Springwound |
| 1950 | 808 | 20 | |
| 1951 | 809 | 20 | |
| 1953 | 811 | 4 | |
| 1954 | 812 | 4 | |
| 1960 | 818 | 12 | |
| 1968 | 832 | 4 | |

# 1985 PRICE LIST FOR BOOKS ON CLOCKS, WATCHES AND RELATED SUBJECTS.

**Prices subject to change without notice.**

## BAROMETERS

| | PRICE |
|---|---|
| Antique Barometers - Banfield, E. | $12.00 |
| Barometers - Bolle, B. | 29.50 |
| English Barometers 1680-1860 - Goodison, N. ★ | 59.50 |

## CATALOGS

| | PRICE |
|---|---|
| American Clocks - From 10 different Clock Co. - 1874 Catalog | $6.95 |
| Ansonia Clock Co. - 1880 Catalog | 6.00 |
| Boston Clock Co. - 1890 Catalog | 4.95 |
| Chauncey Jerome Brass Clocks | 3.95 |
| Chauncey Jerome - 1852 - 1853 Catalog | 5.00 |
| Chelsea Clock Co. - 1911 Catalog | 7.00 |
| Clocks - Catalogue of Imported Clocks, 1911 | 7.50 |
| Fine Imported Clocks - 1915 Catalog | 7.50 |
| Gilbert Clock Co. - 1875 Catalog | 2.95 |
| Gilbert Clock Co. - 1881 Catalog | 6.75 |
| Herschede Hall Clock - 1924 Catalog | 3.00 |
| Herschede Hall Clock - Electrically wound floor & mantel clocks | 4.50 |
| Herschede Mantel Clock - 1923 Catalog | 2.00 |
| Howard Clock Co. - 1858 - Howard & Davis Clocks Catalog | 3.00 |
| Howard Clock Co. - 1874 Catalog | 5.00 |
| Howard Clock Co. - 1889 Catalog - Fine Regulators | 4.50 |
| Howard Clock Co. - 1900 Catalog | 5.00 |
| Howard Clock Co. - 1923 Catalog | 5.00 |
| Howard Clock Co. - Tower & Street Clocks - 1923 Catalog (very good) | 5.95 |
| Ingraham Clock Co. - 1880 Catalog | 5.00 |
| Ingraham Clock Co. - 1885 Supplementary Catalog | 2.00 |
| Ingraham - 1881 - 1886 Catalog | 6.00 |
| Kroeber Clock Co. - 1888 - 1889 Catalog | 8.00 |
| Kroeber Clock Co. - 1898 - 1899 Catalog | 8.00 |
| New Haven Clock Co. - 1880 Catalog | 5.00 |
| New Haven Clock Co. - 1886 Catalog | 8.50 |
| New Haven Clock Co. - 1889-90 Catalog | 8.50 |
| New Haven Clock Co. - 1900 Catalog | 6.95 |
| New Haven Clock Co. - 1906-7 Catalog | 3.00 |
| Self Winding Clock - 1908 Catalog | 16.95 |
| Sessions Clock Co. - 1915 Catalog | 6.00 |
| Seth Thomas Clock Co. - 1863 Catalog | 5.00 |
| Seth Thomas Clock Co. - 1892-93 Catalog | 7.00 |
| Seth Thomas Clock Co. - 1909-10 Catalog ★ | 10.00 |
| Seth Thomas Watches - 1884-1915 Catalog | 7.00 |

**For comprehensive listing of Seth Thomas clocks, see Tran Duy Ly's "Seth Thomas Clocks and Movements, A Guide to Identification and Prices."**

| | |
|---|---|
| South Bend Watch Co. - 1914 Catalog ★ | $10.00 |
| Southern Calendar Clock Co. - "Fashion Magazine" | 5.00 |
| Terry Clock Co. - 1875 Catalog with History of Silas B. Terry | 5.00 |
| Vermont Clock Co. - 1900 Catalog | 5.00 |
| Vienna Regulators of Lenzkirch and Lorenz Bob | 7.00 |
| Waltham Banjo Clocks | 3.50 |
| Waltham Watch Co. - 1885 Material Catalog | 5.00 |

## CATALOGS

| | PRICE |
|---|---|
| Waterbury Clock Co. - 1891 Catalog | $ 7.50 |
| Waterbury Clock Co. - 1893 Catalog | 6.95 |
| Waterbury Clock Co. - 1908 - 9 Catalog | 7.00 |
| Waterbury Ship's Bell Clocks Catalog | 3.00 |
| Welch Clocks - 1880 Catalog | 5.00 |
| Welch Clocks - 1885 Catalog | 5.00 |
| Welch Clocks - 1900 Catalog | 5.95 |
| Western Clock MFG. Co. - 1902 ★ | 5.00 |
| Yale Clocks - Catalog of Miniature Clocks - Catalog 1882 | 4.50 |

**For selected listing of clocks from above catalogs, see Tran Duy Ly's "Clocks, A Guide to Identification and Prices, Volume 1 & 2.**

## CHRONOMETERS

| | PRICE |
|---|---|
| Account of Improvements in Chronometers - Eiffee, J. | $17.00 |
| Construction & Management of Chronometers, Watches & Clocks - Dent, E. | 3.50 |
| Edward John Dent & His Successors - Mercer, V. | 54.00 |
| Edward John Dent, a lecture by Vaudrey Mercer | 2.95 |
| Frodshams - History of Chronometer Makers - Mercer, V. | 54.00 |
| How the Chronometer Went to Sea - May, W.E. | 3.50 |
| Marine Chronometer - Gould, R. | |
| Marine - Und Taschenchronometer - Bertelle, H. - Text in German | 82.00 |
| Mercer Chronometer - Mercer, V. | |
| Paul Phillip Barraud - Jagger, C. | 9.95 |
| Pioneers of Precision Timekeeping | 6.95 |
| Robert Molineau's Astronomical Clocks & Chronometers | 4.95 |
| Ships Chronometers - Whitney, M. - Available soon | - About 80.00 |
| Supplement to John Arnold & Son | 3.50 |

## CLOCKS

| | PRICE |
|---|---|
| A Handbook & Directory of old Scottish clockmakers - Smith, J. | $ 9.95 |
| Alte Uhren/Old Clocks - Muhe & Vogel, Text in German | 55.00 |
| Alte-Wiener Uhren/Old Viennese Clocks - Helich, E. - 1st edition, Hardbound ★ | 35.00 |
| - Reprint, Softbound | 19.95 |
| American Clock - Distin & Bishop - 1st Edition | 40.00 |
| - Reprint | 12.98 |
| American Clocks for the Collector - Tyler, E.J. - Hardbound | 20.75 |
| - Softbound | 13.50 |
| American Clockwork Toys - 1862 - 1900 - Whitton, B. | 25.00 |
| Asprey, Catalog of Skeleton and Decorative Clocks | 7.95 |
| Baird Advertising Clocks - Schiff, L. | 4.95 |
| Banjo Timepiece - Ela, Chip | 29.95 |
| Black Forest Cuckoo Clocks - Kochman, K. | 9.50 |

**For selected listing of Black Forest Cuckoo Clocks, see Tran Duy Ly's "Clocks, A Guide to Identification and Prices, Volume 2."**

★ **Denotes Soon to be Out of Print.**

## CLOCKS

| | PRICE |
|---|---|
| Book of American Clocks - Palmer, B. ★ | $21.95 |
| Boston Clock Club Papers - 1934 - 1940 | 19.95 |
| Bracket Clocks - Deryck Roberts | 26.50 |
| Calendar Clocks - Miller, A&D | 12.95 |
| Canadian Clocks & Clockmakers - Burrows, G. | 25.00 |
| Carriage Clocks - Allix & Bonnert | 59.50 |
| Characteristics of Standard Terry Type 30-hour wooden movement as a guide to identification of movement makers - Taylor, S. - Bulletin | 3.95 |
| Chauncey Jerome 30 hr. brass weight and related movements - Taylor, S. - Bulletin - Revised Edition | 3.95 |
| Chester County Clocks & Their Makers - James, E. | 20.00 |
| Clock Decorating Stencils of the mid-19th Century - American Clock & Watch Museum | 6.95 |
| Clockmaker's Library - Bromley, J. - Pub. at 52.50 | 24.98 |
| Clockmakers of Lancaster County & Their Clocks - Wood, S. | 15.95 |
| Clocks of Shenandoah - Whitney, P. | 10.95 |
| Clockwork Universe, The - German Clocks - Maurice & Mayr- ★ | 55.00 |
| Clocks made in Canada - Pequegnat story - Varkaris, J&C | 28.50 |
| Collecting & Identifying Old Clocks - Harris, H.G. | 12.95 |
| Complete British Clocks - Loomes, B. | 24.00 |
| Complete Clock Book, The - Wallace, H ★ 1973 Revised and Enlarged edition by Wm. B. Jacobs, Jr., 483 pages. | 29.50 |
| Connecticut Clockmakers of The 18th Century - DeLuxe Edition | 9.50 |
| Contribution of Joseph Ives To Conn. Clock Tech. - Roberts, K. | 24.95 |
| Cornish Clocks & Clockmakers - Brown, H.M. | 14.95 |
| Country Clocks & Their London Origins - Loomes, B. | 19.95 |
| Country Life Collector's Pocket Book of Clocks - Brewer, C. | 10.95 |
| Country Life Library of Antique Longcase Clocks - McDonald, J. | 13.95 |
| Dixie Clocks & Clockmakers - Gibbs, J. | 19.95 |
| Dutch Antique Domestic Clocks - Sellink ★ | 59.50 |
| Early English Clocks - Dawson, Drover & Parkes | 99.50 |
| Eli Terry Pillar & Scroll Shelf Clocks - Barr, L. | 3.50 |
| English Dial Clocks - Rose, R. | 44.50 |
| English Domestic Clocks - Cescinsky-Webster | 49.50 |
| English Lantern Clocks - Hana, W.F.J. ★ | 29.50 |
| Escapements, The - Britton, F.J. | 6.95 |
| Evolution of Clockwork & The Clock of Japan - Robertson | 14.95 |
| For Generations to Come - Story of Elias Ingraham - Holly, F. | 8.95 |
| French Clocks in North American Collection - Edey, W. | 15.00 |
| French Clocks The World Over - 5th Edition - Tardy - Text in English | |
| ■Part I: From Their Beginnings to the Louis XV - Louis XVI - Transition .......................... | 85.00 |
| ■Part II: From Louis XVI to the Second Empire .............. | 85.00 |
| ■Part III: Clocks of 1900 and 1925 from the French Provinces, Belgium, the Netherlands and the United Kingdom ..... | 95.00 |
| ■Part IV: Germany, Austria, Switzerland, Russia, Poland, Czechoslovakia, Hungary, Turkey, Denmark, Sweden, Norway, Spain, Asia, and America ............... | 105.00 |
| Geared to the Stars - The evolution of planetariums, orceries and astronomical clocks - King, H.C. | 75.00 |
| Grandfather Clocks - Barder, R. | 31.50 |
| Grandfather Clocks, The - Edwards, E.J. | 45.00 |
| Guinness Book of Clocks - Smith, A. | 19.95 |
| Gustav Becker Story European Clocks - Kochman, K. | 9.75 |

**For selected listing of Gustav Becker Clocks, see Tran Duy Ly's "Clocks, A Guide to Identification and Prices, Volume 2."**

| | |
|---|---|
| Hamburg American Clock Co., The - Kochmann, K. | 11.20 |
| Heman Clark & The Salem Bridge Shelf Clocks - Bailey & Blackwell - Bulletin - Reprint | 3.95 |

★ **Denotes Soon to be Out of Print.**

## CLOCKS

| | PRICE |
|---|---|
| History of the American clock business for the past 60 years & life of Chauncey Jerome | $ 7.50 |
| Hobson's Choice - Hobson, C. | 29.95 |
| Horology Americana - Dworetsky & Dickstein - Published at $30.00 | 14.95 |
| In Quest of Clocks - Ullyett, K. ★ | 35.00 |
| International Dictionary of Clocks - Smith, A. - Published at $45.00 | 24.98 |
| It's About Time - Chamberlain, P. | 50.00 |
| Japanese Clocks - N.H.N. Mody | 22.50 |
| Lantern Clock - White, G. | |
| Longcase Clock - Brutton - 2nd Edition ★ | 35.00 |
| Longcase Clock - Tom Robinson | 69.50 |
| 19th Century American Clocks - Harris, G. | 12.95 |
| Old Clock Book, The - Moore, N. | 19.95 |
| Old English Master Clockmakers & Their Clocks - Cesinsky, H. | 9.50 |
| Pequegnat Story (Canadian Clocks) - Varkaris, J. | 28.95 |
| Revolution in Time - Landes, D. | 8.95 |
| Royal Clocks - Jagger, C. | 110.00 |
| Royal Clocks - 1904 | 10.95 |
| Pennsylvania Clocks & Clockmakers - Eckhardt, G. ★ | 25.00 |
| Seth Thomas Clocks and Movements, A Guide to Identification & Prices - Ly, T.D. - 1985 Edition | 19.95 |
| Shaker Clock Makers - Gibbs & Meader | 5.00 |
| Silversmiths and Clockmakers of Fulton County, Pa. 1785 - 1900 - Nelson, J.H. | 5.00 |
| Simon Willard & His Clocks - Willard, J.W. | 6.95 |
| Six Quaker Clockmakers - Chandlee, E. | 9.00 |
| Skeleton Clocks - Royer-Collard - Revised Edition | 29.50 |
| Some outstanding clocks over 700 years - Lloyd | 69.50 |
| Story of the Pendulum Clock, The - Edward, E.J. | 52.00 |
| The Time Museum - Andrews & Atwood | 4.00 |
| Thomas Cole & Victorian Clockmaking - Hawkins, J.B. | 30.00 |
| Treasury of American Clocks - Palmer, B. | 21.95 |
| Viennese Regulators & Industrial Clock Factories - Ortenburger | 18.50 |
| Warner Collectors Guide to American Clocks - Schorsch, A. | 9.95 |
| Weatherfield Collection - Britten's Old English Clocks ★ | 89.50 |
| Weatherfield Collection of Clocks - A Guide to Dating - Brutton, E. | 49.95 |
| White Dial Clock - Loomes, B. ★ | 25.00 |
| White Dial Clocks - The Complete Guide - Loomes, B. | 32.00 |

## CLOCKS AND WATCHES

| | PRICE |
|---|---|
| A list of Irish Watch & Clock Makers - Fennel, G. | $ 6.95 |
| American Clocks & Watches - 200 Years - Bailey ★ | |
| Britten's Old Clocks & Watches & Their Makers - 3rd Edition | 49.50 |
| Britten's Old Clocks & Watches & Their Makers - 891 pp. - 6th Ed. ★ | 65.00 |
| Buckeye Horology, Ohio Clocks & Watchmakers - Gibbs, J. | 10.00 |
| Clock and Watch Advertising - Edwards - Collector Edition ★ | 25.00 |
| Clock & Watch Pronunciary - Gerschler, M. | 11.95 |
| Clock & Watchmakers in Canada - Langdon, J. | 12.95 |
| Clocks & Watches - An Historical Bibliography - Baillie, G. - Vol. 1 | 40.00 |
| Clocks & Watches - Clutton & Daniels - Published at $52.00 | 24.98 |
| Clocks & Watches - Willsberger, J. ★ | 30.00 |
| Complete Checklist of American Clocks & Watchmakers - Edwards | 5.50 |

★ **Denotes Soon to be Out of Print.**

## CLOCKS & WATCHES

| | PRICE |
|---|---|
| Dictionaire Des Horologers Francaise - Tardy - French text | $105.00 |
| Dictionary of American Clock & Watch Makers - Sposato, K. | 47.00 |
| Die Uhren von A. Lange & Sohne, Glashutte Sachsen - Martin Huber - Deutsch/English | 42.00 |
| Frodshams - History of Chronometer Makers - Mercey | 54.00 |
| History of Clocks & Watches, The - Bruton, E. - Published at $40.00 | 15.95 |
| Introduction To Clock & Watchwork - Hatton, T. | 19.50 |
| Pennsylvania Clocks & Watches, Antique Timepieces and Their Makers - Gibbs, J. | 39.50 |
| The Time Museum - Andrews & Atwood | 4.00 |
| Watch & Clock Encyclopedia - DeCarle, D. - Third Edition | 19.95 |
| Watches & Clocks of Sir David Salomons Collection - Daniel, G. & Markarian - Published at $65.00 | 50.00 |
| Watchmakers & Clockmakers in Russia - Chenakal, V. | 9.00 |
| Watchmakers & Clockmakers of the World - Baille - Volume 1 | 22.00 |
| Watchmakers & Clockmakers of the World - Loomes - Volume 2 | 19.95 |

## ELECTRIC

| | PRICE |
|---|---|
| Digital Electronic Watch - Hyltin, T.M. | $19.95 |
| Electrical & Electronic Clocks & Watches - Weaver, J. | 49.50 |
| Electrical Timekeeping - Aked, C. | 6.75 |
| Electrifying Time - Aked, C. | 11.00 |
| Herschede Hall Clock Catalog - Electrically wound floor & mantel clocks | 4.50 |
| How to Make An Electric Clock - Way, R.B. | 5.85 |
| Self Winding Clock - 1908 Catalog | 16.95 |

**For selected listing of Electric Clocks, see pages 54 - 73 of Tran Duy Ly's "Clocks, A Guide to Identification and Prices, Volume 1."**

## MUSIC BOXES

| | PRICE |
|---|---|
| Cylinder Musical Box Handbook - Webb, G. | $14.95 |
| Encyclopaedia of Automatic Musical Instruments - Bowers, W. | 45.00 |
| History of the Music Box & Mechanical Music - Chapius, A. | 27.50 |
| How to Repair Musical Boxes - Jacot & Son | 2.95 |
| Ideal & Peerless Cylinder Musical Boxes - 1885-6 Catalog | 7.50 |
| Making Musical Miniatures: Music Boxes - Greenhowe | 19.95 |
| Musical Box - A History & Collector's Guide - Ord-Hume | 55.00 |
| Musical Boxes - Silver Anniversary Collection - 1949-74 | 25.00 |
| Official Price Guide to Music Machines - 1984, 3rd Edition | 9.95 |
| Regina Music Box Repair Manual | 2.95 |
| Restoring Musical Boxes - Ord-Hume | 34.00 |
| Stella Music Boxes - Jacot | 2.95 |
| Symphonion, Regina & Polyphone Music Boxes - Hasse - 1894-5 Catalog | 2.95 |
| Swiss Musical Boxes, Musical Novelties & Clocks - Ca. 1890 A. Schellhase Catalog | 4.95 |
| Violano-Virtuoso Service Manual | 2.95 |
| Wood Turning Music Boxes - Jacobson, J.A. | 10.95 |

★ **Denotes Soon to be Out of Print.**

## PRICE GUIDE

| | PRICE |
|---|---|
| A Guide to Watch Holders - Niehaus, J. | $ 4.95 |
| Antique Clocks & Value Guide - Wescot, A. - 1977 Edition | 11.95 |
| Calendar Clocks Price Guide - Miller - 1985 Edition | 4.95 |
| Clock Guide ID and Prices - Book 1 - Miller, R.W. - 1981 Edition | 10.95 |
| Clocks, A Guide to Identification & Prices - Ly, T.D. - Volume 1 - 1984 Edition | 15.00 |
| Clocks, A Guide to Identification & Prices - Ly, T.D. - Volume 2 - 1985 Edition - Available Soon | 17.50 |
| Complete Guide to American Watches - Shugart, C. & Engle, T. | |
| - 1983 Edition | 4.00 |
| - 1984 Edition | 5.00 |
| - 1985 Edition | 10.95 |
| Guide to the Purchase of a Watches - Smith & Sons Ltd. | 14.95 |
| Howard - 1858 - 1903 Watches - Townsend, G. | 8.00 |
| Lux Pendulette Clocks & Price Guide - Nicholson, D. | 12.95 |
| Lyle Antiques & Their Value - Clocks - Curtis, A. - 1983 Edition | 5.95 |
| Main Street Pocket Guide to American Clocks - Schorsch, A. 1983 Edition | 5.95 |
| Official Price Guide to Antique Jewelry - 1985 Edition | 9.95 |
| Official Price Guide to Antique and Modern Dolls - 1985 Edition | 9.95 |
| Official Price Guide to Antiques and other Collectibles - 1985 Edition | 9.95 |
| Official Price Guide to Collectible Toys - 1985 Edition | 9.95 |
| Official Price Guide to Music Machine - 1985 Edition | 9.95 |
| One Hundred Years of Collectible Jewelry - Bake, L. | 9.95 |
| Price Guide to American Pocket Watches - 1985 Edition - Criss, D. | 9.95 |
| Price Guide to Clocks - Volume II 1840-1940 - Shenton - New Edition - Available Soon | 49.50 |
| Price Guide to Jewelry - Poynder, M. | 59.50 |
| Seth Thomas Clocks and Movements - A Guide to Identification and Prices - Ly, T.D. - 1985 Edition | 19.95 |
| Vintage American & European Wrist Watches Price Guide - Planes, P. | 9.95 |

## SUNDIALS

| | PRICE |
|---|---|
| Die Sonneuhr - Rene, R.J., Rohr - Text in German | $55.00 |
| Easy to Make Wooden Sundials - Stoneman, M. | 2.95 |
| From Sundials to Atomic Clocks - Jespersen & Randolph | 5.00 |
| Sonneuhren 1 - Schumacher, Heinz - Text in German | 29.95 |
| Sonneuhren 2 - Peitz, Adolf - Text in German | 25.00 |
| Sonneuhren 3 - Schumacher, Heinz, Peitz, Adolf - Text in German | 29.95 |
| Sundials at Oxford College - Pattenden, P. | 12.95 |
| Sundials at Walls - Daniels, C. | 8.50 |
| Sundials: Their Theory & Construction - Waugh, A. | 4.00 |

## TECHNICAL AND REPAIRING

| | PRICE |
|---|---|
| A Manual of Veneering - Villard, P. | $ 3.50 |
| Advanced Watch & Clock Repair - Harris, H. | 10.95 |
| Art of Painting on Glass - Albinas-Elskus | 12.95 |
| Bench Practices for Watch & Clockmakers - Fried, H. | 9.00 |
| Best of Coleman, The Clockmaker - Hagans, O.R. | 30.00 |
| Britten's Watch & Clockmakers Handbook, Dictionary & Guide - 11th Edition — Hardbound | 29.50 |
| — Paperback | 19.95 |
| Britten's Watch & Clockmakers Handbook, Dictionary & Guide - 16th Edition - Published at $69.50 | 24.98 |
| Care & Keeping of Antiques - Kinard, E. | 6.95 |
| Clock & Watch Escapements - Gazeley, W.J. | 25.00 |
| Clock Cleaning & Repairing - Jones, B.E. | 10.00 |
| Clocks Construction, Maintenance & Repair - Coggins, F. — Hardbound | 18.95 |
| — Softbound | 13.50 |

## TECHNICAL AND REPAIRING

| | PRICE |
|---|---|
| Clock Design and Construction - Penman, L. | $15.95 |
| Clock Question & Answer Book - Baier, Tigner, Whitney | 14.95 |
| Clockmaking & Repairing - Garrard, F.J. Poor Reprint but good for references. | 12.95 |
| Clockmaking for the Amateur - Reeve, C. | 12.75 |
| Clockmaking for the Woodworker - Kadar, W.L. — Hardbound | 16.95 |
| — Softbound | 11.50 |
| Clock Repairing as a Hobby - Kelly, H.C. | 5.95 |
| Compensating Pendulums - Finn, J.L. | 4.95 |
| Complete Handbook of Sand Casting - Ammen, C. | 8.95 |
| Complicated Watches & Their Repair - DeCarle | 5.95 |
| Complicated Watches - Seibel & Hagans | 6.95 |
| Construction & Management of Chronometers, Watches & Clocks - Dent, E. | 3.50 |
| Designing and Building a Grandfather Clock - William, G. | 12.95 |
| Digital Electronic Watch - Hyltin, T.M. | 19.95 |
| Electrical & Electronic Clocks & Watches - Weaver, J. | 49.50 |
| Escapements, The - Britton, F.J. | 6.95 |
| Essence of Clock Repair - Monk, S. | 19.95 |
| Geared to the Stars - The evolution of planetariums, orceries and astronomical clocks - King, H.C. | 75.00 |
| George Graham Classic Escapement - Aydlett, G. | 3.50 |
| Gold Leaf Technical - LeBlanc - Hardbound - New Edition | 18.50 |
| Grandfather Clock Maintenance Manual - Vernon, J. | 11.50 |
| Hairsprings - Manipulating of Old & New Springs | 4.95 |
| Handbook of Clock & Watch Repairs - Harris, H.G. | 8.95 |
| Helpful Information for Watchmakers - Ohlson, O. - 11th Edition | 6.95 |
| Horological Hints & Helps - Britten, F.W. — Hardbound | 25.50 |
| — Paperback | 17.95 |
| Horological Wheel Cutting Engines - Crom, T. | 20.00 |
| Horolovar 400 Day Repair Guide - Terwilliger, C. - 9th Edition | 19.95 |
| How to Build 35 Great Clocks - Daniele, J. | 24.95 |
| How to Collect, Refinish & Restore Antique & Country Furniture - Damato, H. | 12.95 |
| How to Gold Leaf Antiques & Others - Chambers, D. | 9.95 |
| How to Make an Electric Clock - Way, R.B. | 5.85 |
| How to Repair Clock - Smith, E. - Hardbound | 10.95 |
| How to Repair Clock - Smith,E. - Softbound | 7.65 |
| How to Repair Musical Boxes - Jacot & Son | 2.95 |
| How to Train a Cuckoo Clock - Gerschler, M. | 3.98 |
| It's About Time - Chamberlain, P. | 50.00 |
| Jeweled Bearings for Watches - Higginbotham, C. | 4.95 |
| Jewelry, Basic Techniques & Design - Sprintzen, A. — Hardcover | 15.95 |
| — Paperback | 9.95 |
| Jewelry Repair Manual - Hardy, R.A. - 2nd Edition | 9.25 |
| Kemlo's Watch Repair Manual - Kemlo, F. | 9.95 |
| Marine Chronometer - Gould, R. | |
| Modern Clock - Construction Regulating & Repair - Goodrich, W. (This is the best book on the market for repairing American and European clocks; including Electric, 400-Day and Cuckoo clocks.) - 1984 Edition | 19.95 |
| Modern Calendar Watches - Humbert, B. | 27.95 |
| Modern Watch & Clock Repairing - Harris, P.B. | 15.95 |
| Musical Boxes - Silver Anniversary Collection 1949-74 | 25.00 |
| Navy Overhaul Manual for Hamilton Chronometer | 14.95 |
| Ordinance Manual T M 1575 Repair Manual for Pocket Watches, Stop Watches and Clocks | 14.95 |
| Paul Phillip Barraud - Jagger, C. | 9.95 |
| Pioneers of Precision Timekeeping | 6.95 |
| Practical Benchwork for Horologist - Levin, S. ★ | 47.00 |
| Practical Clock Repairing - DeCarle | 19.95 |
| Practical Watch Adjusting & Springing | 17.85 |
| Practical Veneering - Hayward, C. | 6.95 |

## TECHNICAL AND REPAIRING

| | PRICE |
|---|---|
| Practical Watch Adjusting & Springing - DeCarle, D. | 17.85 |
| Practical Watch Repairing - DeCarle, D. | 19.95 |
| Prazision Pendel Uhren - Riefler - Text in German | 35.00 |
| Prazision Pendel Uhren - Erbrich, K. - Text in German (Excellent Pictures) | 67.00 |
| Precision Time Measures - Higginbotham, C.T. | 6.95 |
| Quartz Watch Repair Manual - Zanoni, L. - Volume I | 19.95 |
| Quartz Watch Repair Manual - Zanoni, L. - Volume II | 19.95 |
| Questions & Answers for the Watchmaking Profession - AWI | 5.95 |
| Questions & Answers on Watch & Clockmaking - Elgin Collection | 5.00 |
| Regina Music Box Repair Manual | 2.95 |
| Repairing Antique Pendulum Clocks - Jendritski | 29.95 |
| Repairing and Restoring Antiques - Young, D. | 9.98 |
| Repairing and Restoring Pendulum Clocks - Plewes, J. — Hardbound | 19.95 |
| — Softbound | 12.95 |
| Repairing American Watches - Detent | 5.00 |
| Repairing American Wood Gear Clock Movements - Barlow, H. | 6.95 |
| Repairing Old Clocks & Watches - Whiten, A. | 15.95 |
| Repairing Quartz LED, LCD & Analog Watches - Volume 2 | 36.00 |
| Repairing Quartz LED, LCD & Analog Watches - Volume 3 | 36.00 |
| Repairing Quartz Watches - Fried, H. | 22.95 |
| Restoring Antique Furniture - Johnson, E. | 6.95 |
| Restoring Musical Boxes - Ord-Hume | 34.00 |
| Reverse Paintings on Glass - Mildred Lee Ward Collection | 9.00 |
| Revolution in Time - Landes, D. PB | 8.95 |
| Seth Thomas Clocks and Movements, A Guide to Identification and Prices - Ly, T.D. - 1985 Edition | 19.95 |
| Scientific Timing - Purdom, C. | 5.95 |
| Square Recoil Escapement - Aydlett, G. | 4.50 |
| Staff Making & Pivoting - Hall, E. | 4.95 |
| Staining and Wood Polishing - Hayward, C. | 6.95 |
| Striking Clocks - Baier, J.G. | 7.95 |
| Striking and Chime Clocks, Their Working & Repair - Smith, E. | 19.95 |
| Swiss Watch Repairer's Manual - Jendritski, H. | 29.95 |
| Technical Horology - Wright, T. | 5.95 |
| Timing Manipulations - AWI | 4.50 |
| Tower Clock & How to Make It - Ferson, E. | 6.00 |
| Watch & Clock Encyclopedia - DeCarle, D. | 19.95 |
| Watch & Clock Information, Please - Samelius & Hagens | 30.00 |
| Watch & Clockmaker's Buyer's Guide - AWI | 7.95 |
| Watch & Clockmaking & Repairing - Gazeley, W.J. | 24.95 |
| Watch Cleaning & Repairing - Ries, F. | 4.95 |
| Watch Collector's Handbook - Cutmore, M. | 14.50 |
| Watch Escapement - Fried, H.B. | 8.00 |
| Watch Repairer's Manual - Fried, H.B. | 17.95 |
| Watch Repairing as a Hobby - Fletcher, D.W. | 6.95 |
| Watch Repairing, Cleaning & Adjusting - Garrard, F.J. | 12.95 |
| Watchmaking - Daniels, G. (The best book ever written) — First Edition | 75.00 |
| — Reprint | 47.50 |
| Watchmaking in England 1760-1820 - Weiss, L. (Very good) | 29.95 |
| Wood Finishing - Vanderwalker, F. | 7.95 |
| Wood Finishing & Refinishing - Gibbia, S. - Revised Edition — Hardbound | 19.95 |
| — Softbound | 13.50 |

★ **Denotes Soon to be Out of Print.**

## TOOLS

| | PRICE |
|---|---|
| A Man and His Lathe - Sparey, L.S. | $ 6.95 |
| Amateur's Lathe - Sparey, L. | 11.95 |
| Amateur's Workshop - Bradley, I. | 11.95 |
| Catalog of Tools for Watch & Clockmakers - Wycke, J. | 23.00 |
| Clock Wheel and Pinion Cutting - Wild, J.M. | 3.85 |
| Early Lathes & Machine Tools of Interest - Deith, C.L. | 5.95 |
| Electroplating & Electroforming - Newman, L. | 7.95 |
| Electro-Plating for the Amateur - Warburton | 6.95 |
| Engine Turning 1680-1980 - The Tools & Techniques - Mathews, M. | 87.50 |
| Gear Wheels & Gear Cutting - Marshall, A. | 6.95 |
| Hardening, Tempering & Heat Treatment - Cain, T. | 13.95 |
| Horological Shop Tools - Crom, T. | 115.00 |
| Horological Wheel Cutting Engines - Crom, T. | 20.00 |
| How to Use Staking Tool - K&D | 4.00 |
| How to Work Sheet Metal - Dyer, H. | 6.95 |
| Lathe Accessories - Westbury, E. | 7.50 |
| Metal Turning Lathes - Westbury, E. | 8.95 |
| Metal Working Tools - Bradley, I. | 6.95 |
| Milling Operations in the Lathe - Westbury, E. | 9.95 |
| Precision Machinery - American Watch Tool Co. | 4.50 |
| Shaping Machine & Lathe Tools - Bradley, I. | 6.95 |
| Sharpening Small Tools - Bradley, I. | 6.95 |
| Soldering & Brazing - Turpin, A. | 6.95 |
| Tools for the Trade - Roberts, K. | 24.00 |
| Using the Small Lathe - Mason, L. | 7.95 |
| Watches by Automatic Machinery at Waltham - Marsh, E. | 6.95 |
| Watchmaker and His Lathe - Jendritski, H. | 29.95 |
| Watchmaker Lathe - DeCarle, D. | 19.95 |
| Watchmaker Lathe - Goodrich, W. - Very Good. | 14.95 |
| Welding & Cutting - Oats, J. | 5.95 |
| Welding Skills and Practices - 5th Edition - Giachino & Weeks | 19.95 |
| Wheel & Pinion & How to Determine Their Exact Size - Gribi, T. | 3.00 |
| Woodworking Tools & How to Use Them - Hill, J. | 18.00 |

## TRADE MARKS

| | PRICE |
|---|---|
| Hallmark & Date Letters on Silver, Gold, Platinum - Brutton, E. PB | $ 4.50 |
| International Silver Hallmarks - Tardy, L. | 25.00 |
| Pewter Hallmark (Poincons d'Etain) - Tardy, L. - Text in French. PB. | 18.00 |
| Trademark Index - Clocks & Watches - Kochmann, K. | 12.50 |
| Trade Marks - Watch Cases, Pocket Watches, Precious Stones - Ehrhardt, R. | 10.00 |

## WATCHES

| | PRICE |
|---|---|
| Almost Everything you wanted to know about American watches & didn't know who to ask - Townsend, G. | $ 8.00 |
| American Railroad Watches - Townsend, G. | 8.00 |
| Antique Enamelled Watch - Lewis, M. | $25.00 |
| Antique Watches - Camerer - Cuss | 49.50 |
| Armbanduhren - Wrist Watches - Helmut Kahlert/Richard Muhe/ Gisbert L. Brunner - Text in German | 55.00 |
| Art of Brequet, The - Daniels, G. | |
| Balance Staff and Cylinder - Eigner, P.W. with Excessive or Small Vibrations in Cylinders Watches & Their Corrections by Hillman, B. | 5.95 |
| Cavalcade of Time - Fried, H.B. | 10.00 |
| Collecting & Identifying Old Watches - Harris, H.G. | 13.95 |
| Complete History of Watchmaking in America - Crossman | 12.50 |
| Dollar Watch That Made The Dollar Famous - Townsend, G. | 8.00 |
| Dueber Hampden Story - Gibbs, J. ★ | 6.95 |
| Englishman's Watch, The - Kemp, H. | 35.00 |
| Four Hundred Years of Watchmaking - Benis, A. | 20.00 |
| Helpful Information for Watchmakers - Ohlson, O. - 11th Edition | 6.95 |
| Hours of Love 'Naughty Watches' - Carrera, R. | 62.00 |
| Howard 1858 - 1903 Watches - Townsend, G. | 8.00 |
| It's About Time - Chamberlain, P. | 50.00 |
| Lancashire Watch Co. - 1899 - 1910 | 7.50 |
| Patek Philippe Geneve - Hubert & Banbara | 130.00 |
| Patek Philippe - Vogel, H. | 44.00 |
| Pocket Watch Handbook, The - Cutmore, M. | 19.95 |
| Taschenuhren/Watches - Meis, R. - Text in German | 55.00 |
| Watch Collector's Handbook - Cutmore, M. | 14.50 |
| Watch Companies of America - A History - Dixon, J. | 10.95 |
| Watch Dials of the French Revolution - Paricaldi | 27.50 |
| Watch Escapements - Pellaton, J.C. | 12.95 |
| Watch Factories of America - Past & Present - Abbott, H. | 8.50 |
| Watch Keys - Gunther, N. | 45.00 |
| Watch Repairing, Cleaning & Adjusting - Garrard, F.J. | 12.95 |
| Watches - Clutton & Daniels - 3rd Edition | |
| Watches by Automatic Machinery at Waltham - Marsh, E. | 6.95 |
| Watches, Their History, Decoration and Mechanism - Baillie, G.H. | 45.00 |
| Watches in Color - Good, R. | 6.95 |
| Watches of Fantasy, 1790-1850 - Patrizzi-Sturm | 59.00 |
| Wrist Watches - Kahzert, Muher, Brunner | |

★ **Denotes Soon to be Out of Print.**

*Prices shown are those customarily charged for non-discounted single copies, and will change based on availability and demand. Foreign publications are usually priced based on currency exchange rates in use at the time of purchase by the dealer, and may therefore be higher or lower than indicated by comparison of prices using current exchange rates. Purchase prices will only occasionally be higher than list prices for some difficult to locate books, due to fluctuations in publisher's discount rates.*

*Most mailorder book dealers charge extra for postage and handling, with the usual rate being $1.25 for single copies and $.50 for each additional copy. If UPS is used, this charge is usually a flat $3.50, regardless of the number of copies mailed. If COD is requested, the additional charge is a flat $1.90 for UPS, or a minimum of $1.50 for U.S. Postal Service.*

# INDEX

## ALARM CLOCKS



## BANJO CLOCKS



## CALENDAR CLOCKS



## CALENDAR CLOCKS



## CARRIAGE CLOCKS



## CHIME CLOCKS



## CHIME CLOCKS

## CITY CLOCKS



## CITY CLOCKS



## CRYSTAL REGULATOR CLOCKS



## CRYSTAL REGULATOR CLOCKS

# CRYSTAL REGULATOR CLOCKS



# ELECTRIC CLOCKS



# HALL CLOCKS



# HALL REGULATORS



# HANGING CLOCKS



# HANGING CLOCKS

## HANGING CLOCKS



## MANTLE CLOCKS — METAL CASES



## MANTEL CLOCKS — METAL CASES



## MANTEL CLOCKS — WOOD CASES



## MANTEL CLOCKS — WOOD CASES

## MANTEL CLOCKS — WOOD CASES



## MANTEL CLOCKS — WOOD CASES



## MUSICAL CLOCKS



## NOVELTY CLOCKS



## NOVELTY CLOCKS



## SHIP'S BELL CLOCKS

# PUBLIC CLOCKS

## BRACKET CLOCKS



## POST CLOCKS



## STREET CLOCKS



## TOWER CLOCKS



## TOWER CLOCKS



## REFERENCE DATA